Electronic Devices

Other books for basic electronics courses

DIGITAL ELECTRONICS: CONCEPTS AND APPLICATIONS
  by Terry L. M. Bartelt

DC/AC CIRCUITS: CONCEPTS AND APPLICATIONS
  by Richard Parrett

Electronic Devices
Concepts and Applications

JOHN HENDERSON
Instructor of Electronics Technology
Computer Processing Institute

Prentice Hall, Upper Saddle River, NJ 07458

Editorial/production supervision and
    interior design: Ed Jones
Cover design: Computer Graphic Resources
Manufacturing buyers: Dave Dickey,
    Mary McCartney, and Ed O'Dougherty

Printed in the United States of America

10 9

ISBN   0-13-042656-3

Prentice-Hall International (UK) Limited,London
Prentice-Hall of Australia Pty. Limited, Sydney
Prentice-Hall Canada Inc., Toronto
Prentice-Hall Hispanoamericana, S.A., Mexico
Prentice-Hall of India Private Limited, New Delhi
Prentice-Hall of Japan, Inc., Tokyo
Pearson Education Asia Pte. Ltd., Singapore
Editora Prentice-Hall do Brasil, Ltda., Rio de Janeiro

To my wife, Joannie,
with all my love

Contents

9

Field-Effect Transistors 210

10

Thyristors and Unijunction Devices 251

11

Operational Amplifiers 272

12

Op-Amp Filters 313

13

Oscillators 341

14

555 Timer 377

15

A

B

C

D

Preface

The purpose of this book is to provide the reader with an understanding of semi-conductors. A wide range of devices are covered, both independently and in the context of an actual circuit. After the proper operation of a device is explained, most chapters end with a troubleshooting discussion of possible faults involving the device.

Ease of understanding and convenience were goals in preparing this book. All formulas are supported by worked-through Examples and Practice Problems. Review Questions at the end of each section allow you to monitor your progress, and their answers are found at the end of the chapter. Important formulas are repeated as needed to avoid backtracking. The glossary at the end of each chapter refreshes your understanding of key words. A comprehensive collection of Problems is also located at the end of each chapter. They are grouped according to chapter sections in order to provide a quick and easy means of referencing needed information.

I would like to acknowledge the helpful feedback that I received from students at the Computer Processing Institute. They helped to shape my approach to much of the material in this book. Also, the teaching staff and management of CPI deserve a note of gratitude for their support and encouragement. Finally, a note of thanks to the many reviewers whose criticisms and suggestions have led to what I hope will be a very useful textbook.

John Henderson

Safety

Safety should be a priority whenever working with electricity. A relatively small amount of current (10 mA) can lead to serious injury under the right conditions. The following lists of procedures and tips are provided to ensure a safe approach to the study of electronics.

GOLDEN RULE: ALWAYS ASK YOUR INSTRUCTOR IF YOU ARE IN DOUBT.

ENVIRONMENT

1. The work area should be well lit.
2. A proper fire extinguisher (usually CO_2) should be handy.
3. NO LIQUIDS should be permitted in the work area!

PERSONAL

1. Don't wear metal jewelry as this can lead to a shock.
2. Always wear safety glasses. (Capacitors, diodes, transistors, etc. can explode.)
3. NEVER TOUCH components in a live circuit; turn off the power first. Also note that many components retain a charge even after power is turned off (i.e., capacitors, cathode-ray tubes, etc.).

EQUIPMENT

1. Only use equipment (e.g., meters, power supplies, tools, etc.) that are properly rated and approved.

2. Remember that a circuit with a low voltage at one section may have a very high voltage at another section.

3. Never leave equipment (i.e., circuits, soldering irons) unattended. Always turn off the power when you leave your work area.

4. Do not use adapters to defeat the ground when plugging in equipment.

Electronic
Devices

Semiconductor Construction

When you have completed this chapter, you should be able to:

- Recognize the differences between a conductor, an insulator and a semiconductor.
- Define P-type and N-type materials.
- Measure and observe the effects of forward and reverse biasing on a PN junction device.

INTRODUCTION

We are about to enter a world that lies between the two extremes of a conductor and an insulator. Keep in mind as you consider this subject that the semiconductor has the ability to change its characteristics. One moment it will act like a conductor; at another it will act as an insulator. Because of its efficiency in changing characteristics (resistance), it is used when you need a conductor and when you need an insulator.

Where vacuum tubes were once used, semiconductors are now used. What once took a room full of tubes to accomplish can now be packaged in a small chip or integrated circuit (IC). It is the small yet powerful semiconductor that has revolutionized the electronics industry.

REVIEW QUESTIONS

1. A semiconductor has the ability to change its _characteristics_
2. The semiconductor can appear either as a(n) _insulator_ or as a(n) _conductor_
3. Modern day semiconductors have replaced _vacuum tubes_.

1-1 DIFFERENCES IN THE VALENCE SHELL

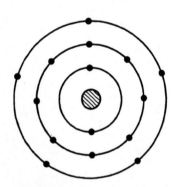

FIGURE 1-1 Phosphorus atom (15).

Electron: A negatively charged particle of an atom.

Proton: A positively charged particle found in the nucleus of an atom.

We begin by finding the differences among insulators, conductors, and semiconductors. First, we review some basic principles and terms of atomic structure.

From the atomic model shown in Fig. 1-1, we recognize some of its features. There are a number of concentric circles representing the "shells" of an atom. Each of these shells contains electrons that in turn revolve around the central portion of the atom called the nucleus. Inside the nucleus, though not shown here, are protons and neutrons. The number of protons is equal to the number of electrons. This is true for any atom.

Remember that **electrons** are negatively charged particles and **protons** are positively charged particles. If there is an equal number of each, then the net charge is neither more positive nor more negative; the atom is, therefore, considered neutral.

There is a third type of particle contained in the atom called the neutron. The **neutron** is a particle having no electrical charge and will be ignored in our discussion.

Examine the outermost shell of the atomic model. This shell is called the "valence shell" and the electrons contained within it are aptly called "valence electrons." There are five valence electrons shown in this example of the phosphorus atom, but the number of valence electrons varies according to the element. However, there is a limit of eight valence electrons. When an atom has eight electrons in its valence shell, that shell is considered full.

It is the differences in the valence shell that are studied in the next few sections. Refer to Appendix A in order to gain a broader view of the many elements as well as additional information on the structure of the atom.

Neutron: A particle having no charge that is found in the nucleus of an atom.
Valence Shell: The outermost shell of an atom.
Valence Electron: An electron occupying the outermost shell of an atom.

REVIEW QUESTIONS

4. Electrons can be found in the _____ of an atom.

5. The central portion of an atom is called the _____ .

6. The nucleus of an atom may contain both _____ and _____ .

7. If an atom has 14 electrons, then it has _____ protons.

8. Electrons are _____ charged particles.

9. Protons are _____ charged particles.

10. The net charge of an atom that has an equal number of electrons and protons is _____ .

11. The outermost shell of an atom is called the _____ shell.

12. Electrons found in the outermost shell are called _____ electrons.

13. The outermost shell can hold a maximum of _____ electrons.

1-2 THE CONDUCTOR

A **conductor** is, by definition, anything that conducts electricity. Yet what is it that makes one material able to conduct better than another? The answer is found in the valence shell of the atom.

Refer to the copper atom of Fig. 1-2. If you look up the atomic number of copper in the periodic table of elements found in Appendix A, you find it to be 29. The atomic number refers to the number of electrons contained in the atom. By referring to the atomic model in Appendix B, the maximum number of electrons allowable in each shell can be seen. By combining these two sets of facts, the model of Fig. 1-2 is derived.

In Fig. 1-2, the 29 electrons are distributed throughout the shells with one in the valence shell. It is the valence electron that is of importance, because it can be freed from its shell, becoming available for current flow. The ease with which the valence electron can do this determines the quality of conductance.

Two factors that effect the valence electron's ability to move are the number of valence electrons and their distance from the nucleus. The fewer the number of valence electrons, the easier it is for them to break free. In the example, there is only one, making it an ideal case. There are three shells between the valence electron and the nucleus, which is helpful toward release. The further the valence shell is from the nucleus the better. This is because of the electrostatic force that the nucleus exerts on the shells.

At this point, we can conclude that copper is a good conductor; however, it is not the best. To illustrate this, see Fig. 1-3, the silver atom. Notice that silver has only one valence electron, which makes it ideal. It also has four shells between it and the nucleus, which makes it an even better choice than copper. However, the high cost of silver versus copper is the reason copper is used for wire.

Conductor: A material having many free electrons and a low resistivity.

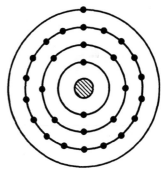

FIGURE 1-2 Copper atom (29).

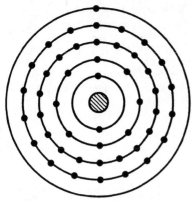

FIGURE 1-3 Silver atom (47).

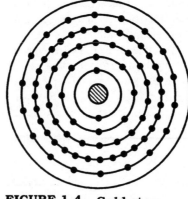

FIGURE 1-4 Gold atom (79).

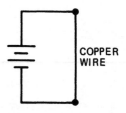

FIGURE 1-5 Conduction in a copper wire. WARNING: DO NOT ATTEMPT THIS.

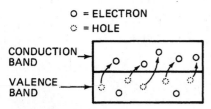

O = ELECTRON
◌ = HOLE

CONDUCTION BAND →
VALENCE BAND →

FIGURE 1-6 Valence electrons freeing themselves from the valence band.

Free Electron: An electron that has left its valence shell.

First Law of Electrostatics: Like charges repel and unlike charges attract.

Hole: Positively charged, yet has no mass.

Speaking of high costs leads us to gold. Figure 1-4 shows the atomic model of gold. Its single valence electron is the furthest from the nucleus of all the metals. Although silver remains the best conductor at room temperature, gold is less susceptible to corrosion and oxidation.

You should now have a sufficient idea of how to determine whether or not a material is a good conductor. Now let us explore how conduction itself occurs. To do this, an outside force (heat, light, chemical, electrical) and a piece of copper wire is needed.

If a wire were connected across the terminals of a battery, as shown in Fig. 1-5 (DO NOT ATTEMPT THIS AS IT WOULD RESULT IN A DIRECT SHORT!), the difference in potential would cause the valence electrons to become "excited." Most of these valence electrons from the billions of copper atoms would take on new energy levels. This simply means that the "potential energy," which is due to electrostatic attraction, and the "kinetic energy," which is due to the motion of the atoms, are increased. The end result is that the valence electron no longer stays in its valence band of energy, but jumps to the conduction energy band, where it is now a **"free electron."** Figure 1-6 shows this movement.

As each of these electrons jump to the conduction band, they, in turn, become available for current flow. From Coulomb's **first law of electrostatics,** "unlike charges attract and like charges repel," free electrons travel toward the positive terminal of the battery.

One final note. As each electron leaves its valence shell, a **"hole,"** or absence of an electron, is left behind. These holes, although having no mass, are considered to have a positive charge. Therefore, the movement of the holes is toward the negative terminal of the battery. This constitutes hole flow, or conventional current flow.

REVIEW QUESTIONS

14. A _____ can conduct electricity.
15. The _____ number reveals the number of electrons an atom contains.
16. Copper has _____ valence electron(s).
17. A conductor has _____ valence electrons.
18. The _____ the distance between the valence electron(s) and the nucleus, the better the conductor.

19. The nucleus exerts an _____ force.

20. Three common metallic elements that have the properties of a conductor are _____ , _____ , and _____ .

21. _____ energy is due to electrostatic attraction.

22. _____ energy is due to motion.

23. When an electron gains enough energy to jump into the conduction band, it is called a _____ _____ .

24. When an electron leaves its valence shell, it leaves a _____ behind.

25. Holes have a _____ charge.

26. Holes flow from _____ to _____ .

1-3   THE INSULATOR

Having examined what makes a material a good conductor and the direction of electron/hole current flow, we now apply the same principles to the insulator.

Insulation in everyday life is shown in the home that is insulated from the cold or heat. The object is to prevent the outer climate from entering the home as well as preventing the inner climate of the home from escaping.

The **insulator** in the field of electronics also attempts to protect a device or circuit from outside interference as well as to prevent current from leaving its conductive path within the insulation. The synthetic rubber or plastic coating on a wire is an example of an insulator.

Insulator: A material having very few free electrons and a high resistivity.

The atomic structures of these compound materials, called polymers, are those of complex molecules. Yet without even discussing a specific material, it is still possible to understand the basic principle of a full valence shell. We stated earlier that a valence shell is considered full once it contains eight electrons; it cannot receive any more electrons. It is also very difficult for it to lose a valence electron because of the large amount of energy required to cause an electron to jump into the conduction band. Figure 1-7 shows a gap labeled as the forbidden band, which does not exist with conductors (Fig. 1-6). Electrons may not travel in this band; therefore, the amount of energy necessary for them to jump from the valence band to the conduction band is much greater.

Because of its difficulty in conducting current, the insulator achieves its purpose, which is to limit current flow. The principle of the full valence shell becomes more important as we discuss semiconductors and the process of doping.

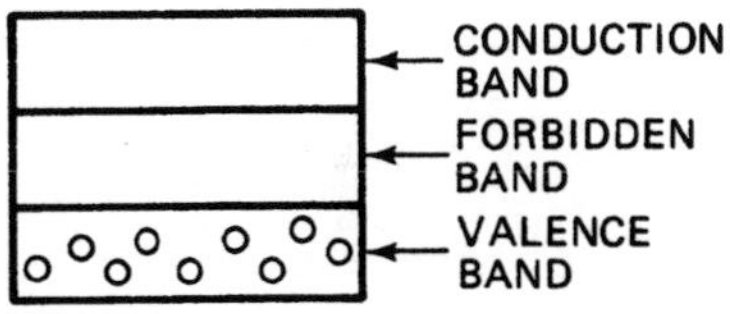

FIGURE 1-7   Energy bands of an atom.

REVIEW QUESTIONS

27. Synthetic rubber or plastic are examples of an _____ .

28. It is difficult for a _____ valence shell to lose an electron.

29. It takes a _____ amount of energy to cause an insulator's valence electron to jump into the conduction band.

30. Electrons may not travel in the _____ band.

1-4   THE SEMICONDUCTOR

Two of the most common semiconductor materials are silicon and germanium. The term **semiconductor** literally means a device that is partly conductor and, therefore, partly insulator. It is the ability to change resistance that makes the semiconductor device unique.

Semiconductor: A material having a conductivity and resistivity between that of a conductor and an insulator.

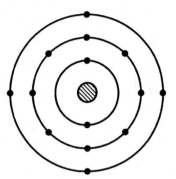

FIGURE 1-8 Silicon atom (14).

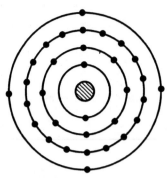

FIGURE 1-9 Germanium atom (32).

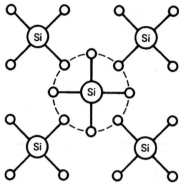

FIGURE 1-10 Covalent bonding of silicon atoms.

Ion: An atom that has either gained or lost an electron.

Covalent Bonding: The sharing of valence electrons among atoms.

Intrinsic: Having no impurities.

Figure 1-8 shows a silicon atom with its four valence electrons. This valence shell falls between a conductor and an insulator in its number of valence electrons. The same is true of germanium, shown in Fig. 1-9.

If we take a number of these semiconductor atoms and join them together, we create what is called a "wafer." The process of joining them is called "bonding," and there are two types of bonding we need to discuss.

The first is called "ionic bonding," which refers to the process of an atom either gaining or losing an electron. If an atom actually gains an electron, it is then called a negative **ion** because it contains one more electron than it does protons, thus giving it an overall net negative charge. The opposite is true if it were to lose an electron in the process of joining itself to another atom. In this case, it is called a positive ion because it now has more protons than electrons.

The second form of bonding is called **"covalent bonding."** This form of bonding is achieved by sharing valence electrons, as shown in Fig. 1-10. Note that only the valence electrons of each atom are shown for simplicity.

When viewing Fig. 1-10, imagine each of the atoms turning as a set of gears that mesh one into another. If we now focus on the center atom, it appears always to have eight valence electrons. This is due to the four surrounding atoms, each sharing one of their valence electrons. The end result is a very stable structure with each atom retaining its four valence electrons.

Keep in mind that this is a tiny part of an extremely large picture. There are billions of silicon or germanium atoms joined together in this crystallike lattice structure.

At this point, the semiconductor material is rather useless because it is **"intrinsic,"** which means it is pure. In other words, there is not a significant number of atoms other than silicon or germanium in the crystal wafer. We are now about to make something useful of this atomic structure through the process of doping.

REVIEW QUESTIONS _____

31. The two most popular semiconductor materials are _____ and _____.

32. A semiconductor can change _____.

33. Semiconductors have _____ valence electrons.

34. A _____ is formed by the joining of semiconductor atoms.

35. The process of joining atoms is called _____.

36. When an atom gains an electron, it is called a _____ _____ .

37. When an atom loses an electron, it is called a _____ _____ .

38. When an atom bonds by sharing valence electrons with another atom, it is called _____ _____ .

39. A pure semiconductor can also be called _____ .

1-5 DOPING

The process of **doping** adds impurities to the silicon wafer, thus creating an extrinsic material. These impurities are atoms other than silicon or germanium and the procedure involves exposing the silicon wafer at high temperatures to a particular gas. The impurities in the gas are absorbed into the exposed portion of silicon.

Doping: The process of adding impurities to a semiconductor.

The purpose of doping is to create either an excess of electrons or an excess of holes in the silicon wafer. By creating an excess of electrons, we cause the material to be negative in nature. Exposing the silicon to an impurity that creates an excess of holes results in a material that is positive in nature.

REVIEW QUESTIONS

40. Adding impurities to a semiconductor wafer is called _____ .

41. Excess _____ and _____ are a result of doping.

1-6 N-TYPE MATERIAL

Let us start with a discussion of how an N-type material is formed. We begin with the intrinsic silicon wafer and dope it with a "pentavalent" element. A pentavalent element is an atom that has five valence electrons. Examples of these are phosphorus, arsenic, and antimony. Arsenic is a commonly used dopant.

When arsenic atoms are added to silicon, the outcome looks like the structure of Fig. 1-11. Notice that arsenic's fifth electron seems to be misplaced. This is because only eight electrons may exist in the valence shell of any atom. When you take atoms with four valence electrons and then add pentavalents, which have five valence electrons, you end up with nine valence electrons, which cannot be accommodated. The effect is that the excess electron becomes a free electron available for current flow.

This brings us to a very subtle point: when an electron breaks away from a covalent bond, it leaves a hole behind. However, if it was not part of the covalent bond, as is the case for the example just given, then no hole is left behind. Figure 1-12 illustrates this point. In this very simple depiction, there are two free elec-

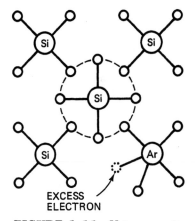

EXCESS ELECTRON

FIGURE 1-11 N-type material.

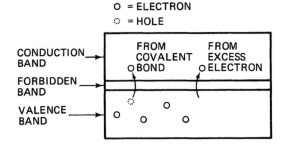

FIGURE 1-12 N-material current carriers.

trons compared to the one hole. This makes electrons in the majority and holes in the minority in the N-type material. As a matter of fact, these are the names given to the two types of current flow. On the one hand, you have "majority carriers" of current, and on the other, "minority carriers" of current. When a device depends on both types of current flow, we call it a "bipolar" device. Consider now the other side of the wafer, P-type material.

REVIEW QUESTIONS

42. An atom with five valence electrons is called a _____ .
43. When arsenic is added to silicon, the excess electron becomes a _____ electron.
44. In N-type materials, the _____ are majority current carriers.
45. In N-type materials, the _____ are minority current carriers.
46. When a device depends on majority and minority current carriers, it is called _____ .

1-7 P-TYPE MATERIAL

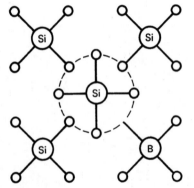

FIGURE 1-13 P-type material.

Again, the starting point is the intrinsic silicon wafer, and through the process of doping, we add some "trivalents." You may have already guessed that trivalents are atoms with three electrons in the valence shell. Included in the list of trivalents are boron, gallium and indium. We select boron since it is a long used dopant. Consider Figure 1-13 as you read the following.

When arsenic is added to a silicon wafer via doping, the end result is that you have silicon with its four valence electrons covalently bonding with boron and its three valence electrons. If you stop and add these, you come up with seven valence electrons. Remember that the valence shell can hold eight. This leaves us with one vacancy, or hole. When another valence electron fills this hole, it in turn leaves a hole behind. By doping the silicon with a number of boron atoms, it is left with a number of excess holes.

Finally, in Fig. 1-14, when a valence electron frees itself from the covalent bond, it leaves a hole behind. In considering the overall picture, there are more holes than conduction band electrons (two to one). Therefore, the holes are in the majority and electrons are in the minority in a P-type material. So holes are the "majority carriers" of current and electrons are the "minority carriers" of current.

Some of you may be thinking that you see more electrons in Fig. 1-14 than holes. That is because you are looking at a stationary picture. All of those elec-

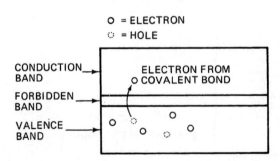

FIGURE 1-14 P-type material current carriers.

trons in the valence band are in constant motion, and as they move, they leave an equal number of holes behind (that are not shown for the sake of clarity). The particles we are concerned with are those that do not have an associated hole or electron in the valence band.

We have discussed both N-type and P-type materials separately. It is now time to combine the two materials and create what is known as a PN junction semiconductor device.

REVIEW QUESTIONS

47. An atom with three valence electrons is called a _____ .
48. When boron is added to silicon, _____ are created.
49. In P-type materials, the _____ are the majority current carriers.
50. In P-type materials, the _____ are the minority current carriers.

1-8   THE PN JUNCTION

The intrinsic silicon wafer is doped half with boron and half with arsenic. This creates an initial condition similar to that shown in Fig. 1-15. There are an excess of holes, or positive charges, on one side and an excess of electrons, or negative charges, on the other. This situation does not last for long because opposite charges attract.

Immediately after the doping process has ended, the excess electrons begin to combine with the excess holes near the center of the wafer. This area where the materials join is called the "junction," from which we get the name PN junction.

As the holes and electrons combine, ions are formed. You recall that an ion is developed when an atom either gains or loses an electron. In this case, the atoms in the N-type material are losing their electrons and the atoms in the P-type material are gaining them. The result is seen in Fig. 1-16, where a wall of positive and negative ions develops.

Keep in mind that when an atom gains an electron, it then has more electrons than protons, giving it an overall net negative charge. The electron lost by the N-type material causes an overall net positive charge. Both sets of ions are very stable, each having a full valence shell. Once they are formed, they do not accept nor lose any more electrons.

This process of electron/hole combination forming ions continues up to a certain point. The point at which it stops is determined by the semiconductor material. For silicon, it continues until a 0.7-volt difference in potential is reached, and for germanium, 0.3 volt. The region where this potential develops as a result of the ions is called the "depletion region" because this area is depleted of current carriers (see Fig. 1-17).

This combination of electrons and holes will not encompass the entire wafer. This is because the excess electrons from the N-type material eventually need too much energy to jump across the depletion region to reach the vacancy (hole) on the other side.

Finally, the 0.7 or 0.3 volt exists internally within the semiconductor. It cannot be measured because you are unable to enter the device and directly measure that region. However, you see the affect of this voltage when a power supply is connected to the device.

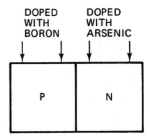

FIGURE 1-15   PN junction.

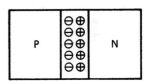

FIGURE 1-16   Formation of ions.

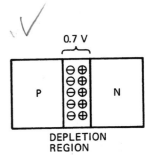

FIGURE 1-17   Depletion region forms.

51. The region where the P-type and N-type materials join is called the PN _____.

52. When an atom gains or loses an electron, it is called an _____ .

53. The area around the PN junction where the ions are formed is called the _____ region.

54. The potential difference of the depletion region for silicon is _____ volt.

55. The potential difference of the depletion region for germanium is _____ volt.

56. The hole–electron combination in the depletion region eventually stops because too much _____ is required for current carriers to cross that region.

1-9 BIASING THE PN JUNCTION _____

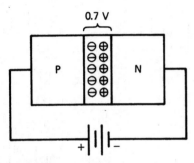

FIGURE 1-18 Forward biasing a PN junction.

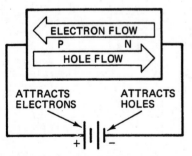

FIGURE 1-19 Two majority-carrier current flows.

Now that we have finished the fabrication of our PN junction, it is time to connect two electrodes (leads or legs) to either end of the wafer. This enables us to attach it to a power supply for proper "biasing."

Biasing simply means applying a voltage that causes a device to operate in a particular way. With our device, we have one of two options, "forward bias" and "reverse bias." This is because there are only two ways that you can connect a power supply to a two-legged device.

"Forward biasing" is illustrated in Fig. 1-18. Observe how the negative terminal of the supply is connected to the N-type material and the positive terminal is connected to the P-type material. This is the way you always connect the power supply in order to forward bias the PN junction.

The effect this has on the PN junction device is the narrowing of the depletion region until finally current begins to flow. This happens because the negative terminal of the power supply repels the majority carriers (electrons) in the N-type material and the positive terminal repels the majority carriers (holes) in the P-type material. Figure 1-19 shows how these two current carriers travel as they finally overcome the 0.7-volt "barrier voltage" of the depletion region.

A very important point needs to be emphasized here and that is that approximately 0.7 volt of the power supply voltage is required to overcome the barrier voltage within the silicon PN junction wafer. (It is 0.3 volt for germanium.) As soon as this voltage is reached, current flows.

Reverse biasing is shown in Fig. 1-20, where the negative terminal of the battery is connected to the P-type material and the positive terminal to the N-type material. This causes quite a different mode of operation in the device. As a matter of fact, it has the effect of turning off the device. In other words, we find that the PN junction device does not conduct.

Once again, we call to mind the first law of electrostatics, which states that opposites attract. Keeping this thought in mind, let us have another look at Fig. 1-20 to see how it applies.

The action in this circuit is the attraction of the electrons in the N-type material to the positive terminal and the attraction of the holes in the P-type material to the negative terminal of the power supply. You can almost imagine the depletion region expanding and this is what happens. The depletion region widens and the barrier voltage increases from 0.7 volt to a value equal to the power-supply voltage.

Although the device is connected in this manner, there is no current flow as far as we are concerned as technicians. In actuality, there may be a very small reverse current flow, but not enough for us to consider.

Of course, there is a limit to the amount of voltage or current any device can withstand and these maximum amounts are found on specification (spec) sheets under the heading of "peak inverse voltage." You will eventually find that each device has its own part number that can be used to look up the specifications. The spec sheets are usually published in large volumes that deal with various semiconductor devices.

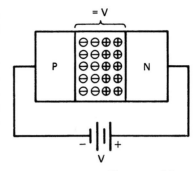

FIGURE 1-20 Reverse biasing a PN junction.

REVIEW QUESTIONS

57. Another name for the leads or legs of a component is _____ .

58. A device can be caused to operate in a certain way through _____ .

59. Forward biasing is achieved when the _____-type material is connected toward the negative terminal of a supply, and the _____-type material is connected toward the positive terminal of a supply.

60. The opposite of forward biasing is _____ biasing.

61. The depletion region becomes _____ as a result of forward biasing.

62. Another name for the potential difference of the depletion region is _____ voltage.

63. _____ biasing has the effect of turning off the device.

64. The first law of electrostatics states that like charges _____ each other and unlike charges _____ each other.

65. The depletion region _____ when reverse biased.

1-10 I–V CURVES

The final topic in this chapter deals with a pictorial representation of the operation of a PN junction device. Figure 1-21 shows the relationship of current and voltage. The designations are as follows:

I_F = forward current
V_F = forward voltage
I_R = reverse current
V_R = reverse voltage

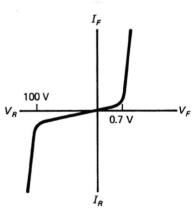

FIGURE 1-21 I–V characteristic curve for a silicon device.

When considering a device's operation in forward bias, we refer to the upper right quadrant of the graph. Here we find that when the forward voltage across the device reaches 0.7 volt, it begins to conduct, as shown by the upward curving line.

Toward the other extreme, which would be reverse biasing the device, refer to the lower left quadrant of the graph. As the reverse voltage is increased slightly past 100 volts, the device begins to conduct. The 100 volts is considered the peak inverse voltage, which is the maximum allowable reverse-biased voltage before the device breaks down.

These characteristics were discussed previously in the sections on forward and reverse biasing. If you understand these topics, then this figure should make sense.

66. The *I–V* curve chart shows that a silicon diode begins to conduct at _____ volt.

67. The maximum reverse voltage allowable before a diode begins to conduct is called the _____ _____ _____ .

SUMMARY

1. The nucleus of an atom contains protons and neutrons.

2. A proton is a positively charged particle.

3. A neutron is an electrically neutral particle.

4. The shells of an atom contain negatively charged particles called electrons.

5. The outermost shell of an atom is called the valence shell and can hold a maximum of eight electrons.

6. Conductors have few electrons in the valence shell (e.g., copper, gold, silver).

7. Conduction occurs when the increased energy level of a valence electron causes it to jump to the conduction energy band and leave a hole behind.

8. Insulators have full valence shells that prevent conduction.

9. Semiconductors, having four electrons in the valence shell, are partly conductors and partly insulators (e.g., silicon and germanium).

10. Semiconductor wafers are formed through covalent bonding (sharing valence electrons).

11. Doping is used to create an excess of electrons or holes in a normally intrinsic (pure) semiconductor wafer.

12. N-type materials are formed by doping a semiconductor wafer with pentavalent (five valence electrons) atoms, thus creating an excess of electrons free for current flow.

13. In N-type materials, electrons are the majority carriers and holes are the minority carriers.

14. P-type materials are formed by doping a semiconductor wafer with trivalent (five valence electrons) atoms, thus creating an excess of holes.

15. In P-type materials, holes are the majority carriers and electrons are the minority carriers.

16. The PN junction is formed at the point where the P-type material and N-type material of the wafer meet.

17. The region surrounding the PN junction is depleted of current carriers by the combination of electrons and holes.

18. The combination continues until a difference in potential of 0.7 volt for silicon or 0.3 volt for germanium is reached.

19. Forward biasing a PN junction increases current flow, whereas reverse biasing a PN junction decreases current flow.

20. The relationship of current and voltage can be viewed on an *I–V* characteristic curve.

GLOSSARY

Conductor: A material having many free electrons and a low resistivity.

Covalent Bonding: The sharing of valence electrons among atoms.

Doping: The process of adding impurities to a semiconductor.

Electron: A negatively charged particle of an atom.

First Law of Electrostatics: Like charges repel and unlike charges attract.

Free Electron: An electron that has left its valence shell.

Hole: Positively charged, yet has no mass.

Insulator: A material having very few free electrons and a high resistivity.

Intrinsic: Having no impurities.

Ion: An atom that has either gained or lost an electron.

Neutron: A particle having no charge that is found in the nucleus of an atom.

Proton: A positively charged particle found in the nucleus of an atom.

Semiconductor: A material having a conductivity and resistivity between that of a conductor and an insulator.

Valence Electron: An electron occupying the outermost shell of an atom.

Valence Shell: The outermost shell of an atom.

PROBLEMS

SECTION 1-1

1-1. Draw the atomic model for a copper atom.

1-2. Label the shells, electrons, nucleus, and protons of the copper atom.

1-3. Explain the importance of the valence shell.

1-4. What is the difference between the electron, proton, and neutron?

1-5. How is the net charge of an atom determined?

SECTION 1-2

1-6. What is the definition of a conductor?

1-7. What is the significance of the atomic number?

1-8. List the factors that affect a valence electron's ability to break free of its orbit.

1-9. What are the characteristics of a good conductor?

1-10. How does conduction occur?

1-11. Explain the concept of the excited electron.

1-12. What is Coulomb's first law of electrostatics.

1-13. What is a hole and how is it formed?

SECTION 1-3

1-14. Explain the concept of insulation.

1-15. Describe the relationship between the valence shell and an insulator.

1-16. How does an insulator differ from a conductor?

1-17. What is the forbidden band?

1-18. What is the definition of a semiconductor?

1-19. Name two common semiconductor materials and list their differences.

1-20. What are the two types of bonding and how do they differ?

1-21. What is meant by the term intrinsic?

SECTION 1-5

1-22. Define doping.

1-23. How does doping occur?

SECTION 1-6

1-24. What is an N-type material?

1-25. Define pentavalent.

1-26. Explain the relationship between a hole and a covalent bond.

1-27. What are the majority and minority current carriers of N-type materials?

SECTION 1-7

1-28. What is a P-type material?

1-29. Define trivalent.

1-30. Give some examples of trivalents.

1-31. What are the majority and minority current carriers of P-type materials?

SECTION 1-8

1-32. Define a PN junction.

1-33. Explain how ions are formed.

1-34. What is meant by the term depletion region?

1-35. What is the difference in potential between germanium and silicon?

SECTION 1-9

1-36. Define biasing.

1-37. What are the differences between forward biasing and reverse biasing?

1-38. Describe how forward and reverse biasing affect the PN junction.

1-39. Explain how current flows when the junction is forward biased.

1-40. Explain why current does not flow when the junction is reverse biased.

SECTION 1-10

1-41. What does an $I-V$ curve demonstrate?

1-42. Define peak inverse voltage.

1-43. Describe how the $I-V$ curve is read.

ANSWERS TO REVIEW QUESTIONS

1. resistance	**2.** conductor, insulator	**3.** vacuum tubes
4. shell	**5.** nucleus	**6.** protons, neutrons
7. 14 protons	**8.** negatively	**9.** positively
10. neutral	**11.** valence	**12.** valence
13. eight	**14.** conductor	**15.** atomic number

16. one	**17.** few	**18.** greater
19. electrostatic	**20.** copper, silver, gold	**21.** potential
22. Kinetic	**23.** free electron	**24.** hole
25. positive	**26.** positive, negative	**27.** insulator
28. full	**29.** large	**30.** forbidden
31. germanium, silicon	**32.** resistance	**33.** four
34. wafer	**35.** bonding	**36.** negative ion
37. positive ion	**38.** covalent bonding	**39.** intrinsic
40. doping	**41.** holes, electrons	**42.** pentavalent
43. free	**44.** electrons	**45.** holes
46. bipolar	**47.** trivalent	**48.** holes
49. holes	**50.** electrons	**51.** junction
52. ion	**53.** depletion	**54.** 0.7 volt
55. 0.3 volt	**56.** energy	**57.** electrode
58. biasing	**59.** N, P	**60.** reverse
61. narrower	**62.** barrier	**63.** Reverse
64. repel, attract	**65.** expands	**66.** 0.7
67. peak inverse voltage		

2

PN Junction Diode

When you have completed this chapter, you should be able to:

- Analyze the process of rectification, including transformer action, filter capacitors, and rectifier diodes.
- Discern the differences between the three types of rectifiers.
- Investigate other applications for the diode such as clippers and clampers.
- Analyze the effect of a faulty diode on circuit operation.

INTRODUCTION

In the previous chapter, the term "PN junction device" was repeatedly used. This is a very broad description. Now we give it the name "diode," which is more specific.

The diode is one of the most common devices in electronics. It comes in many packages and provides different functions. You will eventually learn about light-emitting diodes, zener diodes, and Shockley diodes. But to begin, we discuss the rectifier diode.

2-1 BIASING THE DIODE

Everything discussed in Section 1-9, "Biasing the PN Junction," applies here. We now refer to the PN junction device as a diode and use the schematic symbol instead of the block diagrams of the previous chapter.

Figure 2-1 compares the block diagram and schematic symbol. Notice the names "anode" and "cathode" are used to label the P-type and N-type materials. To forward bias a diode, as shown in Fig. 2-2, the anode (P-type) is connected to the positive terminal of the power supply and the cathode (N-type) to the negative terminal.

If the power supply of Fig. 2-2 were turned on, the action in the circuit would begin with electron current flowing from the negative terminal to the positive terminal. In order for current to flow, however, the **barrier voltage** within the diode must be overcome. Remember that overcoming this barrier requires approximately 0.7 volt for a **silicon diode.** Therefore, approximately 0.7 volt is dropped across the diode and the remaining 9.3 volts is dropped across the resistor. The resistor is the current-limiting device. If it were not there, the diode would attempt to conduct a huge current and be destroyed.

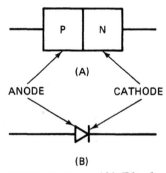

FIGURE 2-1 (A) Block diagram of a diode. (B) Schematic symbol of a diode.

Barrier Voltage: The voltage that must be overcome for a diode to conduct.

Silicon Diode: A semiconductor whose barrier voltage is approximately 0.7 V.

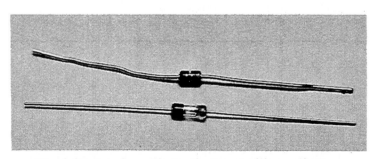

(Top) Germanium diode; (bottom) silicon diode.

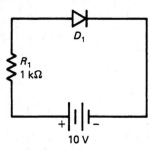

FIGURE 2-2 Forward-biased diode.

Taking the circuit of Fig. 2-2 one step further, we find the actual value of current flowing in this circuit through the use of Ohm's law. If 9.3 volts are dropped across R_1, which is a 1-kΩ resistor, then:

$$I_{R1} = \frac{9.3 \text{ V}}{1 \text{ k}\Omega}$$

$$= 9.3 \text{ mA}$$

This also happens to be the value of the total current since it is a series circuit. So the current through D_1 is 9.3 mA.

What about the power dissipated by each of the components? Calculate these values by applying Watt's law:

$$P = I \times E$$

$$P_{R1} = 9.3 \text{ mA} \times 9.3 \text{ V}$$

$$= 86.5 \text{ mW}$$

and

$$P_{D1} = 9.3 \text{ mA} \times 0.7 \text{ V}$$

$$= 6.51 \text{ mW}$$

These calculations are made with several assumptions. The first is that we are using a silicon diode. This is why the value of 0.7 volt is used. You will find with experience that the voltage of the silicon diode can fall anywhere from 0.5 to 1.0 volts, depending on the current. Also, under lab conditions, the tolerance of resistors and meters must be kept in mind. All of this is mentioned to alleviate some of the questions that may arise when the calculated values do not exactly match the measured values.

Now let us try solving for the current of Fig. 2-3 using a **germanium diode**. To begin, recall that a germanium diode requires approximately 0.3 volt to overcome its potential barrier. With this in mind and the fact that the diode is forward biased, we already have one voltage drop solved. This leaves 4.7 volts to be divided across the resistors. Begin by solving for total current:

Germanium Diode: A semiconductor whose barrier voltage is approximately 0.3 V.

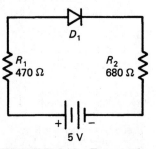

FIGURE 2-3 Forward-biased germanium diode.

$$I_T = \frac{4.7 \text{ V}}{470 \text{ }\Omega + 680 \text{ }\Omega}$$

$$= 4.09 \text{ mA}$$

This is the current that flows through each component.

$$V_{R1} = 4.09 \text{ mA} \times 470 \text{ }\Omega$$

$$= 1.92 \text{ V}$$

and

$$V_{R2} = 4.09 \text{ mA} \times 680 \text{ }\Omega$$

$$= 2.78 \text{ V}$$

Adding V_{R1} and V_{R2} equals 4.7 volts, leaving the remaining 0.3 volt to drop across the diode. Finally, the power of each component:

$$P_{R1} = 4.09 \text{ mA} \times 1.92 \text{ V}$$
$$= 7.85 \text{ mW}$$

and

$$P_{R2} = 4.09 \text{ mA} \times 2.78 \text{ V}$$
$$= 11.37 \text{ mW}$$

and

$$P_{D1} = 4.09 \text{ mA} \times 0.3 \text{ V}$$
$$= 1.23 \text{ mW}$$

ALL DIODES FROM THIS POINT ON ARE SILICON UNLESS OTHERWISE NOTED!

Turning our attention to Fig. 2-4, an example of a reversed-biased diode is shown. In this circuit, the anode (P-type) is connected to the negative terminal and the cathode (N-type) is connected to the positive terminal. Recall how the depletion region is made wider (Section 1-9) under these circumstances. Also, recall that almost no current flows, and for all intent and purposes, the diode acts like an open circuit.

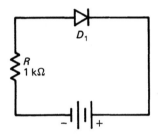

FIGURE 2-4 Reverse-biased diode.

Solving for the values of this circuit is a simple matter. The diode drops the source voltage, which in this case is 10 volts. That leaves 0 volts across the resistor, which means there is no current flowing:

$$I_T = \frac{0 \text{ V}}{1 \text{ k}\Omega}$$
$$= 0 \text{ A}$$

If the power supply of Fig. 2-4 were replaced with a 20-volt supply, then the reverse-biased diode would drop 20 volts and again no current would flow. The conclusion is that a diode does not conduct in reverse bias.

EXAMPLE 2-1

A silicon diode is used in Fig. 2-2 and R_1 is changed to a 3.3-kΩ resistor. Find the value of (a) V_{R1} and (b) P_{D1}.

Solution:

(a) A forward-biased silicon diode drops 0.7 V; therefore, the balance of the supply voltage is seen across the resistor.

$$V_{R1} = 10 \text{ V} - 0.7 \text{ V}$$
$$= 9.3 \text{ V}$$

(b) Finding the power dissipated by the diode requires that we find both its voltage drop and current. The voltage drop of a forward-biased silicon diode is 0.7 V. Current through the diode is determined by the value of the series resistor.

$$I_T = \frac{9.3 \text{ V}}{3.3\text{-k}\Omega}$$

$$= 2.82 \text{ mA}$$

and

$$P_{D1} = 2.82 \text{ mA} \times 0.7 \text{ V}$$

$$= 1.97 \text{ mW}$$ ∎

REVIEW QUESTIONS

1. A _diode_ is a PN junction device.
2. The P-type material of a diode is called the _Anode_.
3. The N-type material of a diode is called the _Cathode_.
4. Current does not flow through a diode until the _barrier_ voltage is overcome.
5. Resistors are used as _Current_-limiting devices.
6. A germanium diode requires _.3_ volts to overcome its potential barrier.
7. A reverse-biased diode acts like a(n) _inductor_.
8. A diode conducts when it is _Forward_ biased.

2-2 DIODE CHARACTERISTICS

You have already encountered several characteristics of the diode during our dc analysis. We will now systematically consider some of the more important points.

These characteristics apply to general-purpose diodes. Table 2-1 illustrates a number of these characteristics.

Conduction: A diode conducts when it is forward biased and receives the necessary voltage to overcome its potential-barrier voltage (0.7 or 0.3 V). It does not conduct in reverse bias.

Maximum Forward Current (I_{FM}): Approximately 1.0 A. This is the average direct current with reference to a forward-biased diode in a half-wave rectifier. A constant dc current at this level may damage the diode if the device's maximum temperature rating is exceeded.

Maximum Forward Voltage (V_{FM}): Approximately 1 V. As you know, the proper voltage drop across a diode is approximately 0.7 V. Forcing the diode much further than this will damage it.

Repetitive Peak Reverse Voltage (V_{RRM}): Ranges from 50 to several hundred volts. This refers to an ac application wherein the diode is forward and reverse biased alternately.

Maximum Repetitive Peak Reverse Current (I_{RRM}): Reverse current that flows through the diode in an ac application.

Reverse Recovery Time (t_{rr}): The time it takes for the diode to switch from a forward-bias to reverse-bias condition. It is measured in nanoseconds (ns).

TABLE 2-1 Diode Characteristics

| Diode Products by Ascending V_{RRM} and t_{rr} (Continued) | | | | | | |

Part No.	V_{RRM} (V)	I_{RRM} (nA)	V_{FM} @ I_F (V)	(mA)	t_{rr} (ns) Max	Package
1N659	60	5000	1.0	6.0		DO-35
BAV18	60	100	1.0	100	50	DO-35
FDLL659	60	5000	1.0	6.0		LL-34
FDH600	65	100	1.0	200	4.0	DO-35
BAV70	70	5000	1.1	50	6.0	TO-236
BAV99	70	2500	1.1	50	6.0	TO-236
BAW56	70	2500	1.1	50	6.0	TO-236
1N457	70	25	1.0	100		DO-35
1N457A	70	25	1.0	100		DO-35
1N462A	70	500	1.0	100		DO-35
FDLL457	70	25	1.0	20		LL-34
FDLL457A	70	25	1.0	100		LL-34
FDLL462A	70	500	1.0	100		LL-34
1N4153	75	50	0.66	20	2.0	DO-35
1N4151	75	50	1.0	50	2.0	DO-35
1N4305	75	100	0.85	10	2.0	DO-35
FDLL4153	75	50	0.88	20	2.0	LL-34
FDLL4151	75	50	1.0	50	2.0	LL-34
FDLL4305	75	100	0.85	10	2.0	LL-34
1N3600	75	100	1.0	200	4.0	DO-35
1N3064	75	100	1.0	10	4.0	DO-35
1N4150	75	100	1.0	200	4.0	DO-35
1N4454	75	100	1.0	10	4.0	DO-35
FDLL3600	75	100	1.0	200	4.0	LL-34
FDLL600	75	100	1.0	200	4.0	LL-34
FDLL3604	75	100	1.0	10	4.0	LL-34
FDLL4150	75	100	1.0	200	4.0	LL-34
FDLL4454	75	100	1.0	10	4.0	LL-34
BAS16	75	1000	1.1	50	6.0	TO-236
FDH1000	75	50	1.0	500	100	DO-35
BA128	75	100	1.0	50		DO-35
BAW76	75	100	1.0	100	2.0	DO-35
1N5194	80	25	1.0	100		DO-35
1N5282	80	100	1.3	500	2.0	DO-35
1N483B	80	25	1.0	100		DO-35
FDLL483B	80	25	1.0	100		LL-34
1N914	100	25	1.0	10	4.0	DO-35
1N914A	100	25	1.0	20	4.0	DO-35
1N914B	100	25	1.0	100	4.0	DO-35
1N916	100	25	1.0	10	4.0	DO-35
1N916A	100	25	1.0	20	4.0	DO-35
1N916B	100	25	1.0	30	4.0	DO-35
1N4148	100	25	1.0	10	4.0	DO-35
1N4149	100	25	1.0	10	4.0	DO-35
1N4446	100	25	1.0	20	4.0	DO-35
1N4447	100	25	1.0	20	4.0	DO-35
1N4448	100	25	1.0	100	4.0	DO-35
1N4449	100	25	1.0	30	4.0	DO-35
FDLL914	100	25	1.0	10	4.0	LL-34
FDLL914A	100	25	1.0	20	4.0	LL-34
FDLL914B	100	25	1.0	100	4.0	LL-34
FDLL916	100	25	1.0	10	4.0	LL-34
FDLL916B	100	25	1.0	30	4.0	LL-34
FDLL4148	100	25	1.0	10	4.0	LL-34
FDLL4149	100	25	1.0	10	4.0	LL-34
FDLL4446	100	25	1.0	20	4.0	LL-34
FDLL4447	100	25	1.0	20	4.0	LL-34

Resistance: One of the unique characteristics of a semiconductor device is its ability to change resistance. This can be demonstrated mathematically as well as through circuit operation. We use the circuit of Fig. 2-5 to prove this point.

If we begin by adjusting the power supply to 5 V and calculate the current, we would find 4.3 mA of total current. Next, calculate the resistance of the diode at this value of source voltage:

$$R_{D1} = \frac{0.7 \text{ V}}{4.3 \text{ mA}}$$

$$= 163 \ \Omega$$

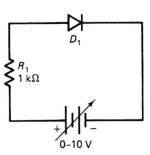

FIGURE 2-5 Circuit with variable power supply.

Using the same circuit with the power supply increased to 10 V gives a new current of 9.3 mA and, therefore, a new diode resistance:

$$R_{D1} = \frac{0.7 \text{ V}}{9.3 \text{ mA}}$$

$$= 75.3 \ \Omega$$

Stop and compare these values, taking careful note that the same diode changed its internal resistance according to the demands of the circuit. The only diode value that remained constant was the voltage across the diode.

REVIEW QUESTIONS

9. A silicon diode requires _____ volt and a germanium diode requires _____ volt for normal conduction.

10. A diode does not conduct in reverse bias. True or false?

11. The _____ _____ _____ rating reveals the maximum voltage a reverse-biased diode can safely handle.

12. A forward-biased diode maintains a constant internal resistance regardless of the level forward current. True or false?

2-3   HALF-WAVE RECTIFIER (UNFILTERED)

Rectify: To change ac to dc.

Half-Wave: Conducts on only one alternation.

Rectification refers to the process of changing ac to dc. AC voltage, by definition, continuously changes voltage and periodically reverses polarity. This means that the diode is forward biased on one alternation and reverse biased on the next. It passes only half of the input waveform **(half-waves).** Remember that the diode only conducts when forward biased. The circuit of Fig. 2-6 shows the direction of electron-current flow and the alternation that caused it. On the next alternation, no current flows through the resistor or diode. This is because the diode is reverse biased.

The ac voltage at the secondary of Fig. 2-6 is periodically changing its polarity. Therefore, the voltage across the diode looks similar to the waveform of Fig. 2-7. The alternation when the diode is reverse biased results in its acting like an open and it drops the total voltage. When forward biased, it drops 0.7 volt, the voltage necessary for it to conduct. The remainder of this alternation's voltage is measured across the resistor.

Consider the circuit of Fig. 2-8 with its values and attempt to calculate some voltages and the current. To begin, we need to find the voltage at the secondary, so we use the turns ratio shown in the schematic. The turns ratio of 10 : 1 indicates that the primary winding has 10 times more turns or coil windings than the second-

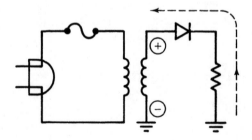

FIGURE 2-6   Half-wave rectifier.

CHAP. 2 / PN JUNCTION DIODE

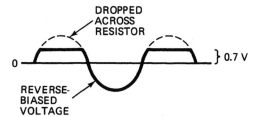

FIGURE 2-7 Voltage across the diode.

ary. The ac voltage is directly related to the turns ratio. The voltage at the primary is 10 times larger than the voltage at the secondary:

$$V_s = \frac{N_s}{N_p} V_{in}$$

where V_s = voltage at the secondary

N_s = number of turns in the secondary

N_p = number of turns in the primary

V_{in} = voltage input to the primary

This gives us

$$V_s = \frac{1}{10} 120 \text{ V}$$

$$= 12 \text{ V}$$

The value of 12 V is a root-mean-square (rms) value and needs to be converted to peak voltage. We can achieve that by

$$V_{s\,pk.} = \frac{12 \text{ V}}{0.707}$$

$$= 17 \text{ V}$$

At this point, it can be predicted that the anode of the diode of Fig. 2-8 alternately receives a negative and positive peak voltage of 17 V. Only a positive voltage on the anode will forward bias it. Finally, you know that the forward-biased diode drops 0.7 V. Putting it all together allows you to calculate the voltage across the load resistor:

$$V_{R1} = V_{s\,pk.} - V_{D1}$$

$$= 17 \text{ V} - 0.7 \text{ V}$$

$$= 16.3 \text{ V pk.}$$

The resistor's voltage is 16.3 V peak.

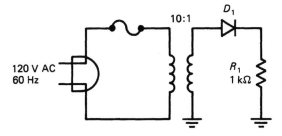

FIGURE 2-8 Half-wave rectifier.

FIGURE 2-9 Resistor wave-form.

This voltage is not the true dc voltage. The actual dc voltage would be an average of the voltages across the resistor. Figure 2-9 shows the actual waveform across the resistor. To find the overall dc value of this waveform, we use the formula

$$V_{dc\,out} = V_{R1\,pk.} \times 0.318$$

Therefore,

$$V_{dc\,out} = 16.3\ V \times 0.318$$
$$= 5.18\ V$$

So the dc output voltage of Fig. 2-8 is 5.18 V. This means that the dc current flowing through the resistor is 5.18 mA.

The reason that 0.318 is used in the calculation of dc out is because we are trying to find the overall voltage of the area within each alternation. The value 0.318 is the reciprocal of pi. An alternate method of finding dc out is to divide $V_{R1\,pk.}$ by pi.

The problem with this circuit is that the output voltage is not very steady. It is actually a pulsating dc voltage, which does not make it too useful. This is a problem with "unfiltered circuits." You will now learn how a capacitor can be used to "smooth," or "filter," the wave across the resistor, causing it to look more like dc.

EXAMPLE 2-2

The input voltage of the circuit of Fig. 2-8 is adjusted to 25 VAC. What is the value of the dc output voltage?

Solution: Begin by finding the voltage at the secondary:

$$V_s = \frac{1}{10}\,25\ V$$
$$= 2.5\ V$$

so that

$$V_{s\,pk.} = \frac{2.5\ V}{0.707}$$
$$= 3.54\ V$$

The forward-biased diode drops 0.7 V.

$$V_{R1} = 3.54\ V - 0.7\ V$$
$$= 2.84\ V$$

The direction of the diode determines the polarity of the output voltage. In this case, the output voltage is positive. For a half-wave rectifier, the dc output voltage is found by

$$V_{dc\,out} = 2.84\ V \times 0.318$$
$$= 903\ mV$$

Practice Problem: If the input voltage of Fig. 2-8 were changed to 40 V and an 8:1 transformer were used, what would be the dc output voltage? *Answer:* 2.03 V.

REVIEW QUESTIONS

13. <u>Rectifier</u> refers to the process of changing ac to dc.

14. A diode remains forward biased on both alternations of an ac signal. True or false?

15. A diode acts like an open when it is <u>Reve</u> biased.

16. A diode requires a (positive or negative) voltage on the anode to forward bias it.

17. The voltage at the secondary of a transformer is (ac or dc) voltage.

18. The output from Fig. 2-8 is (filtered or unfiltered).

2-4 FILTERED HALF-WAVE RECTIFIER

Figure 2-10 is a filtered half-wave rectifier. Notice that the capacitor is in parallel with the output resistor. Remember that a characteristic of capacitors is that they oppose changes in voltage. With this in mind, consider the operation of the circuit.

The procedure for finding the peak voltage at the secondary is the same as with the unfiltered circuit. The voltage at the secondary is 17 V. Also, the peak voltage across the resistor remains 16.3 V.

Figure 2-11 shows the main difference between this and the previous circuit. As we examine the waveform, we see that the capacitor's ability to maintain a charge allows it to sustain the voltage from one peak to the next. The end result is an almost constant dc level. We say *almost* because there is actually a small amount of "ripple." It is this ripple voltage that we investigate at this point.

The dc output voltage is almost a constant 16.3 V. The slight variation in this voltage is taken into consideration by finding the ripple voltage. The formula for this is

$$V_{rip(p-p)} = \frac{I}{fC}$$

Current

―――――――――

Freq × Capacitance

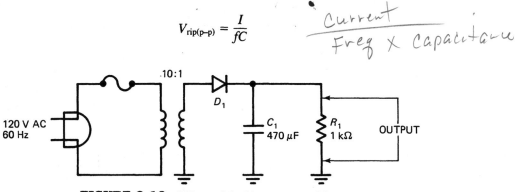

FIGURE 2-10 Filtered half-wave rectifier.

16.3 V

FIGURE 2-11 Filtered output waveform.

where $V_{rip(p-p)}$ = peak-to-peak voltage ripple

I = peak resistor current

f = ripple frequency

C = capacitance

The peak resistor current can be found by using the values of 16.3 V and 1 kΩ. The ripple frequency in the half-wave rectifier is equal to the input frequency. And the capacitance equals the value of the filter capacitor. Therefore,

$$V_{rip(p-p)} = \frac{16.3 \text{ mA}}{60 \text{ Hz} \times 470 \text{ } \mu\text{F}}$$

$$= 578 \text{ mV}$$

Figure 2-12 shows a magnified view of the ripple waveform. This illustration shows the 578-mV p–p signal. Considering this and Fig. 2-11, we can calculate that the dc voltage actually fluctuates between 16.3 and 15.7 V. The latter is found by simply subtracting the 578 mV from the 16.3 V.

The average value of the dc voltage output is found by subtracting the peak value of the ripple voltage from the peak output voltage:

$$V_{dc \text{ out}} = V_{R1 \text{ pk.}} - V_{rip \text{ pk.}}$$

$$= 16.3 \text{ V} - 289 \text{ mV}$$

Therefore,

$$V_{dc \text{ out}} = 16 \text{ V}$$

There is not a great difference between the peak voltage across the resistor and the final calculated dc voltage output. This is the whole purpose of **filtering.** The ideal is to have absolutely no ripple voltage.

The size of the filter capacitor has a direct affect upon the amount of ripple voltage. The larger the capacitance, the smaller the ripple voltage because the *RC* time constant is increased and, therefore, the capacitor does not discharge as quickly. Also, frequency affects the ripple voltage. The higher the frequency, the smaller the ripple voltage because the higher (quicker) frequency does not allow the capacitor time to discharge. And, finally, the load current affects the ripple voltage. The lower the load current, the smaller the ripple voltage because the capacitor does not discharge as much current. All of these relationships can be examined by substituting values into the formula for ripple voltage and then calculating its effect on the dc output.

The diode in the circuit of Fig. 2-13 must be able to withstand a peak reverse voltage, PRV (or peak inverse voltage, PIV), of at least 33.3 V. When the negative peak is applied to the anode in the circuit (reverse biasing it), a positive potential of approximately 16.3 V still exists at the cathode. The difference in potential between these two is 33.3 V.

The PRV is twice the secondary voltage. This is a disadvantage since it places a greater strain on the diode. Another disadvantage of this circuit is that it

16.3 V → 578 mV

FIGURE 2-12 Magnified view of the ripple waveform.

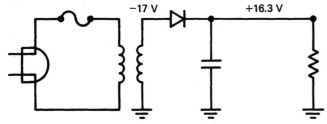

FIGURE 2-13 The PRV of a diode is almost double the secondary voltage.

is difficult to filter. Its only advantage is simplicity of design in that it uses one diode.

EXAMPLE 2-3

Determine the output of Fig. 2-10 if the input voltage were changed to 35 VAC.

Solution: The rms voltage at the secondary is equal to 3.5 V. The peak value of this voltage is 4.95 V. This produces a peak current of 4.25 mA through the resistor. The ripple voltage for this circuit can now be found.

$$V_{rip(p-p)} = \frac{4.25 \text{ mA}}{60 \text{ Hz} \times 470 \text{ } \mu\text{F}}$$
$$= 151 \text{ mV}$$

so

$$V_{rip\ pk.} = 75.5 \text{ mV}$$

Therefore,

$$V_{dc\ out} = 4.25 \text{ V} - 75.5 \text{ mV}$$
$$= 4.17 \text{ V} \qquad \blacksquare$$

Practice Problem: What would be the dc output voltage if the filter capacitor of Fig. 2-10 were changed to a value of 100 μF? *Answer:* 14.9 V.

REVIEW QUESTIONS

19. Capacitors oppose changes in _____ .
20. The small fluctuation in dc output is called _____ voltage.
21. Ripple frequency in a half-wave rectifier is always equal to the primary input frequency. True or false?
22. A filter capacitor (increases or decreases) ripple voltage.
23. The larger the capacitance, the larger the ripple voltage. True or false?
24. A filter capacitor _____ the PRV of a diode in a half wave-rectifier.

2-5 UNFILTERED FULL-WAVE RECTIFIER

In order to overcome the difficulty in filtering a half-wave rectifier, a full-wave rectifier can be used. This circuit requires two changes in the configuration of the last circuit. The first is the use of another diode and the second is the requirement

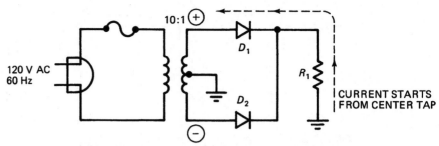

FIGURE 2-14 Full-wave rectifier (unfiltered).

of a center-tapped transformer. These changes are combined to form the circuit of Figure 2-14.

Using Fig. 2-14 as your guide, follow the path of current during the displayed alternation. It should be obvious that D_1 is conducting because it is forward biased and D_2 is not conducting because it is not forward biased. As the polarity at the secondary switches on the next alternation, D_2 conducts. Both diodes together pass the **full-wave.** Keep in mind that current starts at the center tap and always flows upward through the load resistor.

If you were to place an oscilloscope across the resistor, the waveform of Fig. 2-15 would appear. Now comparing this with the waveform of the half-wave shown in Fig. 2-9, we see the vast difference. We see that this waveform is easier to filter because there is less open area between peaks. But let us continue our examination of the unfiltered circuit.

For the circuit of Fig. 2-16, begin by calculating the voltage at the secondary:

$$V_s = \frac{1}{10} (120 \text{ V})$$

$$= 12 \text{ V}$$

Now because the secondary has a center tap, the 12 V are split in half, resulting in 6 V from center tap to either end. The peak value of this center-tapped voltage is

$$V_s \text{ (center tap)} = \frac{6 \text{ V}}{0.707}$$

$$= 8.49 \text{ V pk.}$$

Earlier it was stated that only one diode at a time conducts. Each diode receives alternately +8.49 V on its anode, causing it to conduct. When the diode conducts, it drops 0.7 V, allowing the remaining 7.79 V pk. to drop across the load resistor.

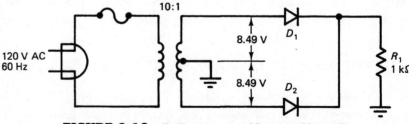

FIGURE 2-16 Full-wave rectifier (unfiltered).

Full Wave: Conducts on both alternations.

FIGURE 2-15 Resistor's waveform.

This brings us to the point of determining the actual dc voltage output across the load resistor. The formula for determining this is

$$V_{dc\ out} = V_{R1\ pk.} \times 0.636$$
$$= 7.79\ V \times 0.636$$
$$= 4.95\ V$$

Once again this formula takes into consideration the unfiltered output by taking an average of the overall voltage. Observe that this voltage is considerably less than the dc output of the unfiltered half-wave, because the center-tapped transformer cuts the secondary voltage in half. This is considered a disadvantage of the full-wave rectifier because of the greater expense of center-tapped trans-formers. The advantage is that the full-wave rectifier is easier to filter.

EXAMPLE 2-4

The transformer of Fig. 2-16 is substituted by one that has a 8 : 1 turns ratio. Determine the dc output voltage.

Solution: The change in the transformer turns ratio means a change in the voltage at the secondary.

$$V_s = \frac{1}{8}\ 120\ V$$
$$= 15\ V$$

The center-tap transformer splits this rms voltage.

$$V_s\ (center\ tap) = 7.5\ V$$

and

$$V_{s\ pk.}\ (center\ tap) = \frac{7.5\ V}{0.707}$$
$$= 10.6\ V\ pk.$$

Since only one diode conducts on each alternation, we subtract the voltage drop of one diode. This leaves us with the peak voltage at the load.

$$V_{R1} = 10.6\ V\ pk. - 0.7\ V$$
$$= 9.9\ V\ pk.$$

Finally, the dc output voltage is the average of the full-wave output signal.

$$V_{dc\ out} = 9.9\ V\ pk. \times 0.636$$
$$= 6.3\ V \qquad \blacksquare$$

Practice Problem: What would be the dc output voltage in the circuit of Fig. 2-16 if the input voltage were 10 V and the transformer had a turns ratio of 1 : 4? *Answer:* 17.5 V.

REVIEW QUESTIONS

25. Both diodes conduct on the same alternation for a full-wave rectifier. True or false?

26. A _____ transformer is necessary for the construction of a full-wave rectifier.

27. The output waveform of both a half-wave and a full-wave rectifier look the same. True or false?

2-6 FILTERED FULL-WAVE RECTIFIER

Filtering involves the addition of a filter capacitor to the circuit. The schematic of Fig. 2-17 includes this change along with a reversal of the diodes. By turning the diodes around, the polarity of the output voltage becomes negative. The operation of the circuit is essentially the same except that current flows downward through the load resistance toward ground.

The voltage at the secondary is 12 V, which is split in half due to the center tap. The peak value from center tap to either end is 6 V/0.707 = 8.49 V. The diodes each continue to drop 0.7 V, so the peak voltage across the load resistor is −7.79 V. This is the same value as calculated in the previous circuit. The only difference is that the polarity has changed.

A filter capacitor has the effect of increasing the dc output voltage. This is due to its ability to maintain a charge. The waveform of Fig. 2-18 illustrates this point. By comparing this waveform to the waveform of Fig. 2-11, we see that it is much easier to filter the full-wave rectifier due to the shorter time intervals between peaks.

To calculate the dc output of Fig. 2-17, it is necessary to calculate the small ripple voltage that remains after filtering. This is accomplished by the same formula:

$$V_{\text{rip(p–p)}} = \frac{I}{fC}$$

where $I = -7.79 \text{ V}/1 \text{ k}\Omega = -7.79 \text{ mA}$

f = twice the input frequency = 120 Hz

C = value of the capacitor = 470 μF

Therefore,

$$V_{\text{rip(p–p)}} = \frac{-7.79 \text{ mA}}{120 \text{ Hz} \times 470 \text{ } \mu\text{F}}$$

$$= -138 \text{ mV}$$

As we can see, the value of the ripple voltage is quite low. It should be emphasized that with a full-wave rectifier, we always double the input frequency in order to get the ripple frequency because both alternations are passed. The

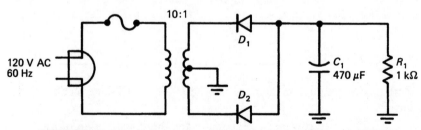

FIGURE 2-17 Full-wave rectifier with negative output.

0 ── EFFECT OF FILTER CAPACITOR **FIGURE 2-18** Resistor waveform.

ripple voltage in this example causes the dc output to fluctuate between -7.79 and -7.65 V. This means that the average dc output is

$$V_{\text{dc out}} = V_{R\,\text{pk.}} - V_{\text{rip pk.}}$$
$$= -7.79 \text{ V} - (-69 \text{ mV})$$
$$= -7.72 \text{ V}$$

The peak reverse voltage across the nonconducting diode is approximately the same as the secondary peak voltage (17 V). The nonconducting diode has $+8.49$ V at its cathode and -7.72 V at its anode. The difference between these two voltages is 16.21 V.

Before leaving this circuit, take note that in order to create a negative dc output, *both* the diodes and the capacitor are reversed. Forgetting to reverse the capacitor is dangerous and results in the destruction of the capacitor and other components.

EXAMPLE 2-5

The circuit of Fig. 2-17 is changed so that the transformer now has a 5:1 turns ratio. Determine the dc output voltage.

Solution: The voltage at the secondary with reference to the center tap is

$$V_s = \frac{1}{5}\,120$$
$$= 24 \text{ V}$$
$$V_s \text{ (center tap)} = 12 \text{ V}$$

and

$$V_{s\,\text{pk.}} \text{ (center tap)} = \frac{12 \text{ V}}{0.707}$$
$$= 17 \text{ V pk.}$$

The diodes drop 0.7 V on each alternation. The remainder of the peak voltage is passed on to the resistors. Polarity is determined by the direction of the diodes and capacitor.

$$V_{R1} = 17 \text{ V pk.} - 0.7 \text{ V}$$
$$= -16.3 \text{ V pk.}$$

Next, the ripple voltage is determined.

$$V_{\text{rip(p–p)}} = \frac{-16.3 \text{ V}/1 \text{ k}\Omega}{120 \text{ Hz} \times 470 \text{ }\mu\text{F}}$$
$$= -289 \text{ mV}$$

and

$$V_{\text{rip pk.}} = \frac{-289 \text{ mV}}{2}$$
$$= -145 \text{ mV}$$

Finally, the dc output is the difference between the peak voltage across the load resistor and the peak ripple voltage.

$$V_{dc\ out} = -16.3 \text{ V pk.} - (-145 \text{ mV pk.})$$
$$= -16.2 \text{ V} \qquad \blacksquare$$

Practice Problem: What would be the ripple voltage for the circuit of Fig. 2-17 if the input frequency were increased to 10 kHz?   *Answer:* $-829 \ \mu\text{V p-p.}$

REVIEW QUESTIONS

28. The polarity of the dc output of a rectifier can be changed by simply reversing the diodes and filter capacitor. True or false?

29. A full-wave rectifier is easier to filter than a half-wave rectifier. True or false?

30. The ripple frequency of a full-wave rectifier is equal to _____ the value of the primary input frequency.

31. The PRV of any diode in a full-wave rectifier is approximately (equal to or double) the secondary center-tapped voltage.

2-7   (UNFILTERED) FULL-WAVE BRIDGE RECTIFIER

The rectifier that we are most likely to encounter is called the full-wave bridge rectifier. In fact, it is so common that it comes in IC (integrated-circuit) form. The circuit of Fig. 2-19 illustrates the bridge rectifier (it is understood to be full wave). Notice the elimination of the center-tap transformer.

The operation of the circuit can be understood by examining one alternation. In Fig. 2-19, the alternation shown causes diodes D_1 and D_3 to conduct. When the next alternation occurs, diodes D_2 and D_4 conduct. This process continues with two diodes conducting on each alternation. The end result is that the full wave is passed on to the load.

The waveform across the load resistor is similar to that of Fig. 2-20. The only difference between this waveform and the waveform of Fig. 2-15 is that it is approximately twice the voltage. This is because no center tap is used here.

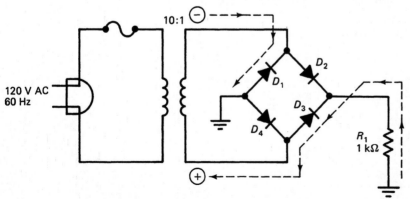

FIGURE 2-19 Unfiltered bridge rectifier.

FIGURE 2-20 Resistor waveform.

Full-wave bridge rectifiers in a single package.

Begin the calculations by determining the voltage at the secondary. This is accomplished by the same formula as that used earlier:

$$V_s = \frac{1}{10} \, 120 \text{ V}$$
$$= 12 \text{ V}$$
$$V_{s \text{ pk.}} = \frac{12 \text{ V}}{0.707} = 17 \text{ V}$$

Because two diodes conduct on every alternation, it is necessary to subtract 1.4 V from the peak secondary voltage. This includes 0.7 V for each diode. Therefore,

$$V_{R1 \text{ pk.}} = 17 \text{ V} - 1.4 \text{ V} = 15.6 \text{V}$$

Finally, to find the dc or average of this peak voltage, we use the formula

$$V_{\text{dc out}} = V_{R1 \text{ pk.}} \times 0.636$$
$$= 9.92 \text{ V}$$

That is all there is to determining the dc output voltage. This brings us to the filtered version of the bridge rectifier.

EXAMPLE 2—6

The transformer in Fig. 2-21 is replaced by one that has a 6 : 1 turns ratio. Determine the dc output voltage.

Solution: Begin by determining the peak voltage at the secondary.

$$V_s = \frac{1}{6} \, 120 \text{ V}$$
$$= 20 \text{ V}$$
$$V_{s \text{ pk.}} = \frac{20 \text{ V}}{0.707}$$
$$= 28.3 \text{ V pk.}$$

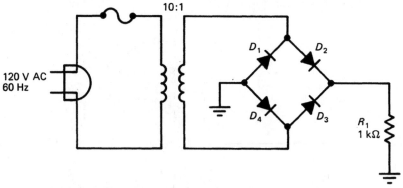

FIGURE 2-21 Unfiltered bridge rectifier.

Two diodes conduct on each alternation; therefore, it is necessary to subtract the value of both diode voltage drops.

$$V_{R1\,pk.} = 28.3 \text{ V pk.} - 1.4 \text{ V}$$
$$= 26.9 \text{ V pk.}$$

The dc average of the full-wave output is determined by

$$V_{dc\,out} = 26.9 \text{ V pk.} \times 0.636$$
$$= 17.1 \text{ V} \qquad \blacksquare$$

Practice Problem: How would you modify the circuit of Fig. 2-21 in order to obtain a negative dc output voltage? *Answer:* Reverse all the diodes.

REVIEW QUESTIONS

32. The full-wave bridge rectifier eliminates the need for a center-tapped transformer. True or false?

33. How many diodes conduct on each alternation for a full-wave bridge rectifier? _____

34. A full-wave bridge rectifier provides a larger dc output voltage than a full-wave rectifier using a center-tapped transformer with the same turns ratio. True or false?

2-8 FILTERED FULL-WAVE BRIDGE RECTIFIER

The previous examples have shown us that the addition of a filter capacitor causes an increase in the dc output voltage. The output also becomes more steady, or smooth. The end result is a much more useful voltage.

In Fig. 2-22, an example of a filtered bridge rectifier is shown. It is very important to realize how current flows in order to determine the correct way to connect a filter capacitor. In this circuit, the current is flowing from ground upward through the load resistor. This results in a positive output voltage with reference to ground.

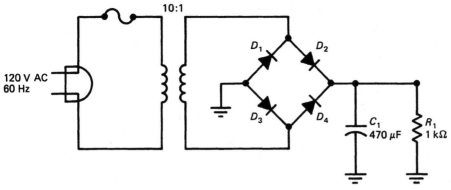

FIGURE 2-22 Filtered bridge rectifier.

Retrace the steps to find $V_{R1\,pk.}$:

$$V_s = \frac{1}{10}\,120\text{ V} = 12\text{ V}$$

$$V_{s\,pk.} = \frac{12\text{ V}}{0.707} = 17\text{ V}$$

$$V_{R1\,pk.} = 17\text{ V} - 1.4\text{ V} = 15.6\text{ V}$$

At this point, it is necessary to recall that a small ripple voltage exists and is determined by the formula:

$$V_{rip(p-p)} = \frac{I}{fC}$$

Therefore,

$$V_{rip(p-p)} = \frac{15.6\text{ mA}}{120\text{ Hz} \times 470\ \mu\text{F}}$$

$$= 277\text{ mV}$$

$$V_{rip\,pk.} = 138\text{ mV}$$

So the output voltage fluctuates between 15.6 and 15.3 V. The actual dc output is found by

$$V_{dc\,out} = V_{R1} - V_{rip\,pk.}$$

$$= 15.6\text{ V} - 138\text{ mV}$$

$$= 15.5\text{ V}$$

The peak reverse voltage that any one diode receives is approximately equal to the value of the secondary voltage. Let us use diode D_4 as an example. When it is reverse biased, the voltage at its cathode is $+15.5$ V and the voltage at its anode is -0.7 V (the drop across D_3). The difference of these two voltages results in a peak reverse voltage of 16.2 V.

EXAMPLE 2-7

The circuit of Fig. 2-22 is modified such that the transformer now has a 4 : 1 turns ratio. Determine the dc output voltage.

Solution: Start by determining the peak voltage at the secondary of the transformer.

$$V_s = \frac{1}{4} \, 120 \text{ V}$$

$$= 30 \text{ V}$$

and

$$V_{s \text{ pk.}} = \frac{30 \text{ V}}{0.707}$$

$$= 42.4 \text{ V pk.}$$

Since two diodes conduct on each alternation, the peak voltage at the output is equal to the peak voltage at the secondary minus the diode voltage drops.

$$V_{R1 \text{ pk.}} = 42.4 \text{ V pk.} - 1.4 \text{ V}$$

$$= 41 \text{ V pk.}$$

The filter capacitor attempts to maintain an output equal to the peak voltage. However, a small ripple voltage does exist.

$$V_{\text{rip(p--p)}} = \frac{41 \text{ V pk.}/1 \text{ k}\Omega}{120 \text{ Hz} \times 470 \text{ } \mu\text{F}}$$

$$= 727 \text{ mV p--p}$$

Therefore,

$$V_{\text{rip pk.}} = \frac{727 \text{ mV p--p}}{2}$$

$$= 364 \text{ mV pk.}$$

Finally, the dc output voltage is equal to the small difference between the peak voltage at the output and this peak ripple voltage.

$$V_{\text{dc out}} = 41 \text{ V pk.} - 364 \text{ mV pk.}$$

$$= 40.6 \text{ V} \qquad \blacksquare$$

Practice Problem: What would be the dc output in the circuit of Fig. 2-22 if the input voltage is 25 V and a 1 : 1 transformer is used? *Answer:* 33.7 V.

REVIEW QUESTIONS

35. _____ is used as a reference point for measuring voltages.
36. A full-wave bridge rectifier is easier to filter than a half-wave rectifier. True or false?
37. The PRV across any diode in a filtered full-wave bridge rectifier is approximately double the peak secondary voltage. True or false?

2-9 CLIPPERS AND CLAMPERS

Clippers Another use of diodes is for the purpose of "clipping" off a portion of a wave. The clipper looks very similar to the half-wave unfiltered rectifier. The difference is in the purpose or intention of the circuit. We are not concerned with

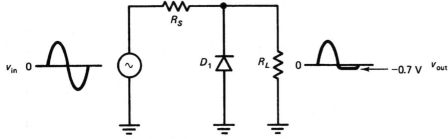

FIGURE 2-23 Negative clipper.

creating dc as much as altering the waveform. In other words, the goal is to clip off the negative alternation. This circuit is also called a limiter. The circuit shown in Fig. 2-23 is considered a negative limiter because it limits the negative alternation to −0.7 V (the diode's voltage).

The circuit of Fig. 2-23 conducts on one alternation only. The output is comprised of positive alternations and a small portion of the negative alternation. The negative alternations are clipped due to the forward-biased diode. Resistor R_s is a series current-limiting resistor. The purpose of the circuit is to clip off the negative peak.

The amount of the negative alternation to be clipped can be varied by the addition of a dc power supply, as shown in Fig. 2-24. Here a −5-V supply is used to increase the clipping level to −5.7 V. All voltages beyond −5.7 V are clipped.

EXAMPLE 2-8 _____

The input signal for Fig. 2-24 is 20 V p–p. Explain (a) the output signal and (b) how to create a positive clipper.

Solution:

(a) The full 10-V positive alternation is passed on to the load due to the reverse-biased diode. Anything below −5.7 V is clipped on the negative alternation. In this example, −4.3 V is clipped off the negative peak.

(b) A positive clipper can be created by reversing both the diode and the dc supply. The result is the clipping of all voltages above +5.7 V on the positive alternation. The full −10-V alternation is passed on to the load. ∎

Clampers   The clamper circuit (also called a dc restorer) is used to change the dc level of the output. The circuit of Fig. 2-25 shows the positive clamper changing the dc level from essentially zero to almost +10 V. This is done with the aid of the capacitor.

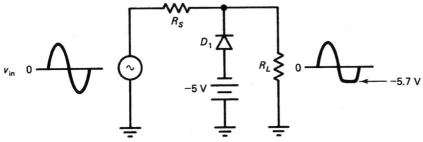

FIGURE 2-24 Negatively biased clipper.

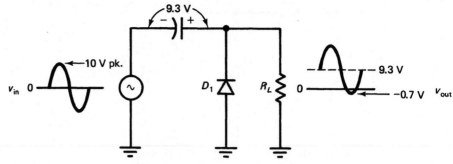

FIGURE 2-25 Positive clamper.

The operation of the clamper circuit can be understood by considering each of the two possible alternations. On the negative alternation, the diode is forward biased and acts as a closed switch. The diode's voltage at this time is -0.7 V with reference to ground. This means that the capacitor charges to the remaining -9.3 V of that same alternation.

When the positive alternation comes along, the capacitor maintains its charge because of the RC time constant involved. Keep in mind that the diode is reversed biased on this alternation and acts as an open switch. View the load resistor as being in parallel with the capacitor (9.3 V) and the ac supply (10 V pk.). This produces 19.3 V peak across the load resistor. On the next alternation, the diode conducts, causing -0.7 V to develop across the load resistor. The capacitor is providing a dc reference of 9.3 V on which a 20-V peak-to-peak ac signal is riding.

Proper design of the clamper circuit requires that the RC time constant be much longer than the period of the input frequency. Improper design causes a larger portion of the negative alternation to swing below the zero reference line. The RC time constant should be a minimum of 10 times larger than that period of the input frequency. This prevents any significant discharge of the capacitor through the resistor when the diode is not conducting.

EXAMPLE 2-9

Explain the output of the circuit of Fig. 2-25 if a 10-V p–p signal were applied to the input.

Solution: The capacitor charges to -4.3 V on the negative alternation. The positive alternation is aided by the voltage across the capacitor and produces a positive 9.3-V peak output voltage. The capacitor maintains its charge due to the large RC time constant. A series opposing condition exists when the negative alternation returns. The -4.3 V of the capacitor is subtracted from the -5 V pk. and the result is -0.7-V pk. at the output. ∎

REVIEW QUESTIONS

38. Another name for a clipper circuit is _____ .
39. A positive clipper clips the (positive or negative) alternation of the input signal.
40. It is the (forward- or reverse-) biased condition of the diode that aids in clipping.
41. How can the clipping level be adjusted?

42. Another name for a clamper circuit is _____ .

43. The capacitor in a clamper circuit ideally charges to the _____ voltage of the input signal.

44. The reason the capacitor maintains its charge is due to the _____ _____ _____ .

45. A negative clamper circuit causes the majority of the output signal to ride (above or below) the zero reference line.

2-10 TROUBLESHOOTING

The final consideration is troubleshooting a bridge rectifier. To begin, examine the method of using an ohmmeter to determine if a diode has been damaged. The majority of problems that we encounter are due to either an open or a short.

Ohmmeter Test of a Diode There are two ohmmeter measurements that are used to determine if a diode is good. The diode's "forward resistance" is measured by placing the common (black) lead of the ohmmeter on the diode's cathode and the positive (red) lead on the anode. The ohmmeter forward biases the diode and reduces the depletion region, which, in turn, reduces the diode's resistance. Figure 2-26 shows the connection of the ohmmeter. The range switch of the meter should be set to the diode range (if available) or to the highest range. This should result in some registered value on the meter. The specific value is unimportant.

By reversing the leads of the meter, we can measure the diode's "reverse resistance." This reading should be much larger than the forward resistance, possibly infinite. The key is to have a low to high ratio when comparing forward resistance to reverse resistance. This means that the diode passes the test and appears to be in good working order.

If the diode measures infinite in both directions, it has developed an open internally. On the other hand, if zero ohms are measured, the diode is shorted internally. A diode may be damaged by subjecting it to a current or voltage beyond its maximum rated values. Another cause of damage might be that the normal life expectancy has been reached. Under any of these circumstances, the diode should be replaced.

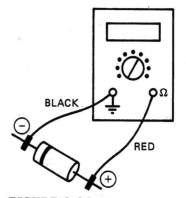

FIGURE 2-26 Measuring a diode's forward resistance.

Effects of a Bad Diode The question is how would an open or shorted diode affect a circuit such as the bridge rectifier? We examine each of these circumstances, including the affects of an open or shorted filter capacitor.

If the circuit of Fig. 2-27 has an open in D_1 (simulated by the missing diode), the output voltage is slightly reduced. This is because the bridge rectifier operates as a half-wave rectifier with diodes D_2 and D_4 conducting. There is no conduction on the other alternation because no current can flow through D_1. This also decreases the ripple frequency to the value of the input frequency, which can be measured with an oscilloscope. The change in dc output voltage is slight because the filter capacitor performs its job of opposing changes in voltage. This same theory applies to the opening of any of the other diodes.

If diode D_2 of Fig. 2-28 were to short (simulated by a piece of wire), there would be a tremendous increase in secondary current. Increasing the secondary current causes an increase in primary current. The end result is the destruction of D_3 and a blown fuse.

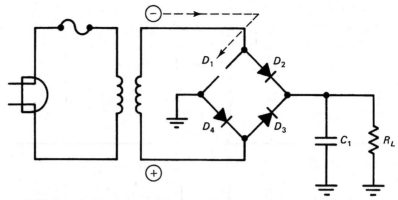

FIGURE 2-27 An open prevents current flow on one alternation.

Effects of a Bad Capacitor A problem in the filter capacitor has a drastic effect on the output voltage. For instance, if C_1 were to open, it would no longer maintain a charge. This has the same effect as removing it completely from the circuit. What we are left with is an unfiltered bridge rectifier. By comparing the results of Sections 2-7 and 2-8, we can see both visually and mathematically that the dc output is greatly changed. On an oscilloscope, the difference is very noticeable. There is an excessive amount of ripple.

We can check an electrolytic capacitor out of the circuit with an ohmmeter. We should see the resistance reading of the meter gradually increase as the capacitor is measured in each direction. Caution should be exercised when testing low-voltage tantalum capacitors.

The final possibility we discuss is the shorted filter capacitor. In this case, the output voltage is reduced to zero volts. This is because the output is in parallel to the filter capacitor, which measures zero volts. The shorted capacitor would most likely cause the fuse to blow or the destruction of one or more diodes. This is due to the tremendous increase in current. This problem is evident on an oscilloscope, which would show no voltage, and on an ohmmeter, which would show zero ohms.

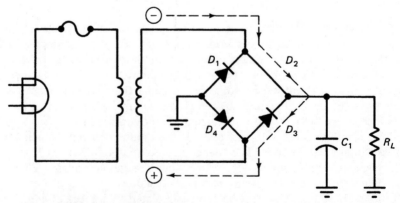

FIGURE 2-28 A short diverts current from the load resistor on one alternation.

46. The two most likely causes of a problem in a circuit are the _____ and the _____ .

47. A diode's forward resistance can be measured by connecting the ohm-meter's common lead to the _____ and the positive lead to the _____ .

48. The resistance ratio of a good diode is a _____ forward resistance to a _____ reverse resistance.

49. A measurement of zero ohms across a diode indicates it has a(n) _____ .

50. Excessive current can destroy a diode. True or false?

51. The (open or shorted) diode can be simulated by removing a diode from the circuit.

52. A full-wave rectifier is turned into a half-wave rectifier when a (diode or capacitor) opens.

53. Which component has a greater effect on the output voltage, an open diode or a shorted filter capacitor?

54. A shorted capacitor measures _____ ohms.

55. Ohmmeter measurements should be made (in or out) of the circuit.

SUMMARY

1. The anode of a diode is composed of P-type material.

2. The cathode of a diode is composed of N-type material.

3. A forward-biased silicon diode drops 0.7 V.

4. A reverse-biased silicon diode acts like an open switch.

5. Diodes are rated for maximum forward and reverse currents and voltages.

6. The internal resistance of a diode changes with varied circuit conditions.

7. A rectifier converts ac to dc.

8. A half-wave rectifier uses one diode that conducts on one alternation of the ac input signal.

9. The dc output voltage for an unfiltered *half-wave rectifier* is found by

$$V_{dc\ out} = V_{R1\ pk.} \times 0.318$$

10. A filter capacitor is used to maintain a constant dc level.

11. The small fluctuation in dc voltage that remains after filtering is called ripple voltage and is found by

$$V_{rip(p-p)} = \frac{I}{fC}$$

12. The dc output voltage for a *filtered half-wave rectifier* is found by

$$V_{dc\ out} = V_{R1\ pk.} - V_{rip\ pk.}$$

13. The peak reverse voltage (PRV) or peak inverse voltage (PIV) is equal to twice the value of the secondary voltage for the *filtered half-wave rectifier*.

14. A *full-wave rectifier* uses a two-diode arrangement in which one diode conducts on each alternation of the input signal. A center-tap transformer is necessary.

15. The dc output voltage of an unfiltered *full-wave rectifier* can be found by

$$V_{\text{dc out}} = V_{R1\,\text{pk.}} \times 0.636$$

16. The dc output voltage of a *filtered full-wave rectifier* is found by

$$V_{\text{dc out}} = V_{R1\,\text{pk.}} - V_{\text{rip pk.}}$$

17. The PRV across the nonconducting diode of a *filtered full-wave rectifier* is equal to twice the value of the center-tap voltage.

18. A *full-wave bridge rectifier* uses a four-diode arrangement in which two diodes conduct on each alternation of the input signal. A center-tap transformer is not necessary.

19. The dc output voltage of an unfiltered *full-wave bridge rectifier* is found by

$$V_{\text{dc out}} = V_{R1\,\text{pk.}} \times 0.636$$

20. The dc output voltage of a *filtered full-wave bridge rectifier* is found by

$$V_{\text{dc out}} = V_{R1\,\text{pk.}} - V_{\text{rip pk.}}$$

21. The PRV that any one diode experiences in a *filtered full-wave rectifier* is approximately equal to the value of the secondary voltage.

22. A clipper or limiter circuit is used to clip or limit a portion of the input signal.

23. A clamper circuit is used to change the dc level of the output signal.

24. The forward resistance of a diode is measured by placing the positive lead of the ohmmeter on the anode and the common lead on the cathode while using the diode range or highest resistance range of the meter.

25. The forward-resistance measurement should register some value of resistance.

26. The reverse resistance of a diode is measured by placing the positive lead of the ohmmeter on the cathode and the common lead on the anode while using the diode range or highest resistance range of the meter.

27. The measured value for reverse resistance should be higher than the reading for forward resistance (possibly infinite).

28. The resistance measurement of an electrolytic capacitor should show a steadily increasing resistance.

29. A measured resistance of zero indicates a shorted component.

30. A measured resistance of infinity in both directions indicates an open component.

31. Open diodes in a rectifier reduce the output voltage to zero.

32. Open capacitors increase the output ripple voltage.

GLOSSARY

Barrier Voltage: The voltage that must be overcome for a diode to conduct.

Filtering: Producing a steadier or more constant voltage level, or eliminating unwanted frequencies.

Full Wave: Conducts on both alternations.

Germanium Diode: A semiconductor whose barrier voltage is approximately 0.3 V.

Half-Wave: Conducts on only one alternation.

Rectify: To change ac to dc.

Ripple: Fluctuation in a dc voltage that remains after filtering.

Silicon Diode: A semiconductor whose barrier voltage is approximately 0.7 V.

PROBLEMS

SECTION 2-1

2-1. Draw and label the schematic symbol for a diode.

2-2. What is meant by the term barrier voltage?

2-3. Create a circuit in which the diode is forward biased and show the path of current flow.

2-4. Create a circuit in which the diode is reverse biased and show the path of current flow.

SECTION 2-2

2-5. List and explain maximum forward current, maximum forward voltage, and peak reverse voltage.

2-6. How does the resistance of a diode vary?

2-7. What is required in order for a diode to conduct?

SECTION 2-3

2-8. Define rectification.

2-9. Explain current flow in a half-wave rectifier.

2-10. How is the value of dc voltage determined in an unfiltered half-wave rectifier?

2-11. What is the dc output voltage of an unfiltered half-wave rectifier whose peak secondary voltage is 21 V?

2-12. Draw the output waveform for Problem 11.

SECTION 2-4

2-13. How does a filtered half-wave rectifier differ from an unfiltered half-wave rectifier?

2-14. Explain the purpose of the filter capacitor.

2-15. What is ripple voltage?

2-16. What is ripple frequency?

2-17. How does the size of a filter capacitor affect the dc output voltage?

2-18. How is the peak reverse voltage determined in a filtered half-wave rectifier?

2-19. What is the dc output voltage across a 2-kΩ load resistance in a filtered half-wave rectifier that has a peak secondary voltage of 25 V at 60 Hz and a 100-μF filter capacitor?

SECTION 2-5

2-20. How does the design of a full-wave rectifier differ from that of a half-wave rectifier?

2-21. Why is the rectifier named full wave?

2-22. How does a center-tapped transformer affect the secondary voltage?

2-23. What is the dc output voltage across a 2-kΩ load resistance in an unfiltered full-wave rectifier that has a peak secondary voltage of 25 V at 60 Hz?

2-24. Draw the output waveform of an unfiltered full-wave rectifier.

SECTION 2-6

2-25. Explain any differences between ripple voltage and frequency in the filtered full-wave rectifier as compared to the filtered half-wave rectifier.

2-26. Draw the output waveform of a filtered full-wave rectifier.

2-27. How is the peak reverse voltage of the diode determined in a filtered full-wave rectifier?

2-28. What is the peak voltage at the secondary of Fig. 2-29?

2-29. What is the ripple frequency of Fig. 2-29?

2-30. What is the dc output voltage of Fig. 2-29?

2-31. What is the dc output voltage across a 2-kΩ load resistance in a filtered full-wave rectifier that has a peak secondary voltage of 25 V at 60 Hz and a 100-μF filter capacitor?

SECTION 2-7

2-32. How does the bridge rectifier differ from the full-wave rectifier of Section 2-6?

2-33. Explain the current path through the bridge rectifier.

2-34. What are the advantages of avoiding the use of a center-tapped transformer?

2-35. What is the dc output voltage across a 2-kΩ load resistance in an unfiltered bridge rectifier that has a peak secondary voltage of 25 V at 60 Hz?

SECTION 2-8

The following problems refer to the circuit of Fig. 2-30.

2-36. Trace the current path.

2-37. What is the peak reverse voltage?

2-38. Is the output positive or negative?

2-39. What is the value of the ripple frequency?

2-40. What is the ripple voltage?

2-41. Determine the dc output voltage.

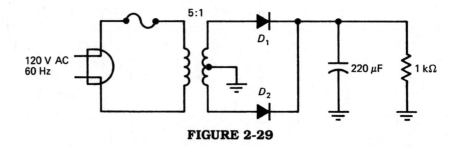

FIGURE 2-29

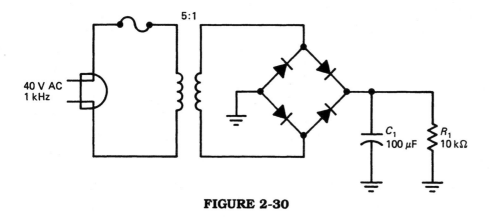

FIGURE 2-30

SECTION 2-9

2-42. What is the purpose of a clipper circuit?

2-43. Draw the schematic for a positive limiter.

2-44. How can the value of the positive limiter be varied?

2-45. What is the purpose of a clamper circuit?

2-46. Draw and explain the operation of a positive clamper.

2-47. How does the concept of RC time constants relate to clamper circuits?

SECTION 2-10

2-48. List the proper forward and reverse resistance of a diode.

2-49. What voltage is measured across an open and shorted diode?

2-50. What effect does an open diode have in the circuit of Fig. 2-30?

2-51. What effect does a shorted diode have in the circuit of Fig. 2-30?

2-52. What effect does an open filter capacitor have in the circuit of Fig. 2-30?

2-53. What effect does a shorted filter capacitor have in the circuit of Fig. 2-30?

2-54. List the steps to take if zero volts were measured at the output of the circuit of Fig. 2-30.

ANSWERS TO REVIEW QUESTIONS

1. diode	**2.** anode	**3.** cathode
4. barrier	**5.** current	**6.** 0.3
7. open	**8.** forward	**9.** 0.7, 0.3
10. True	**11.** peak reverse	**12.** False
13. Rectification	voltage	**14.** False
15. reverse	**16.** positive	**17.** ac
18. unfiltered	**19.** voltage	**20.** ripple
21. True	**22.** decrease	**23.** False
24. doubles	**25.** False	**26.** center-tapped
27. False	**28.** True	**29.** True
30. twice	**31.** double	**32.** True
33. two	**34.** True	**35.** Ground

36. True	**37.** False	**38.** limiter
39. positive	**40.** forward	**41.** Add a dc supply
42. dc restorer	**43.** peak	**44.** *RC* time constant
45. below	**46.** open, short	**47.** cathode, anode
48. low, high	**49.** short	**50.** True
51. open	**52.** diode	**53.** filter capacitor
54. zero	**55.** out	

3

Other Diode Types

When you have completed this chapter, you should be able to:

- Observe, measure, and calculate the regulating characteristics of the zener diode.
- Analyze diodes that emit light such as the light-emitting diode (LED), seven-segment display, and laser diode.
- Analyze how light affects diodes such as the photodiode.

Up to this point, we have studied one type of diode, the PN junction rectifier diode. It is designed to operate in forward bias and begins to conduct when the barrier potential of 0.7 or 0.3 V, depending on the semiconductor material, is exceeded. When the PN junction rectifier diode is reverse biased, it acts as an open switch.

There are a variety of diodes whose construction and operation differ from that of the rectifier diode. Each of these diodes fills a special purpose within the realm of electronics.

3-1 ZENER-DIODE CHARACTERISTICS

Zener Diode: A PN junction diode that is normally operated in reverse bias. It maintains a regular or constant voltage.

Reverse Breakdown Voltage: The reverse voltage necessary to cause a component to conduct.

The **zener diode** is a silicon PN junction diode. It differs from the rectifier diode in the level of doping. It is heavily doped and has a larger depletion region. The result is that a larger zener **reverse breakdown voltage** is necessary to cause conduction. It can range anywhere from a few volts to as much as several hundred volts.

The schematic symbol and the physical package are illustrated in Fig. 3-1. Notice that the cathode in the schematic symbol has a Z shape to it. This is what differentiates it from other diodes. The physical package is not much different from that of a rectifier diode. The cathode lead is marked with a solid band. Only a familiarity with the part number itself allows us to distinguish it as a zener diode.

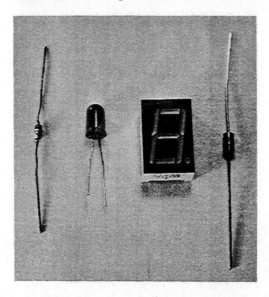

(Left to right) Zener diode, LED, 7-segment display, Schottky diode.

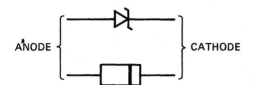

ANODE CATHODE

FIGURE 3-1 Zener diode.

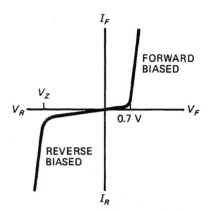

FIGURE 3-2 Characteristic *I–V* curve.

By examining the characteristic *I–V* curve of Fig. 3-2, we can see the unique operating mode of the zener diode. First, a description of the various labels:

I_F = forward current
V_F = forward voltage
I_R = reverse current
V_R = reverse voltage
V_Z = reverse breakdown voltage

The upper right quadrant of the figure shows that the zener diode acts like an ordinary rectifier diode in forward bias. Notice that once a **forward voltage** (V_F) of 0.7 V is applied, the diode begins to conduct, as illustrated by the upward sloping line. This is called **forward current** (I_F).

In the lower left corner of Fig. 3-2, the unique characteristics of the zener diode are brought out. As reverse voltage is applied to the zener diode, it eventually reaches a point where the diode breaks down and allows current to flow. This breakdown voltage does not harm the diode. As a matter of fact, this is the desired operating mode of the zener diode. The **reverse voltage** (V_R) at this point on the curve remains almost constant (or regular) even as **reverse current** (I_R) increases. This is the reason why the zener diode is sometimes referred to as the zener **regulator.** It provides a constant or regulated voltage. The exact voltage can be determined from the specification sheet of the diode.

We may conclude that the zener diode can operate in either forward or reverse bias. In forward bias, it behaves like a conventional diode. In reverse bias, it conducts at the value of its zener breakdown voltage. This breakdown voltage is determined by the internal construction of the diode.

Forward Voltage: The voltage developed across a forward-biased diode.

Forward Current: The current that flows through a forward-biased diode.

Reverse Voltage: The voltage developed across a reverse-biased diode.

Reverse Current: The current that flows through a reverse-biased diode.

Regulator: A circuit or component that maintains a nearly constant or regular voltage.

REVIEW QUESTIONS

1. A zener diode has a cathode and anode. True or false?
2. A zener diode only conducts when forward biased. True or false?
3. As reverse current increases, reverse voltage _____ .
4. A zener diode acts like a conventional diode when _____ biased.
5. What is the range of reverse zener breakdown voltages?

3-2 ZENER-DIODE BIASING

The zener diode is designed to conduct in reverse bias. It is in reverse bias that we can use that unique feature called the zener reverse breakdown voltage. Reverse biasing a zener diode is much like reverse biasing any other diode. The cathode is connected toward the positive side of the source voltage and the anode is connected toward the negative side. Figure 3-3 illustrates a reverse-biased zener diode.

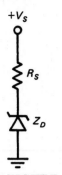

FIGURE 3-3 Reverse-biased zener diode.

In order to discuss Fig. 3-3 in greater detail, it is necessary to assign some values. Assume the zener diode has a rating of 5 V, the series resistor $R_S = 1$ kΩ, and the source voltage $V_S = 12$ V. The circuit is a simple series circuit and Ohm's law can be applied.

First of all, since the source voltage is greater than the voltage rating of the zener diode, the diode conducts. The zener diode drops 5 V and the remainder of the source voltage is across the series resistor. A formula expressing this statement is

$$V_{R_S} = V_S - V_{ZD}$$

Therefore,

$$V_{R_S} = 12\text{ V} - 5\text{ V} = 7\text{ V}$$

The next step is to determine the amount of current flowing. In order to determine the value of current, we need the values of both voltage and resistance. The only component of which both voltage and resistance is known is the series resistor. Therefore, apply Ohm's law using those values.

$$I_{R_S} = \frac{V_{R_S}}{R_S}$$

$$= \frac{7\text{ V}}{1\text{ kΩ}} = 7\text{ mA}$$

Because this is a series circuit, the current through the zener diode is the same as the current through the resistor. Therefore,

$$I_{ZD} = I_{R_S}$$

$$= 7\text{ mA}$$

At this point, the currents and voltage drops of the components are known. The power dissipated can be found by applying Watt's law.

$$P = I \times V$$

$$P_{R_S} = I_{R_S} \times V_{R_S}$$

$$= 7\text{ mA} \times 7\text{ V}$$

$$= 49\text{ mW}$$

and

$$P_{ZD} = I_{ZD} \times V_{ZD}$$

$$= 7\text{ mA} \times 5\text{ V}$$

$$= 35\text{ mW}$$

All of these calculations are based on a source voltage of 12 V. The regulator aspect of the zener diode can be observed by reconfiguring the circuit values of Fig. 3-3 so that the source voltage equals 9 V. The same value resistor and zener diode are used; $R_S = 1$ kΩ and $V_Z = 5$ V.

The source voltage is greater than the zener breakdown voltage rating; therefore, the diode conducts.

$$V_Z = 5 \text{ V}$$

Therefore,

$$V_{R_S} = V_S - V_Z$$
$$= 9 \text{ V} - 5 \text{ V}$$
$$= 4 \text{ V}$$

and

$$I_{R_S} = \frac{V_{R_S}}{R_S}$$
$$= \frac{4 \text{ V}}{1 \text{ k}\Omega}$$
$$= 4 \text{ mA}$$

so

$$I_{ZD} = I_{R_S}$$
$$= 4 \text{ mA}$$

Finally,

$$P_{R_S} = I_{R_S} \times V_{R_S}$$
$$= 4 \text{ mA} \times 4 \text{ V}$$
$$= 16 \text{ mW}$$

and

$$P_{ZD} = I_{ZD} \times V_{ZD}$$
$$= 4 \text{ mA} \times 5 \text{ V}$$
$$= 20 \text{ mW}$$

In considering these calculations, notice how the zener diode maintains a "regulated" zener voltage of 5 V even though the current has changed from the earlier example. Now we can understand why it has the name zener regulator diode.

EXAMPLE 3-1

The values in Fig. 3-3 are changed so that $+V_S = 12$ V, $R_S = 4.7$ kΩ, and $V_Z = 3.3$ V. What is the value of the zener current?

Solution: The zener current is determined by the value of the series resistor. A reverse-biased zener diode drops its zener rated voltage (provided a minimum zener current is maintained). The balance of the source voltage is dropped across the series resistor.

$$V_{R_S} = 12 \text{ V} - 3.3 \text{ V}$$
$$= 8.7 \text{ V}$$

$$I_{R_S} = \frac{8.7 \text{ V}}{4.7 \text{ k}\Omega}$$
$$= 1.85 \text{ mA}$$

$$I_Z = I_{R_S} = 1.85 \text{ mA} \qquad \blacksquare$$

Practice Problem: If the values in Fig. 3-3 are $V_S = 6$ V, $R_S = 2.7$ kΩ, and $V_Z = 3.3$ V, then what are V_Z, V_{R_S}, I_{ZD}, and P_{ZD}? *Answer:* $V_Z = 3.3$ V, $V_{R_S} = 2.7$ V, $I_{ZD} = 1$ mA, and $P_{ZD} = 3.3$ mW.

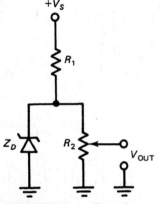

FIGURE 3-4 Regulated variable output.

Variable Zener Regulator Supply The circuit in Fig. 3-4 allows us to vary the output voltage. The source voltage V_S can be tapped off a filtered full-wave bridge rectifier.

The values of Figure 3.4 for this example are $V_S = 20$ V, $R_1 = 2$ kΩ, $R_2 = 10$ kΩ, and $Z_D = 15$ V. The wiper of potentiometer R_2 is adjusted half way.

The circuit is essentially a series–parallel circuit. R_2 is in parallel with the zener diode and consequently drops the same voltage, which in this example is 15 V.

$$V_{R2} = V_{ZD}$$

As a result, V_{out} can vary from 0–15 V, depending on the position of the wiper. The exact value can be determined by using the voltage-divider formula on the potentiometer. It is

$$V_{\text{out}} = \frac{R_W}{R_2} V_{R2}$$

where R_W is the resistance from the wiper to ground. In this example, the wiper is adjusted half way; therefore, the resistance from the wiper to ground is half of 10 kΩ, which is 5 kΩ. So

$$V_{\text{out}} = \frac{5 \text{ k}\Omega}{10 \text{ k}\Omega} (15 \text{ V})$$
$$= 7.5 \text{ V}$$

This assumes that no current is drawn through the output.

Keep in mind that the zener diode maintains a constant voltage of 15 V regardless of the position of the wiper on the potentiometer. It continues to regulate the 15 V as long as the zener current is at or above its minimum level.

This circuit has limitations with regards to the size of the load resistance that can be used. If the load resistance is too small, it will draw most of the current, causing the current through the zener diode to drop below the level necessary for conduction.

EXAMPLE 3-2

The values of Fig. 3-4 are changed so that $+V_S = 15$ V, $R_1 = 330$, $R_2 = 5$ kΩ, and $V_Z = 10$ V. Determine the output voltage when the potentiometer is adjusted half way.

Solution: The zener diode determines the total voltage range of the output. A potentiometer is used to isolate a portion of this voltage. If the 5-kΩ potentiometer is adjusted half way, then the resistance from the wiper to ground would be 2.5 kΩ. The voltage-divider formula is used to determine the portion of the zener voltage developed at the output.

$$V_{out} = \frac{2.5 \text{ k}\Omega}{5 \text{ k}\Omega} (10 \text{ V})$$
$$= 5 \text{ V} \qquad \blacksquare$$

Practice Problem: What would be V_{out} in the circuit of Fig. 3-4 if the wiper were adjusted three-fourths of the way up and all values remained the same as in the previous example? *Answer:* 11.25 V.

REVIEW QUESTIONS

6. When reverse biasing a zener diode, the (anode or cathode) is connected toward the positive terminal of the supply.

7. A zener diode rated at 5 V does not conduct when reverse biased in a circuit powered by a 3-V power supply. True or false?

8. The reverse zener voltage of a conducting diode increases if the supply voltage is increased. True or false?

9. A series resistor is usually used to limit current flow. True or false?

10. The zener diode is sometimes called a _____ diode because it maintains a nearly constant voltage under varied conditions.

3-3 LIGHT-EMITTING DIODES

In most cases, the little colored lights that are observed on stereos, computers, answering machines, etc., are **light-emitting diodes.** The abbreviated name is LED and LEDs come in various shapes and sizes.

The diodes previously discussed were constructed of silicon or germanium and they dissipated energy in the form of heat. LEDs, on the other hand, are constructed from gallium arsenide or gallium phosphide and dissipate energy in the form of light and heat. The light is emitted when conduction-band electrons recombine with holes in the valence shell of the atoms.

The schematic symbol for an LED is shown in Fig. 3-5. It differs from an ordinary diode symbol because of the pair of arrows pointing away from the diode. These arrows are meant to symbolize light being emitted by the diode. The diode retains its names of anode for the P-type region and cathode for the N-type region.

The anode and cathode of an LED can be determined in two ways. Figure 3-6 shows that the longer lead of the diode is generally the anode. The second method involves locating the notched or squared edge of the LED. The lead closest to this edge is generally the cathode.

Light-Emitting Diode (LED): A light-emitting diode emits energy in the form of light. It is operated in forward bias.

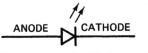

FIGURE 3-5 Light-emitting diode (LED).

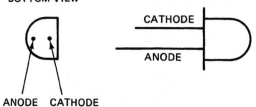

FIGURE 3-6 Determining the anode and cathode on an LED.

An LED should be forward biased for proper operation. Figure 3-7 illustrates a properly biased LED with a resistor to limit current. A forward-biased diode drops approximately 2 V. The remaining 3 V in this circuit is dropped across the series resistor. This means that the calculated current in this circuit is 3 mA using Ohm's law:

$$I_R = \frac{V_R}{R}$$

$$= \frac{3\ V}{1\ k\Omega}$$

$$= 3\ mA$$

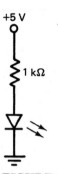

FIGURE 3-7 Properly biased LED.

and

$$I_D = I_R$$

The brightness of the LED is determined by the amount of current. The greater the current, the brighter the LED. There are limitations and an LED can be damaged by excessive current. Typical maximum forward currents vary from 30 to 100 mA.

LEDs can also be damaged if improperly biased. The maximum reverse voltage that an LED can withstand is about 4 V.

EXAMPLE 3-3 _____

The resistor in Fig. 3-7 is increased to 5 kΩ. Will the LED light become dimmer or brighter?

Solution: Increasing the resistance decreases the current flowing through the LED. A decrease in current causes the LED to become dimmer. ■

FIGURE 3-8 Two-color LED.

Two-Color LED: A diode that is capable of emitting different colors, depending on which two leads are forward biased.

Three-Terminal Two-Color LED The **two-color LED** is a three-terminal device, as illustrated in Fig. 3-8. The longest lead is the cathode and the remaining two leads are the anodes.

When leads *R* and *C* are forward biased, the LED emits red light, and when leads *G* and *C* are forward biased, the LED emits green light. The forward voltage for this LED remains at approximately 2 V, while the maximum forward current is only about 10 mA.

Tricolor LED: A diode that has only two leads yet can emit three different colors, depending on the direction of current flow.

Tricolor LED The **tristate LED** emits red, green, or yellow light, depending on operating conditions. It looks similar to the conventional LED. It has two leads and each of these leads acts as both anode and cathode. As dc current flows through it in one direction, the LED lights red. If dc current passes through it in the opposite direction, it lights green. Finally, an ac current causes it to emit yellow light. The brightness of all three colors is equal.

Blinking LED The **blinking LED** is actually a combination of an oscillator and an LED in one package. Having an anode and cathode lead, it looks the same as an ordinary LED. The blinking occurs at a frequency of about 3 Hz when the diode is supplied with a dc forward voltage of 5 V. It conducts about 20 mA of current when on and 0.9 mA when off.

Blinking LED: A diode that blinks at a rate of about 3 Hz when forward biased.

REVIEW QUESTIONS

11. What does the acronym LED represent?
12. An LED should be (forward or reverse) biased.
13. How does the schematic symbol of an LED differ from that of an ordinary diode?
14. How can the anode of an LED be determined?
15. How can the two-color LED of Fig. 3-8 be made to emit green light?
16. What is the effect of applying ac to a tricolor LED?
17. What voltage is necessary to cause the blinking LED to blink and what is the approximate frequency?

3-4 SEVEN-SEGMENT DISPLAY

The **seven-segment** display is composed of seven rectangular LEDs. They are commonly used in digital clocks, radios, stereos, and other appliances to display numbers or letters. Figure 3-9 shows an example. Note that the seven LED segments are labeled "*a*" through "*g*." Each of these segments is controlled through one of the display leads.

Seven-segment displays come in two types, common-cathode and common-anode. The common-cathode version means that all the cathodes of the diodes are tied together (in common), as in Fig. 3-10. This allows any segment to be lit by

Seven-Segment Display: An arrangement of seven LEDs in a single IC that allows numbers and letters to be displayed.

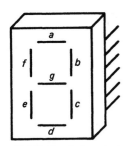

FIGURE 3-9 Seven-segment display.

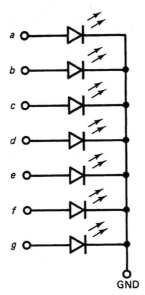

FIGURE 3-10 Schematic diagram of a common-cathode seven-segment display.

forward biasing that particular LED. To light the number 7, for example, segments *a*, *b*, and *c* must be forward biased. The cathodes are tied to ground and 5 V are applied to the anodes of these segments to light them. Of course, limiting resistors should be used to keep the current under 20 mA. The segments require a forward voltage of about 2 V in order to light.

The common-anode seven-segment display has all of its anodes tied together. In this case, the common anodes are tied to +5 V and ground is used to light the individual segments. Again, current-limiting resistors are needed.

Usually, the seven-segment display is used with a seven-segment display decoder, which accepts digital logic levels and converts them to the proper number or letter for display. Refer to a digital textbook for greater detail on the uses and applications of seven-segment displays.

REVIEW QUESTIONS

18. The seven-segment display is composed of seven _____ .
19. (+5 V or ground) is applied to the segment inputs to light the individual segments on a common-cathode seven-segment display.
20. Which segments should be lit to form the number 3?

3-5 PHOTODIODES

Photodiode: A diode whose internal resistance is inversely proportional to the amount of light. It is operated in reverse bias.

Footcandle: An illumination equal to that of a candle burning at a distance of 1 foot.

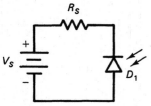

FIGURE 3-11 Properly biased photodiode.

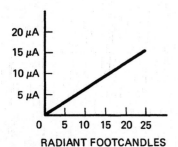

FIGURE 3-12 Illumination specifications.

The **photodiode** is a PN junction device. The schematic symbol is similar to that of the LED with the exception that the arrows are pointing inward. This indicates that the diode is affected by light rather than emitting it. The diode has a glass lens through which light can pass. There is an inverse relationship between light and the internal resistance of the diode. As light increases, the internal resistance decreases.

Proper operation is achieved through reverse biasing the diode. Figure 3-11 shows a properly biased photodiode. Since the internal resistance of the diode is determined by the amount of light, we use Fig. 3-12 to estimate current flow.

The values in Fig. 3-11 are $V_S = 10$ V, $R_S = 470$ kΩ, and there is an illumination of 15 **footcandles.** This means that current flowing in the circuit is approximately 10 μA based on the curve of Fig. 3-12. With this information, we can calculate the voltage across the series resistor using Ohm's law.

$$V_{RS} = I_R \times R_S$$
$$= 10 \ \mu A \times 470 \ k\Omega$$
$$= 4.7 \ V$$

The balance of the source voltage must be across the photodiode. Therefore,

$$V_D = V_S - V_{R_S}$$
$$= 10 \ V - 4.7 \ V$$
$$= 5.3 \ V$$

One can see by examining Fig. 3-12 that as light increases, current increases. Conversely, as it gets darker, current decreases. In absolute darkness, the only current flowing is an extremely small amount due to thermal leakage.

CHAP. 3 / OTHER DIODE TYPES

EXAMPLE 3-4

The circuit of Fig. 3-11 contains the following values: $V_S = 9$ V, $R_S = 330$ kΩ, and the characteristics of D_1 are illustrated in Fig. 3-12. The circuit is initially exposed to light whose illumination started at 25 radiant footcandles and is reduced to 2 radiant footcandles. What are the values of voltage during these different light exposures?

Solution: Voltage across the diode increases as light decreases. The voltage across the diode is calculated as the remaining source voltage after the voltage of the series resistor is taken into consideration. At 25 radiant footcandles, the current is estimated as 15 μA; therefore, the voltage across the diode is

$$V_{D1} = 9 \text{ V} - (15 \text{ } \mu\text{A} \times 330 \text{ k}\Omega)$$
$$= 4.05 \text{ V}$$

As the light decreases to 2 radiant footcandles, so does the current. Current can be estimated as 2 μA at this light exposure.

$$V_{D1} = 9 \text{ V} - (2 \text{ } \mu\text{A} \times 330 \text{ k}\Omega)$$
$$= 8.34 \text{ V} \qquad \blacksquare$$

Practice Problem: If the values of Fig. 3-11 are $V_S = 9$V, $R_S = 1$ MΩ, and there is an illumination of 5 radiant footcandles, what is V_D? *Answer:* 4 V.

REVIEW QUESTIONS

21. How does the diode symbol of the photodiode differ from that of the LED?
22. The photodiode should be (forward or reverse) biased.
23. The internal resistance of a photodiode is (directly or inversely) related to light.
24. The voltage across the photodiode (increases or decreases) as light increases.

3-6 LASER DIODES

The use of lasers is becoming more and more commonplace. The advent of compact disc players and optical disk drives has brought this technology into the hands of the consumer. Laser is the acronym for light amplification by stimulated emission of radiation.

The **laser diode** has a PN junction formed by doping gallium arsenide. The electron–hole recombination causes photons to be released. Photons are tiny packets of light energy. A reflective material within the device causes the photons to bounce back and forth. The photon activity becomes so intense that at some point, a beam of laser light penetrates a partially reflective area of the device. Figure 3-13 illustrates this process.

Laser light has some unique characteristics. The first and foremost is that laser light is coherent. Coherent means that there is no phase difference between the light waves. The second characteristic is that laser light is monochromatic, which means that it is of one color and wavelength. Finally, laser light is collimated; the emitted light waves are parallel to each other.

Laser Diode: A diode that emits a coherent, monochromatic, and collimated light when it receives a large burst of forward current.

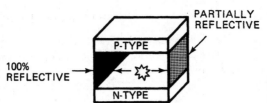

FIGURE 3-13 Photon action within a laser diode.

The schematic symbol for a laser diode is similar to that of an LED. The laser diode is activated by short bursts of a large forward current (5 to 10 A). The beam can be focused using an external lens. A special filter or lens is necessary to view the laser beam. DANGER: Extreme caution should be exercised because serious eye injury can result!

REVIEW QUESTIONS

25. The laser diode should be _____ biased.
26. What does the acronym laser represent?
27. What is the name given to the packets of light emitted by a laser diode?

3-7 PIN AND SCHOTTKY DIODES

PIN Diode: A P-type intrinsic N-type diode having attenuating and switching capabilities.

The **PIN diode** is so named because the heavily doped P-type and N-type regions are separated by an intrinsic (undoped) silicon region. Figure 3-14(A) shows a block diagram of the diodes construction.

When the diode is forward biased, it displays little resistance. If it is unbiased or reverse biased, it develops a very high resistance.

Figure 3.14(B) shows a simple application of the PIN diode (D_1). A positive dc control voltage forward biases the PIN diode, causing it to conduct. Electron

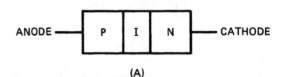

(A)

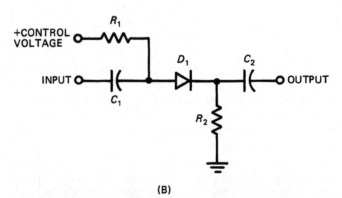

(B)

FIGURE 3-14 (A) Block diagram of a PIN diode. (B) PIN diode used as a switch.

dc current flows from ground through R_2 and D_1 and continues up through R_1. When D_1 is forward biased, it has very little resistance. The input signal can pass through the low resistance of the diode to the output. If the positive control voltage is removed, the PIN diode is no longer forward biased and its resistance greatly increases. This prevents the input signal from reaching the output. The PIN diode acts as a dc-controlled RF switch. It is faster and quieter than a relay.

The **Schottky diode** is shown in Fig. 3-15. The block diagram shows that a metal is joined to the N-type material. This metal is normally platinum, gold, or silver. Metal is used to eliminate the depletion region, and conduction can occur at about 0.3 V when forward biased.

Because the Schottky diode does not retain a charge like conventional diodes, it can be used in higher-frequency applications. It can easily switch at speeds of over 300 MHz, which makes it desirable in high-speed digital applications. It is also used in low-voltage rectification applications such as the 5-V output of a computer's switching power supply.

Schottky Diode: A metal–semiconductor junction diode possessing high-frequency rectifying and switching capabilities.

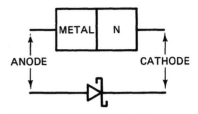

FIGURE 3-15 Schottky diode.

REVIEW QUESTIONS

28. What does the "I" in PIN diode represent?
29. A PIN diode conducts in reverse bias. True or false?
30. What amount of voltage does a forward-biased Schottky diode drop?
31. The Schottky diode is mainly used in switching applications under 3000 Hz. True or false?

SUMMARY

1. The zener diode is a heavily doped PN junction diode.
2. The zener diode acts like an ordinary rectifier diode when forward biased.
3. A zener diode is normally reversed biased and requires a reverse breakdown voltage (V_Z) in order to conduct.
4. Zener diodes are normally used to maintain a specific or regular voltage at some point in a circuit.
5. A zener diode does require a certain level of reverse current in order to maintain its reverse breakdown voltage.
6. An LED is a light-emitting diode.
7. The LED is constructed from gallium arsenide or gallium phosphide and dissipates energy in the form of light.
8. The anode and cathode of an LED must be forward biased in order to emit light.
9. Care must be taken to limit the current of an LED.
10. The two-color LED emits red or green, depending on which two of its three leads are forward biased.
11. The tricolor LED emits red, green, or yellow light, depending on which way current flows through it. It emits a yellow light if ac is applied.
12. The blinking LED is a combination of an oscillator and an LED in one package. It generally blinks at a frequency of 3 Hz.
13. A seven-segment display is composed of seven LEDs.

14. Segments in a seven-segment display are labeled *"a"* through *"g."*
15. The common-anode seven-segment display has its anodes tied together and uses zero volts at one of the cathodes to light a segment.
16. The common-cathode seven-segment display requires a positive voltage at one of the anodes in order to light that segment.
17. The photodiode is a PN junction device.
18. Light causes the internal resistance of a photodiode to decrease and, therefore, allows current to flow.
19. Laser is an acronym for light amplification by stimulated emission of radiation.
20. The laser diode emits laser light when a large forward current is applied.
21. The laser beam can be focused using an external lens.
22. Caution should be exercised when using laser diodes.
23. PIN diodes have heavily doped P-type and N-type regions separated by an intrinsic (undoped) silicon region.
24. Forward-biased PIN diodes display very little resistance, whereas reverse-biased PIN diodes provide extremely large resistances.
25. The Schottky diode is constructed by joining metal and N-type material.
26. The Schottky diode conducts at about 0.3 V when forward biased.
27. Schottky diodes can be used in high-frequency applications.

GLOSSARY

Blinking LED: A diode that blinks at a rate of about 3 Hz when forward biased.

Footcandle: An illumination equal to that of a candle burning at a distance of 1 foot.

Forward Current: The current that flows through a forward-biased diode.

Forward Voltage: The voltage developed across a forward-biased diode.

Laser Diode: A diode that emits a coherent, monochromatic, and collimated light when it receives a large burst of forward current.

Light-Emitting Diode (LED): A light-emitting diode emits energy in the form of light. It is operated in forward bias.

Photodiode: A diode whose internal resistance is inversely proportional to the amount of light. It is operated in reverse bias.

PIN Diode: A P-type intrinsic N-type diode having attenuating and switching capabilities.

Regulator: A circuit or component that maintains a nearly constant or regular voltage.

Reverse Breakdown Voltage: The reverse voltage necessary to cause a component to conduct.

Reverse Current: The current that flows through a reverse-biased diode.

Reverse Voltage: The voltage developed across a reverse-biased diode.

Schottky Diode: A metal–semiconductor junction diode possessing high-frequency rectifying and switching capabilities.

Seven-Segment Display: An arrangement of seven LEDs in a single IC that allows numbers and letters to be displayed.

Tricolor LED: A diode that has only two leads yet can emit three different colors, depending on the direction of current flow.

Two-Color LED: A diode that is capable of emitting different colors, depending on which two leads are forward biased.

Zener Diode: A PN junction diode that is normally operated in reverse bias. It maintains a regular or constant voltage.

PROBLEMS

SECTION 3-1

3-1. Draw and label the schematic symbol of a zener diode.

3-2. How does the zener diode differ from the ordinary rectifier diode?

3-3. Describe the forward-bias operation of the zener diode in terms of the I–V characteristic curve.

3-4. Describe the reverse-bias operation of the zener diode in terms of the I–V characteristic curve.

SECTION 3-2

3-5. What type of biasing is normally used with zener diodes?

3-6. Describe the operation of the circuit of Fig. 3-3.

3-7. In the circuit of Fig. 3-3, $V_S = 10$ V, $R_S = 5$ kΩ, and $V_Z = 4$ V. What is the value of V_{RS}?

3-8. In the circuit of Fig. 3-3, $V_S = 6$ V, $R_S = 2$ kΩ, and $V_Z = 3$ V. What is the value of I_{ZD}?

3-9. In the circuit of Fig. 3-4, $V_S = 15$ V, $R_1 = 1$ kΩ, $R_2 = 10$ kΩ, and $V_Z = 10$ V. What is V_{out} when the wiper is adjusted half way?

3-10. In the circuit of Fig. 3-4, $V_S = 15$ V, $R_1 = 1$ kΩ, $R_2 = 10$ kΩ, and $V_Z = 10$ V. What is the value of I_{ZD}?

SECTION 3-3

3-11. Draw and label the schematic symbol for an LED.

3-12. How does the construction and energy of an LED differ from that of a silicon diode?

3-13. What type of biasing should be used to light an LED?

3-14. What effect does reducing the resistance in the circuit of Fig. 3-7 have on the LED?

3-15. Illustrate how the two-color LED of Fig. 3-8 could be connected to a single supply, using a SPDT switch to choose between red or green.

3-16. Explain the operation of the tricolor LED and the blinking LED.

SECTION 3-4

3-17. Draw and label the block diagram of the seven-segment display.

3-18. List five letters that can be formed on a seven-segment display and the segments necessary to form them.

3-19. List five letters that cannot be formed on a seven-segment display.

3-20. What is the difference in operation between a common-anode and a common-cathode seven-segment display?

SECTION 3-5

3-21. Draw and label the schematic symbol for a photodiode.

3-22. Explain the relationship between light and the internal resistance of a photodiode.

3-23. In the circuit of Fig. 3-11, $V_S = 9$ V, $R_S = 330$ kΩ, and there is an illumination of 20 radiant footcandles. What is V_D? (Use the curve in Fig. 3-12.)

3-24. In the circuit of Fig. 3-11, $V_S = 9$ V, $R_S = 330$ kΩ, and there is an illumination of 10 radiant footcandles. What is V_{RS}? (Use the curve in Fig. 3-12.)

SECTION 3-6

3-25. What does the acronym laser represent?

3-26. What are three characteristics of laser light?

3-27. Explain the operation of a laser diode.

SECTION 3-7

3-28. Draw and label the schematic symbols for a PIN and a Schottky diode.

3-29. How does dc current flow through the circuit of Fig. 3-14(B)?

3-30. Draw a full-wave rectifier using Schottky diodes.

3-31. List some of the unique characteristics of both the PIN and Schottky diode as compared to conventional diodes.

ANSWERS TO REVIEW QUESTIONS

1.	True	**2.**	False
3.	remains the same	**4.**	forward
5.	Few to several hundred	**6.**	cathode
7.	True	**8.**	False
9.	True	**10.**	regulator
11.	light-emitting diode	**12.**	forward
13.	By the arrows indicating light	**14.**	By length or location
15.	Forward bias $C-G$	**16.**	Yellow light
17.	5 V, 3 Hz	**18.**	LEDs
19.	+5 V	**20.**	a, b, c, d, and g
21.	By the arrows pointing inward	**22.**	reverse
23.	inversely	**24.**	decreases
25.	forward	**26.**	light amplification by the stimulated emission of radiation
27.	Photons		
28.	Intrinsic	**29.**	False
30.	0.3 V	**31.**	False

4

Bipolar
Junction
Transistors

When you have completed this chapter, you should be able to:

- Analyze the basic construction of a transistor.
- Bias a transistor properly.
- Recognize the concepts of saturation and cutoff.
- Apply different biasing methods.
- Analyze the effects produced by a variety of faults in a circuit.

INTRODUCTION

BJT: Abbreviation for bipolar junction transistor, a PN junction device that relies on both majority and minority carriers for current flow.

The **bipolar junction transistor (BJT)** was first developed about 1949. It is so named because it depends on the flow of both minority and majority carriers (bipolar). You are already familiar with the concept of the PN junction. The term transistor is an acronym for "transfer resistor."

4-1 BJT CONSTRUCTION

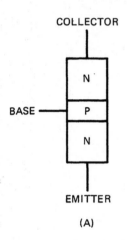

(A)

(B)   (C)

FIGURE 4-1 (A) Block diagram and (B) NPN schematic symbol. (C) PNP schematic symbol.

The block diagram of Fig. 4-1(A) shows three regions: collector (C), emitter (E), and base (B). Each of these regions is doped with either a P-type or N-type material. The emitter is heavily doped and its function is to "emit" current carriers into the base. The base is lightly doped and is a very thin area. It passes most of the emitter-current carriers to the collector. The collector is intermediately doped and its purpose is to "collect" the current carriers passed on from the base. The collector is the largest of the three regions.

As we can see from the block diagram, two PN junctions are formed. One junction is created where the base and emitter meet. The other junction is formed where the collector and base regions meet. Most times these regions are referred to by the designations B–E junction and C–B junction. These junctions play a vital role in the proper biasing of a transistor.

The schematic symbols of Figs. 4-1(B) and 4-1(C) show the two types of BJT. In Fig. 4-1(B), an NPN transistor is represented, and in Fig. 4-1(C), a PNP transis-

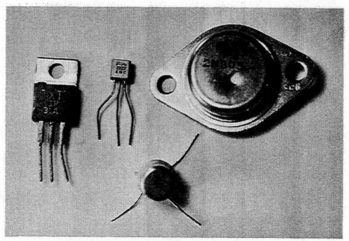

Various transistor packages.

tor is represented. The difference between these two symbols is in the direction of the arrow of the emitter. The arrow points to the N-type material. This is the key to distinguishing between the two in a schematic diagram.

The previous information we studied about the PN junction can be applied to the BJT. Review the material of Section 1-8, "The PN Junction," to refresh your memory.

NPN and PNP bipolar junction transistors.

REVIEW QUESTIONS

1. The letters BJT represent _bipolar Junction transistor_.
2. A device that relies on both majority and minority carriers is called _BJT_.
3. The name "transistor" is an acronym for _Transfer Resistor_.
4. The three regions of a BJT are _Collector_, _Base_, and _Emitter_.
5. The _Emitter_ region is heavily doped.
6. The _Base_ region is lightly doped.
7. If the emitter is composed of N-type material, then the BJT is a (NPN or PNP). _NPN_
8. Which leg of a BJT schematic symbol is marked with an arrow? _Emitter_

4-2 BIASING THE JUNCTIONS

We treat each junction separately in discussing the topic of **biasing.** Keep in mind, however, that both junctions must be properly biased for the BJT to work effectively.

In Fig. 4-2(A), the *B–E* junction is forward biased. This is achieved by connecting the P-type base to the positive terminal of the supply and the N-type emitter to the negative terminal via ground. The resistor is necessary because a forward-biased PN junction only drops a forward voltage of 0.7 V. The remainder

Biasing: Designing a circuit for certain operating voltages.

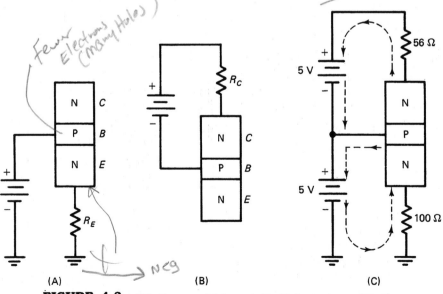

(A)   (B)   (C)

FIGURE 4-2 (A) Forward-biased *B–E* junction. (B) Reverse-biased *C–B* junction. (C) Properly biased BJT.

of the power-supply voltage is dropped across the resistor, which, in turn, determines the limit of emitter-current flow.

In Fig. 4-2(B), the C–E junction is reverse biased. In this figure, a second power supply is used to provide a positive potential to the collector with reference to the base.

Combining the two figures results in the properly biased BJT of Fig. 4-2(C). In order to understand current flow within the transistor, some generalizations are required. The emitter receives 100% of the transistor current, approximately 3% flows out of the base, and the remaining 97% flows out of the collector. This can be stated as

$$I_E = I_C + I_B$$

Because $I_E = 100\%$ and $I_C = 97\%$, it can be stated that emitter current (I_E) and collector current (I_C) are almost the same. Armed with this approximation, consider the values of the circuit of Fig. 4-2(C).

Begin by solving the B–E loop within the circuit. The 5-V supply is tied across the forward-biased base–emitter junction as well as the emitter resistor (R_E). The base–emitter voltage is 0.7 V. This means that the remaining 4.3 V must be dropped across the emitter resistor. Using Ohm's law, we arrive at an emitter current of 43 mA.

Since the emitter current and collector current are almost the same, use the 43 mA for the collector current. This means that 43 mA flows through the 56-Ω resistor (R_C), resulting in a voltage drop of 2.4 V. The remaining 2.6 V of the 5-V supply is found across the C–B junction.

The path of the current flow is indicated in the circuit of Fig. 4-2(C); 100% flows into the emitter and 3% branches off the base, leaving the remaining 97% to flow out of the collector. It is important to understand that there can be no current flow within the transistor unless there is base current. Base current is the result of a properly biased B–E junction. The BJT is a current-controlled device.

REVIEW QUESTIONS

9. How many junctions does a BJT have? *2 (BE & CB)*
10. Which junction is normally forward biased? *BE*
11. The base–emitter junction normally drops ___*.7*___ V.
12. 100% of a BJT's current enters through the ___*Emitt*___.
13. $I_C = I_B + I_E$. True or false? *False*
14. A BJT does not operate without base current. True or false? *True*
15. A BJT is a (voltage- or current-) controlled device. *current*

4-3 BASE BIAS

DC Analysis   There are several configurations that can be used in biasing a transistor. One of the most basic is the base bias. Biasing is necessary to set the transistor for proper operation.

Figure 4-3 shows an example of a base-bias circuit. The first detail we notice is that only one power supply is used. It is much more practical as far as cost is concerned to be able to run a transistor circuit with one power supply rather than two. We see when examining the circuit closely that the collector still receives its reverse bias (positive) through R_C and the N-type material of the emitter receives its forward bias (negative) through ground.

The electron-current flow in the circuit is shown traveling from ground (100%) into the emitter (I_E) and then splitting into base current (I_B) and collector current (I_C). We will be using Ohm's law to determine the various voltage drops and currents. The ratio of collector current to base current is called beta (β). **Beta** is another name for current gain and is a specification included with every transistor. It will become an important part of our calculations. Beta stated as a formula is

$$\text{beta } (\beta) = \frac{I_C}{I_B}$$

Assume that the beta rating of the BJT in Fig. 4-3 is 134. This information is used in the formula to find collector current:

$$I_C = \frac{V_{CC} - V_{BE}}{R_B/\beta}$$

where I_C = collector current

V_{CC} = power-supply voltage

V_{BE} = base–emitter voltage (0.7 V)

R_B = base resistance

β = beta

Substituting the values, we find

$$I_C = \frac{10 \text{ V} - 0.7 \text{ V}}{56 \text{ k}\Omega/134}$$

$$= \frac{9.3 \text{ V}}{418}$$

$$= 22.3 \text{ mA}$$

This is the current flowing out of the collector through resistor R_C. We find the base current by rearranging the formula for beta to read

$$I_B = \frac{I_C}{\beta}$$

Therefore,

$$I_B = \frac{22.3 \text{ mA}}{134}$$

$$= 166 \ \mu\text{A}$$

Finally, the emitter current can be found by adding the collector current and base current.

$$I_E = I_C + I_B$$

$$= 22.3 \text{ mA} + 166 \ \mu\text{A}$$

$$= 22.5 \text{ mA}$$

Beta: The ratio of collector current to base current.

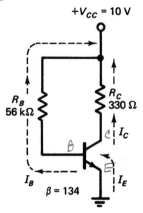

FIGURE 4-3 Base-bias circuit.

Now that the values of these three currents are known, it is possible to solve for the various voltage drops in the circuit. For instance, the voltage across R_B can be found by using Ohm's law:

$$V_{R_B} = I_B \times R_B$$
$$= 166 \; \mu A \times 56 \; k\Omega$$
$$= 9.3 \; V$$

Recall that the B–E junction drops 0.7 V because it is forward biased. In Fig. 4-4, we clearly see how the sum of the voltage drops on the left side of the circuit equals the source voltage.

Using Ohm's law once again, we can solve for the voltage drop across R_C:

$$V_{R_C} = I_C \times R_C$$
$$= 22.3 \; mA \times 330 \; \Omega$$
$$= 7.36 \; V$$

We solve for the collector–emitter voltage drop by using Kirchhoff's law. If 7.36 V are developed across R_C, then the balance of the source voltage must appear from collector to emitter, as shown in Fig. 4-5.

$$V_{CE} = V_{CC} - V_{R_C}$$
$$= 10 \; V - 7.36 \; V$$
$$= 2.64 \; V$$

We will find that in the study of transistors, the collector–emitter resistance varies. The operation of the transistor is dependent on the biasing resistors of the circuit. Remember that one of the unique characteristics about semiconductors is the ability to change resistance. And as the C–E resistance changes, so does its voltage drop.

The voltage at the collector may be solved by using the previous values. First, realize that the voltage at the collector (V_C) is measured with reference to ground. From the voltages of Fig. 4-5, it is evident that the only voltage between the collector and ground is the collector-to-emitter voltage (V_{CE}). Therefore, $V_C = V_{CE}$ in this example.

Another way of approaching voltage V_C is through the formula

$$V_C = V_{CC} - V_{RC}$$

The value of this formula is that it can be applied to all biasing arrangements.

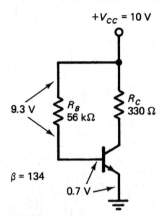

+V_{CC} = 10 V

9.3 V R_B 56 kΩ R_C 330 Ω

β = 134

0.7 V

FIGURE 4-4

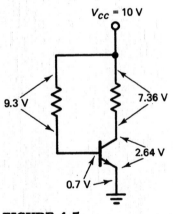

V_{CC} = 10 V

9.3 V   7.36 V

2.64 V

0.7 V

FIGURE 4-5

EXAMPLE 4-1

Consider the circuit of Fig. 4-3. If the circuit values are changed so that R_B = 100 kΩ, R_C = 510 Ω, β = 110, and V_{CC} = 12 V, what are the values of (a) I_B and (b) V_C?

Solution:
(a) The base current is found by determining the collector current and then using the value of beta.

$$I_C = \frac{12\ \text{V} - 0.7\ \text{V}}{100\ \text{k}\Omega/110}$$

$$= 12.4\ \text{mA}$$

$$I_B = \frac{12.4\ \text{mA}}{110}$$

$$= 113\ \mu\text{A}$$

(b) Knowing the value of collector current from the solution of part (a) allows us to find the voltage drop across the collector resistor. This voltage is subtracted from the source voltage to find V_C.

$$V_{R_C} = 12.4\ \text{mA} \times 510\ \Omega$$

$$= 6.32\ \text{V}$$

$$V_C = 12\ \text{V} - 6.32\ \text{V}$$

$$= 5.68\ \text{V} \qquad \blacksquare$$

Practice Problem: If the values of Fig. 4-3 are changed so that $V_{CC} = 15$ V, $R_B = 120$ kΩ, $R_C = 500\ \Omega$, and beta = 110, what is the value of V_C?   *Answer:*   8.45 V.

REVIEW QUESTIONS

16.   A BJT is __biased__ in order to set it up for proper operation.
17.   In a base-bias circuit the *B–E* junction is __Forward__ biased.
18.   The ratio of collector to base current is called __Beta__ .
19.   Base current can be found by (dividing or multiplying) the collector current ~~Dividing~~ by beta.
20.   A BJT's collector-to-emitter resistance can vary. True or false?

4-4   SATURATION AND CUTOFF

One of the uses of a transistor is as a switch. By definition, a switch has two modes of operation: it is either on or off. The BJT can be made to operate as a switch by running it in either saturation or cutoff.

Saturation can best be understood through the analogy of a sponge that can hold no more water—it is saturated. When a transistor reaches a point where it can conduct no more current, it is said to be saturated. **Saturation** can result from either decreased resistance or increased voltage. Let us suppose that the base resistor (R_B) of Fig. 4-4 is decreased to 41 kΩ. This drastically affects the various currents and voltages. As a matter of fact, if we repeat the steps of Section 4-3 using the new R_B value of 41 kΩ, we end up with:

> *Saturation:* The condition of a BJT that results in maximum current flow.

$$I_C = 30.4\ \text{mA}$$
$$I_B = 227\ \mu\text{A}$$
$$I_E = 30.6\ \text{mA}$$
$$V_{R_B} = 9.3\ \text{V}$$
$$V_{BE} = 0.7\ \text{V}$$
$$V_{R_C} = 10\ \text{V}$$
$$V_{CE} = 0\ \text{V}$$

Active @ 0.7v

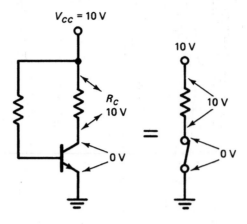

FIGURE 4-6 Saturated transistor compared to a switch.

The power-supply voltage is 10 V; therefore, the sum of V_{R_c} and V_{CE} cannot exceed 10 V.

When a transistor saturates, V_{CE} is reduced to zero volts and the collector current is at its maximum. The condition of the BJT is like that of a closed switch (see Fig. 4-6). The actual value of the saturation current is found by realizing that the total supply voltage is dropped across R_C; therefore,

$$I_{C(SAT)} = \frac{10\text{ V}}{330\ \Omega}$$

$$= 30.3\text{ mA}$$

The saturated base current can be found by dividing $I_{C(SAT)}$ by beta. Then the actual saturated emitter current can be found by adding these two saturation currents.

Cutoff: The condition of a BJT that results in no current flow.

Cutoff is the direct opposite of saturation. When a transistor is cut off, there is no current flow (I_E, I_C, or I_B) and the $C–E$ acts like an open switch. In Fig. 4-7, the voltage drops vividly illustrate the condition of the circuit. Since there is no current flow, there are no voltages across the various resistors.

Table 4-1 shows three possible operating conditions for the base bias circuit of Fig. 4-3.

The applications using a BJT as a switch are many in the area of digital electronics.

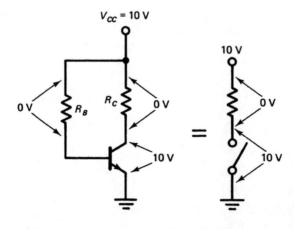

FIGURE 4-7 Cut-off transistor compared to a switch.

TABLE 4-1 Base Bias

	ACTIVE	SATURATION	CUTOFF
I_B	166 μA	227 μA	0 A
I_C	22.3 mA	30.4 mA	0 A
I_E	22.5 mA	30.6 mA	0 A
V_{RB}	9.3 V	9.3 V	0 V
V_{RC}	7.36 V	10 V	0 V
V_{CE}	2.64 V	0 V	10 V
V_C	2.64 V	0 V	10 V

EXAMPLE 4-2

Consider the base-bias circuit of Fig. 4-7. If the circuit values are $R_C = 510\ \Omega$, $\beta = 110$, and $V_{CC} = 12$ V, what are the saturated values of (a) V_{CE} and (b) I_B?

Solution:

(a) The voltage from collector to emitter for a saturated transistor is ideally zero volts.

(b) Maximum current flows through a saturated transistor. If zero volts are dropped across the collector–emitter in Fig. 4-7, then all of V_{CC} is found across the collector resistor. This enables us to find the saturated collector current, which will be used along with beta to determine the saturated base current.

$$I_{C(SAT)} = \frac{12\ \text{V}}{510\ \Omega}$$

$$= 23.5\ \text{mA}$$

$$I_{B(SAT)} = \frac{23.5\ \text{mA}}{110}$$

$$= 214\ \mu\text{A} \qquad \blacksquare$$

REVIEW QUESTIONS

21. I_C is maximum at (saturation or cutoff).

22. V_{CE} is approximately zero volts at (saturation or cutoff).

23. I_B equals _____ A at cutoff.

24. The C–E acts as a closed switch at _____ .

25. The C–E acts more like an (open or closed) switch at cutoff.

26. I_E is maximum at cutoff. True or false?

4-5 EMITTER-FEEDBACK BIAS

DC Analysis   All biasing arrangements have a common goal of providing stability for circuit operation. Eventually, an ac signal is applied to these transistor circuits. In order to prevent unwanted distortion of that signal, certain dc voltages must be maintained throughout the circuit. It is these dc voltages that are examined next. One common factor that affects biasing arrangements is heat. The heat that is produced from normal operation of a circuit can affect the beta of a

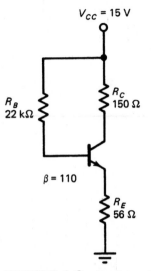

$V_{CC} = 15$ V

R_B
22 kΩ

R_C
150 Ω

β = 110

R_E
56 Ω

FIGURE 4-8 Emitter-feedback bias circuit.

BJT. As heat increases, beta increases. Let us examine how the emitter-feedback bias handles this problem.

First, a dc analysis is made of this circuit. In Fig. 4-8, we see an example of an emitter-feedback bias circuit. To determine the collector current, we must modify the previous formula to reflect the change in the circuit. Since we have added a resistor (R_E), the new formula is

$$I_C = \frac{V_{CC} - V_{BE}}{R_E + (R_B/\beta)}$$

Therefore, using the values of Fig. 4-8,

$$I_C = \frac{15 \text{ V} - 0.7 \text{ V}}{56 \text{ Ω} + (22 \text{ kΩ}/110)}$$

$$= 55.9 \text{ mA}$$

and

$$I_B = \frac{I_C}{\beta}$$

$$= 508 \text{ μA}$$

Therefore,

$$I_E = I_C + I_B$$

$$= 56.4 \text{ mA}$$

With this information and Ohm's law, the following voltage drops are easily solved. It is important to realize the path of the currents in order to understand the circuit's operation. Figure 4-9 shows the paths of electron current flow along with the voltage drops.

$V_{CC} = 15$ V

11.2 V R_B

R_C 8.39 V

3.45 V

0.7 V

R_E 3.16 V

FIGURE 4-9

$$V_{R_B} = I_B \times R_B$$

$$V_{BE} = 0.7 \text{ V}$$

$$V_{R_C} = I_C \times R_C$$

$$V_{R_E} = I_E \times R_E$$

$$V_{CE} = V_{CC} - (V_{R_C} + V_{R_E})$$

so

$$V_{R_B} = 11.2 \text{ V}$$

$$V_{BE} = 0.7 \text{ V}$$

$$V_{R_C} = 8.39 \text{ V}$$

$$V_{R_E} = 3.16 \text{ V}$$

$$V_{CE} = 3.45 \text{ V}$$

The results can be verified by summing the voltage drops of V_{R_C}, V_{R_E}, and V_{CE} to see if they equal the source voltage. Because the calculations are rounded to the third digit, there may be a slight difference. Also, summing the voltages of V_{R_B}, V_{BE}, and V_{R_E} should give the same result.

To find the voltage at each leg of the BJT with reference to ground, we simply add the voltage drops between the two points. For example, the voltage at the emitter with reference to ground is

$$V_E = 3.16 \text{ V}$$

Notice that this is the voltage drop across R_E, because V_{R_E} is the only voltage between ground and the emitter. The voltage at the collector is found by adding V_{R_E} and V_{CE}:

$$V_C = 6.61 \text{ V}$$

V_{BE} is not included when finding V_C, but rather when finding V_B. The voltage at the base in this circuit includes V_{R_E} and V_{BE}, which results in

$$V_B = 3.86 \text{ V}$$

Practice Problem: If the values of Fig. 4-8 are changed so that $V_{CC} = 10$ V, $R_B = 47$ kΩ, $R_C = 200$ Ω, $R_E = 110$ Ω, and beta = 150, what are the values of V_C and I_B? *Answer:* $V_C = 5.61$ V and $I_B = 146$ μA.

Stability of Bias

After having determined all of the dc voltages, what would happen if heat caused the beta of the BJT to increase? Realize that beta affects the total transistor. Consequently, if beta increases, I_C will attempt to increase, which means I_E also tries to increase. However, if I_E should try to increase, then V_{R_E} must also increase. Keep in mind that the sum of V_{R_B}, V_{BE}, and V_{R_E} must always equal the source voltage. Since V_{BE} is a constant (0.7 V), an increase in V_{R_E} means a decrease in V_{R_B}. The effect of V_{R_B} decreasing is a decrease in base current, which is directly related to I_C through beta. If I_B causes I_C to decrease, then I_E must decrease. What began as an increase in current resulted in a stable current due to the biasing circuit. In other words, the effect of heat on beta is offset, or compensated for, by means of the biasing arrangement. This is an example of **thermal stability.**

Thermal Stability: A condition that is unaffected by changes in temperature.

Saturation

Up to this point, the discussion was concerned with the operation of the emitter-feedback circuit in the active region. Yet if saturation or cutoff were to occur, the voltage and current values within the circuit would change. If the emitter-feedback circuit discussed in Fig. 4-8 were to saturate, the following changes would occur.

First, the collector-to-emitter voltage is reduced to zero. This means that the source voltage must split between V_{R_C} and V_{R_E}. Maximum current is flowing and can be approximated as follows:

$$I_C = \frac{V_{CC}}{R_C + R_E}$$

$$= \frac{15 \text{ V}}{150 \ \Omega + 56 \ \Omega}$$

$$= 72.8 \text{ mA}$$

$$I_E = 72.8 \text{ mA}$$

$$I_B = 662 \ \mu\text{A}$$

Because an approximation is being used, I_C and I_E are equal. Finally, the saturated base current is found by dividing I_C by the value of beta.

Saturation can occur when there is either a decrease in resistance or an increase in voltage. This can happen accidentally as a result of a short or an open, or it may be intentional as in the case of digital applications.

Calculate the approximate voltage drops for some of the resistors. The voltage across R_E and R_C are discovered by using the saturated currents and Ohm's law.

$$V_{R_C} = I_{C(SAT)} \times R_C$$
$$= 10.9 \text{ V}$$

and

$$V_{R_E} = 4.1 \text{ V}$$
$$V_{CE} = 0 \text{ V}$$

An attempt to find the voltage across the base resistor would result in an erroneous figure. This is because to assume saturation is to assume a change in the original circuit. In this circuit, saturation could be caused by decreasing the base resistance. The actual value of resistance can be found by determining V_{R_B} and I_B. The value of I_B has already been found, leaving V_{R_B} to be solved. The following formula is based on the fact that the sum of V_{R_E}, V_{BE}, and V_{R_E} equals V_{CC}.

$$V_{R_B} = V_{CC} - (V_{R_E} + V_{B_E})$$
$$= 15 \text{ V} - (4.1 \text{ V} + 0.7 \text{ V})$$
$$= 10.2 \text{ V}$$

Therefore,

$$R_B = \frac{V_{R_B}}{I_B}$$
$$= \frac{10.2 \text{ V}}{662 \ \mu\text{A}}$$
$$= 15.4 \text{ k}\Omega$$

If R_B were decreased from 22 to 15.4 kΩ or less, the transistor would saturate. Though we have maintained a V_{BE} of 0.7 V during saturation, it should be noted that in reality V_{BE} would be affected. But because the voltage is so small to begin with, the change is too negligible to concern us.

Cutoff   The other extreme is cutoff, when no current flows within the transistor. The cause may be an open inside the BJT or in a connected component or possibly on the printed circuit board itself. If there is no base current, then there can be no collector or emitter current. Also, if there is no emitter current, then there can be no base or collector current. Our starting point is to assume an open at point A in the circuit of Fig. 4-10.

Since the open occurs on the base, there is no path for base current to flow. The outcome is that the transistor shuts down, causing its collector-to-emitter region to appear as an open. The source voltage is dropped across the collector-to-emitter region and there are no other voltage drops. This is a result of the absence of current flow.

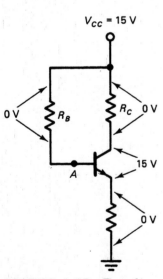

FIGURE 4-10   Results of cutoff.

TABLE 4-2 Emitter-Feedback Bias

	ACTIVE	SATURATION	CUTOFF
I_B	508 μA	662 μA	0 A
I_C	55.9 mA	72.8 mA	0 A
I_E	56.4 mA	72.8 mA	0 A
V_{R_B}	11.2 V	10.2 V	0 V
V_{R_C}	8.39 V	10.9 V	0 V
V_{R_E}	3.16 V	4.1 V	0 V
V_{CE}	3.45 V	0 V	15 V
V_E	3.16 V	4.1 V	0 V
V_B	3.86 V	4.8 V	0 V
V_C	6.61 V	4.8 V	15 V

Table 4-2 shows three possible operating conditions for the emitter-feedback bias circuit of Fig. 4-8.

EXAMPLE 4-3 _____

The values of Fig. 4-8 are replaced with R_B = 150 kΩ, R_C = 850 Ω, R_E = 100 Ω, β = 100, and V_{CC} = 12 V. Determine the voltage at the collector under (a) active, (b) saturated, and (c) cutoff operating conditions.

Solution:

(a) Start by applying the formula for finding collector current. This current is then used to find the voltage drop across R_C. The balance of V_{CC} is found at the collector.

$$I_C = \frac{12 \text{ V} - 0.7 \text{ V}}{100 \text{ } \Omega + (150 \text{ k}\Omega/100)}$$

$$= 7.06 \text{ mA}$$

$$V_{R_C} = 7.06 \text{ mA} \times 850 \text{ } \Omega$$

$$= 6 \text{ V}$$

$$V_C = 12 \text{ V} - 6 \text{ V}$$

$$= 6 \text{ V}$$

(b) The V_{CE} of a saturated circuit is equal to 0 V. Therefore, the V_{CC} must be divided between R_C and R_E. By finding the voltage across R_C and applying the formula for finding V_C, we can find the saturated voltage at the collector.

$$I_{C(\text{SAT})} = \frac{12 \text{ V}}{850 \text{ } \Omega + 100 \text{ } \Omega}$$

$$= 12.6 \text{ mA}$$

$$V_{R_C(\text{SAT})} = 12.6 \text{ mA} \times 850 \text{ } \Omega$$

$$= 10.7 \text{ V}$$

$$V_{C(\text{SAT})} = 12 \text{ V} - 10.7 \text{ V}$$

$$= 1.3 \text{ V}$$

(c) A transistor that ceases to conduct is considered to be cut off. The V_{CE} of a cut-off transistor is equal to the source voltage. Since no current flows, there are no other voltage drops. The voltage at the collector includes V_{CE}; therefore, V_C equals the source voltage.

$$V_{C(CUTOFF)} = 12 \text{ V} \qquad \blacksquare$$

REVIEW QUESTIONS

27. How does the emitter-feedback bias circuit differ physically from the base bias circuit?
28. The sum of V_{R_B} and V_{R_E} equals V_{CC}. True or false?
29. As I_C increases, V_C decreases. True or false?
30. As I_B increases, I_C increases. True or false?
31. As V_{R_E} increases, V_{R_B} increases. True or false?
32. At saturation, (maximum or minimum) current flows.
33. At cutoff, V_{CE} equals (zero or V_{CC}).
34. Heat causes beta to (increase or decrease).

4-6   COLLECTOR-FEEDBACK BIAS

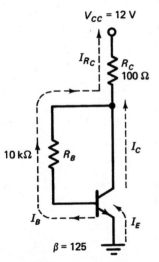

FIGURE 4-11   Collector-feedback bias circuit.

Another circuit arrangement that provides a method of biasing is the collector-feedback bias (also known as "self-bias"). An example of the circuit is shown in Fig. 4-11. There is no longer an emitter resistor, and the base resistor is tied to the collector.

DC Analysis   This difference in circuit configuration brings with it a change in the formula for determining collector current.

$$I_C = \frac{V_{CC} - V_{BE}}{R_C + (R_B/\beta)}$$

Applying the circuit values of Fig. 4-11 to this formula yields a collector current of

$$I_C = \frac{12 \text{ V} - 0.7 \text{ V}}{100 \ \Omega + (10 \text{ k}\Omega/125)}$$

$$= 62.8 \text{ mA}$$

Therefore,

$$I_B = \frac{I_C}{\beta}$$

$$= 502 \ \mu\text{A}$$

and

$$I_E = I_C + I_B$$

$$= 63.3 \text{ mA}$$

If we closely examine the paths of current throughout the circuit of Fig. 4-11, we notice that I_C and I_B join to form the current through the collector resistor. The current through R_C can be calculated as

$$I_{R_C} = I_C + I_B$$
$$= 63.3 \text{ mA}$$

which is the value of the total current that initially flows into the emitter of the transistor.

By using these current values, the following voltage drops can be found:

$$V_{R_C} = I_{R_C} \times R_C$$
$$= 6.33 \text{ V}$$

$$V_{CE} = V_{CC} - V_{R_C}$$
$$= 5.67 \text{ V}$$

$$V_{R_B} = I_B \times R_B$$
$$= 5.02 \text{ V}$$

$$V_{BE} = 0.7 \text{ V}$$

Notice how $V_{R_C} + V_{CE}$ equal the source voltage. Also, V_{BE}, V_{R_B}, and V_{R_C} add up to the source voltage. A closer examination reveals that V_{R_C} is included in both summing processes. This means that V_{R_C} is the compensating factor should beta increase due to heat.

The final voltages to be solved are those at the collector, emitter, and base with reference to ground. This involves recognizing the voltage drops that lay between a specific leg and ground.

$$V_E = 0 \text{ V}$$
$$V_B = 0.7 \text{ V}$$
$$V_C = 5.67 \text{ V}$$

Once again, the operation of a specific bias should result in offsetting thermal (heat) instability. Should heat cause beta to increase, the collector current attempts to increase. An increase in collector current means an increase in the voltage across the collector resistor. If V_{R_C} increases, then the voltage across the base resistor must decrease. This results in lowering the base current, which, in turn, lowers the collector current. So the conclusion is as currents I_C, I_E, and I_B attempt to increase, the collector resistor offsets them.

Practice Problem: If the values of Fig. 4-11 are changed so that $V_{CC} = 15$ V, $R_C = 270 \ \Omega$, $R_B = 18 \text{ k}\Omega$, and beta = 135, what are the values of V_C and I_B? *Answer:* $V_C = 5.43$ V and $I_B = 263 \ \mu\text{A}$.

Saturation   The collector-feedback bias circuit is impossible to saturate. In Fig. 4-12, the base resistor is deliberately shorted by means of a piece of wire. The outcome is that the collector resistor provides the function of the base resistor. As we can see, the sum of V_{R_C} and V_{BE} equals the source voltage. This leaves a voltage of 0.7 V across the collector-to-emitter region.

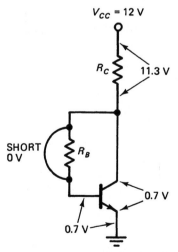

FIGURE 4-12   Attempted saturation of the collector-feedback bias circuit.

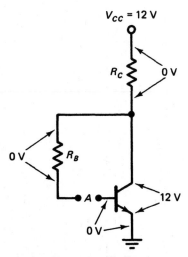

$V_{CC} = 12$ V

R_C

0 V

0 V

R_B

A

12 V

0 V

FIGURE 4-13 Collector feedback in cutoff.

Cutoff The collector-feedback circuit is as susceptible to cutoff as any other biasing arrangement. If the base or emitter current were reduced to zero amperes, then the transistor would be inoperative. The collector-to-emitter region would act as an open switch.

In Fig. 4-13, an open has been introduced at point A. The effect is no base, emitter, or collector current. This in turn produces the voltage drops shown. We are concentrating on the base because in digital applications, this point can be easily controlled. Zero volts applied to the base-to-emitter junction causes cutoff. This is examined in a later section.

Table 4-3 shows three possible operating conditions for the collector-feedback bias circuit of Fig. 4-11.

EXAMPLE 4-4

The values of Fig. 4-11 are replaced with $R_B = 75$ kΩ, $R_C = 470$ Ω, $\beta = 125$, and $V_{CC} = 10$ V. Determine the voltage at the collector under (a) active, (b) saturated, and (c) cutoff operating conditions.

Solution:

(a) The active region of operation for a transistor lies somewhere between cutoff and saturation. First, the collector current is calculated and used to find the voltage across R_C. Next, the voltage across the resistor is subtracted from the source voltage and the difference is equal to V_C.

$$I_C = \frac{10 \text{ V} - 0.7 \text{ V}}{470 \text{ }\Omega + (75 \text{ k}\Omega/125)}$$

$$= 8.69 \text{ mA}$$

$$V_{R_C} = 8.69 \text{ mA} \times 470$$

$$= 4.08 \text{ V}$$

$$V_C = 10 \text{ V} - 4.08 \text{ V}$$

$$= 5.92 \text{ V}$$

(b) At saturation, V_{CE} is equal to 0 V. Since the only voltage between the collector and ground is V_{CE}, the voltage at the collector also equals 0 V.

$$V_{C(\text{SAT})} = 0 \text{ V}$$

(c) $V_{CE} = V_{CC}$ when the transistor is cut off. Once again, V_{CE} is the only voltage drop between the collector and ground in this circuit and is, therefore, equivalent to V_C.

$$V_{C(\text{SAT})} = 10 \text{ V} \qquad \blacksquare$$

TABLE 4-3 Collector-Feedback Bias

	ACTIVE	SATURATION	CUTOFF
I_B	502 μA	904 μA	0 A
I_C	62.8 mA	112 mA	0 A
I_E	63.3 mA	113 mA	0 A
V_{R_B}	5.02 V	0 V	0 V
V_{R_C}	6.33 V	11.3 V	0 V
V_{CE}	5.67 V	0.7 V	12 V
V_E	0 V	0 V	0 V
V_B	0.7 V	0.7 V	0 V
V_C	5.67 V	0.7 V	12 V

35. I_{R_C} equals I_B in the collector-feedback bias circuit. True or false?
36. _____ instability indicates fluctuations in circuit operation due to heat.
37. As I_E increases, I_C (increases or decreases).
38. If R_C were to open, the transistor would (cut off or saturate).
39. What is the voltage at the base of a properly operating BJT in the collector-feedback bias arrangement?

4-7 EMITTER BIAS

This circuit differs from the others in that it uses two separate power supplies to bias the transistor. The use of two supplies provides a greater range of operating voltages. Up to this point, only positive dc voltages were available in the circuit, and now the range has been extended to include negative voltages.

To study the operation of this circuit (see Fig. 4-14), we divide it into two loops. Next, we make an assumption with regard to voltage across the base resistor. If the circuit is properly designed, V_{R_B} will be much less than V_{BE}. Therefore, in calculations, treat it as if it were zero volts.

Beginning with loop A, there are essentially two places for voltage to drop. V_{BE} is 0.7 V as usual. This means that the remaining 9.3 V of the negative supply is across the emitter resistor. The emitter current is

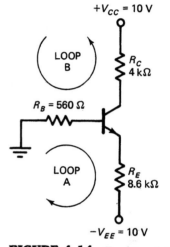

FIGURE 4-14 Emitter-bias circuit.

$$I_E = \frac{V_{R_E}}{R_E}$$

$$= \frac{9.3 \text{ V}}{8.6 \text{ k}\Omega}$$

$$= 1.08 \text{ mA}$$

Since the collector current approximately equals the emitter current, the value of 1.08 mA is used for it. Therefore, the voltage across the collector resistor is

$$V_{R_C} = I_C \times R_C$$

$$= 1.08 \text{ mA} \times 4 \text{ k}\Omega$$

$$= 4.33 \text{ V}$$

In loop B, there are two places for voltage to develop. One is V_{R_C} and the other is V_{BC}, which must drop the remainder of V_{CC}. Therefore,

$$V_{BC} = V_{CC} - V_{R_C}$$

$$= 10 \text{ V} - 4.33 \text{ V}$$

$$= 5.67 \text{ V}$$

It can be stated that the voltage at the collector with reference to ground is 5.67 V and the voltage at the emitter is −0.7 V. The voltage between collector and emitter must equal the difference between these two voltages:

$$V_{CE} = 5.67 \text{ V} - (-0.7 \text{ V})$$

$$= 5.67 \text{ V} + 0.7 \text{ V}$$

$$= 6.37 \text{ V}$$

If we reflect upon V_{R_C}, V_{CE}, and V_{R_E}, it becomes evident that their sum is 20 V. This is as it should be, for the circuit is composed of a positive and negative supply whose total range spans 20 V.

EXAMPLE 4-5

The values of the emitter-bias circuit of Fig. 4-14 are changed so that $R_C = 2.7$ kΩ, $R_E = 4.7$ kΩ, $R_B = 330$ Ω, $+V_{CC} = 12$ V, and $-V_{EE} = 12$ V. Determine the voltage at the collector.

Solution: The voltage at the collector is measured with reference to ground. Since V_{R_B} is assumed to equal 0 V, the only other voltage that lies between the collector and ground is V_{CB} (collector–base voltage). We use the same formula as in the previous examples to find V_C and, therefore, V_{CB}.

Begin by solving for the emitter current. Remember that V_{BE} is 0.7 V and the balance of $-V_{EE}$ is found across R_E.

$$I_E = \frac{11.3 \text{ V}}{4.7 \text{ k}\Omega}$$

$$= 2.4 \text{ mA}$$

$$I_C = I_E = 2.4 \text{ mA}$$

$$V_{R_C} = 2.4 \text{ mA} \times 2.7 \text{ k}\Omega$$

$$= 6.48 \text{ V}$$

$$V_C = 12 \text{ V} - 6.48 \text{ V}$$

$$= 5.52 \text{ V} \qquad \blacksquare$$

Practice Problem: If the values of Fig. 4-14 were changed so that $V_{CC} = 15$ V, $-V_{EE} = 15$ V, $R_C = 12$ kΩ, $R_E = 22$ kΩ, and $R_B = 680$ Ω, what are the values of V_{CE} and I_C? *Answer:* $V_{CE} = 7.9$ V and $I_C = 650$ µA.

REVIEW QUESTIONS

40. The emitter-bias circuit uses both a positive and negative supply. True or false?

41. I_E and I_C are considered approximately equal in the emitter-bias circuit. True or false?

42. The voltage across R_B in a properly biased emitter-bias circuit is _____ .

43. What is the formula used to find V_{CE} in an emitter-bias circuit?

4-8 VOLTAGE-DIVIDER BIAS

DC Analysis One of the more popular forms of biasing is voltage-divider bias. It is a very stable circuit design. We will notice that beta is not a consideration in any of the calculations. This means that if beta increases due to heat, it has little effect on the dc operating voltages of the circuit. Therefore, the voltage-divider bias has a high degree of **thermal stability.**

Thermal Stability: A condition that is unaffected by changes in temperature.

Figure 4-15 is an example of the voltage-divider bias circuit. Its name is derived from the voltage divider formed by resistors R_1 and R_2. The voltages across these two resistors are solved for the same as with any series circuit. Begin by determining the current through both these resistors:

$$I_{R1+R2} = \frac{V_{CC}}{R_1 + R_2}$$

$$= \frac{15 \text{ V}}{10 \text{ k}\Omega + 7.5 \text{ k}\Omega}$$

$$= 857 \ \mu\text{A}$$

In Fig. 4-15, the base current joins with the current through R_2. It may seem that this should be taken into consideration in determining R_1's current. However, if the circuit is properly designed, the base current should not have much of an effect on the base voltage (V_B). The rule of thumb in designing a voltage divider is that I_{R2} should be about one-tenth the size of the collector current, or 10 times the base current.

Now that the current through each of these resistors is known, the next step is to determine their individual voltage drops:

$$V_{R1} = I_{R1} \times R_1$$

$$= 857 \ \mu\text{A} \times 10 \text{ k}\Omega$$

$$= 8.57 \text{ V}$$

and

$$V_{R2} = I_{R2} \times R_2$$

$$= 857 \ \mu\text{A} \times 7.5 \text{ k}\Omega$$

$$= 6.43 \text{ V}$$

Focusing on R_2 of Fig. 4-15, we notice that it is separated from R_E by the base-to-emitter junction of the BJT. Because V_{BE} is 0.7 V and V_{R_E} is at a more negative potential, the value of V_{R_E} is found by

$$V_{R_E} = V_{R2} - V_{BE}$$

$$= 6.43 \text{ V} - 0.7 \text{ V}$$

$$= 5.73 \text{ V}$$

Therefore,

$$I_E = \frac{V_{R_E}}{R_E}$$

$$= \frac{5.73 \text{ V}}{680 \ \Omega}$$

$$= 8.42 \text{ mA}$$

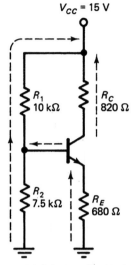

FIGURE 4-15 Voltage-divider bias circuit.

And I_C is approximately equal to I_E:

$$I_C = 8.42 \text{ mA}$$

and

$$V_{R_C} = I_C \times R_C$$
$$= 8.42 \text{ mA} \times 820 \ \Omega$$
$$= 6.91 \text{ V}$$

Lastly,

$$V_{CE} = V_{CC} - (V_{R_C} + V_{R_E})$$
$$= 15 \text{ V} - (6.91 \text{ V} + 5.73 \text{ V})$$
$$= 2.36 \text{ V}$$

The steps for solving many of these values should look familiar. The voltage-divider bias is probably the most straightforward value to solve for since there are no formulas besides Ohm's law. The voltages on the left and right sides of the BJT add up to the source voltage. The base-to-emitter voltage is not included in either of the summing processes.

The voltages at the various legs of the transistor can be determined by recognizing the individual voltage drops within the circuit. For instance, the voltage at the base is equivalent to the voltage across R_2:

$$V_B = V_{R2}$$
$$= 6.43 \text{ V}$$

and the voltage at the emitter is the same as V_{R_E}:

$$V_E = V_{R_E}$$
$$= 5.73 \text{ V}$$

Finally, the voltage at the collector is the sum of V_{R_E} and V_{CE}:

$$V_C = V_{R_E} + V_{CE}$$
$$= 8.09 \text{ V}$$

Practice Problem: If the values of Fig. 4-15 were changed to $V_{CC} = 20$ V, $R_C = 470 \ \Omega$, $R_E = 220 \ \Omega$, $R_1 = 6.8$ kΩ, and $R_2 = 2.5$ kΩ, what is the value of V_C? *Answer:* $V_C = 10$ V.

Saturation   As with the other circuits, saturation causes the transistor to act like a closed switch. The collector-to-emitter voltage decreases to zero volts. This leaves the supply voltage to be dropped across R_C and R_E. The saturated collector current of the circuit of Fig. 4-15 is found by

$$I_C = \frac{V_{CC}}{R_C + R_E}$$
$$= \frac{15 \text{ V}}{820 \ \Omega + 680 \ \Omega}$$
$$= 10 \text{ mA}$$

Therefore,

$$I_E = 10 \text{ mA}$$

and

$$V_{R_C} = I_C \times R_C$$
$$= 10 \text{ mA} \times 820 \ \Omega$$
$$= 8.2 \text{ V}$$

$$V_{CE} = 0 \text{ V}$$

$$V_{R_E} = I_E \times R_E$$
$$= 10 \text{ mA} \times 680 \ \Omega$$
$$= 6.8 \text{ V}$$

It was stated earlier that saturation must have a cause. One way in which a saturated condition is created is through an increase in resistor R_2 or a decrease of resistor R_1. A higher resistance value for R_2 results in a larger base voltage, which increases the emitter voltage and current. On the other hand, the base current travels through R_1; a decrease in resistance causes an increase in current.

Cutoff The characteristics of cutoff are also the same as discussed earlier. The collector-to-emitter region acts like an open switch, dropping the source voltage. This leaves zero volts for both V_{R_C} and V_{R_E}. There is no current flow (I_E, I_C, or I_B). Finally, the voltage divider formed by R_1 and R_2 is unaffected. The source voltage continues to split between these two resistances.

Table 4-4 shows three possible operating conditions for the voltage-divider bias circuit of Fig. 4-15.

EXAMPLE 4-6 _____

The values of the voltage-divider bias circuit of Fig. 4-15 are changed so that $R_1 = 47 \text{ k}\Omega$, $R_2 = 27 \text{ k}\Omega$, $R_E = 220 \ \Omega$, $R_C = 390 \ \Omega$, and $V_{CC} = 10 \text{ V}$. Determine the voltage at the collector under (a) active, (b) saturated, and (c) cutoff operating conditions.

TABLE 4-4 Voltage-Divider Bias

	ACTIVE	SATURATION	CUTOFF
I_C	8.42 mA	10 mA	0 A
I_E	8.42 mA	10 mA	0 A
V_{R1}	8.57 V	*	8.57 V
V_{R2}	6.43 V	*	6.43 V
V_{R_E}	5.73 V	6.8 V	0 V
V_{R_C}	6.91 V	8.2 V	0 V
V_{CE}	2.36 V	0 V	15 V
V_E	5.73 V	6.8 V	0 V
V_B	6.43 V	7.5 V	*6.43 V
V_C	8.09 V	6.8 V	15 V

* This component involved in a fault

Solution:

(a) Begin with the voltage divider of R_1 and R_2. Find the voltage at the base that allows the voltage at the emitter to be found. The emitter current is then calculated (which is equal to I_C) and is used to find the voltage across R_C. Finally, the voltage at the collector is the difference between V_{R_C} and V_{CC}.

$$I_{R1+R2} = \frac{10 \text{ V}}{47 \text{ k}\Omega + 27 \text{ k}\Omega}$$
$$= 135 \text{ } \mu\text{A}$$

$$V_{R2} = 135 \text{ } \mu\text{A} \times 27 \text{ k}\Omega$$
$$= 3.65 \text{ V}$$

$$V_{R_E} = 3.65 \text{ V} - 0.7 \text{ V}$$
$$= 2.95 \text{ V}$$

$$I_E = \frac{2.95 \text{ V}}{220 \text{ } \Omega}$$
$$= 13.4 \text{ mA}$$

$$I_C = I_E = 13.4 \text{ mA}$$

$$V_{R_C} = 13.4 \text{ mA} \times 390 \text{ } \Omega$$
$$= 5.23 \text{ V}$$

$$V_C = 10 \text{ V} - 5.23 \text{ V}$$
$$= 4.77 \text{ V}$$

(b) If the voltage divider were to saturate, V_{CE} would equal 0 V. This leaves V_{CC} to be divided between R_C and R_E. The voltage at the collector is equal to the difference between V_{CC} and V_{R_C}.

$$I_{C(SAT)} = \frac{10 \text{ V}}{220 \text{ } \Omega + 390 \text{ } \Omega}$$
$$= 16.4 \text{ mA}$$

$$V_{R_C(SAT)} = 16.4 \text{ mA} \times 390 \text{ } \Omega$$
$$= 6.4 \text{ V}$$

$$V_{C(SAT)} = 10 \text{ V} - 6.4 \text{ V}$$
$$= 3.6 \text{ V}$$

(c) The V_{CE} of the transistor increases to V_{CC} when it is cut off. Since this voltage lies between the collector and ground, $V_{C(CUTOFF)}$ is equal to the source voltage.

$$V_{C(CUTOFF)} = 10 \text{ V} \qquad \blacksquare$$

REVIEW QUESTIONS

44. Will a change in beta affect the voltage-divider bias from the perspective of the formulas discussed?

45. Which resistors form the voltage divider in Fig. 4-15?

46. The voltage across R_2 and R_E in Fig. 4-15 are (directly or inversely) related.

47. The voltage at the collector and the voltage at the base in Fig. 4-15 are (directly or inversely) related.

48. What is the saturation voltage from the collector to the emitter in a voltage-divider bias circuit?

49. What is the voltage at the collector in a voltage-divider bias circuit at cutoff?

4-9 TROUBLESHOOTING

The topic of discussion at this point is troubleshooting a transistor circuit. The voltage-divider bias circuit of Fig. 4-16 is used. A series of "what if" scenarios will demonstrate the outcome of a variety of troubles. The goal is to understand why certain voltages increase, decrease, or remain the same.

Start by calculating the circuit values under normal operating conditions. Then consider each of the following "faults" and the effects they have on the various voltages.

Results of an Open in R_2

V_{R1} = decreases
V_{R_C} = increases
V_{R_E} = increases
V_{CE} = decreases

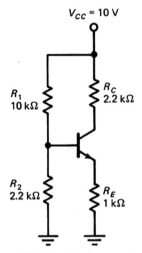

FIGURE 4-16 Voltage-divider bias circuit.

When R_2 opens, it is no longer part of the circuit. The circuit actually changes into an emitter-feedback bias circuit (see Fig. 4-8). The emitter current increases to the point of saturating the transistor. A standard beta value of 100 can be used to calculate the drastic change in circuit conditions. Saturation is the cause of V_{CE} decreasing (0 V). And if I_E increases, then I_C must increase, causing V_{R_C} to increase. Finally, an increase in V_{R_E} due to the increase in I_E means a decrease in V_{R1} (see Section 4-5, "Emitter-Feedback Bias").

Results of an Open in R_1

V_{R2} = decreases to zero
V_{R_C} = decreases to zero
V_{R_E} = decreases to zero
V_{CE} = increases to V_{CC}

When R_1 opens, the path for both base current and the voltage-divider current is taken away. The result is a transistor operating at cutoff. The voltages that decrease measure zero volts and the voltages that increase, including V_{R1}, show the source voltage.

Results of an Open in R_E

V_{R1} = remains the same
V_{R2} = remains the same
V_{R_C} = decreases to zero
V_{CE} = decreases to approximately zero

The results in this circuit seem rather odd at first. But we must remember that opening R_E prevents emitter current. This results in a cut-off transistor whose collector-to-emitter region acts like an open switch. Viewing the circuit as having both an open R_E and open C–E leg may help to explain why the individual components show a decrease in voltage. The cut-off transistor acts as though it were not there, so the voltage divider (R_1 and R_2) is unaffected.

Results of Shorting R_2

V_{R1} = increases to V_{CC}

V_{R_E} = decreases to zero

V_{R_C} = decreases to zero

V_{CE} = increases to V_{CC}

A short always drops a potential of zero volts. If V_{R2} is zero, then the source voltage must be across R_1, which explains the increase. Also, if V_{R2} is zero volts, then V_{R_E} must be zero volts (we cannot subtract 0.7 V from zero). And if V_{R_E} is zero, the emitter current is zero ampere. No emitter current causes cutoff, which explains the increase in V_{CE} (to source voltage) and also the decrease in V_{R_C} (to 0 V).

Results of Shorting R_E

V_{R1} = increases

V_{R2} = decreases to approximately 0.7 V

V_{R_C} = increases

V_{CE} = decreases to approximately zero

It is important to note the relationship between V_{R2} and V_{R_E}. Shorting R_E reduces V_{R_E} to zero volts, which means that V_{R2} is 0.7 V higher in potential (or $V_{R2} = 0.7$ V). The decrease in V_{R2} increases V_{R1}. The shorting of R_E (no resistance) causes a greater emitter current, which saturates the transistor. Saturation explains why V_{CE} decreases, and the larger I_E ($= I_C$) explains why V_{R_C} increases.

Static Test of a Transistor by Using an Ohmmeter
Transistors are PN junction semiconductor devices. This means that they react essentially the same as diodes when tested with an ohmmeter. The difference is that there are two junctions to test. A transistor should be removed from the circuit before testing to prevent other components from interfering.

Once the BJT is removed from the circuit, the forward resistance of the B–E junction is measured by placing the positive lead to the P-type material and the common lead to the N-type material. We should read some resistance value.

By reversing the leads, we can measure the reverse resistance of the B–E junction. This reading should be much higher or possibly infinite. If zero ohms is measured in either direction or infinity is measured in both directions, we have a bad transistor.

The same procedure should be followed for the C–B junction. The key is to identify the legs properly and realize whether we are working with a PNP or NPN transistor.

Finally, any resistance reading from C–E (in either direction) should be suspect.

50. What value of resistance is measured across a short?
51. What value of resistance is measured across an open?
52. What value of voltage is measured across a short?
53. What value of voltage is measured across an open?
54. A short generally (increases or decreases) current flow along the path in which it occurs.
55. An open generally (increases or decreases) current flow along the path in which it occurs.
56. If R_1 opens in the circuit of Fig. 4-16, V_C (increases, decreases, or remains the same).
57. If R_2 shorts in the circuit of Fig. 4-16, V_C (increases, decreases, or remains the same).
58. If R_E opens in the circuit of Fig. 4-16, V_C (increases, decreases, or remains the same).
59. If R_C shorts in the circuit of Fig. 4-16, V_C (increases, decreases, or remains the same).
60. The forward resistance of the $B-E$ junction should measure infinite. True or false?
61. A measurement of zero ohms at the $B-C$ junction indicates a bad transistor. True or false?

SUMMARY

1. BJT stands for bipolar junction transistor.
2. Bipolar means that it depends on both majority and minority carriers for current flow.
3. Transistor is an acronym for "transfer resistor."
4. A BJT is composed of three regions: the emitter, base, and collector.
5. The emitter is heavily doped and emits electrons into the base.
6. The base is lightly doped and passes the emitted electrons to the collector.
7. The collector is intermediately doped and collects the electrons.
8. NPN refers to an N-type emitter, P-type base, and N-type collector.
9. PNP refers to a P-type emitter, N-type base, and P-type collector.
10. A BJT has two PN junctions: the $B-E$ (base–emitter) and $B-C$ (base–collector).
11. The $B-E$ junction is normally forward biased and the $B-C$ is normally reverse biased.
12. $I_E = I_C + I_B$ $(I_E \cong I_C)$
13. Electron current flows into the emitter of an NPN BJT and out of the base and collector.
14. The operation of the BJT is controlled by the base current.

15. The current gain of a BJT is stated in terms of beta.

$$\beta = \frac{I_C}{I_B}$$

16. Important formulas for *base bias* are

$$I_C = \frac{V_{CC} - V_{BE}}{R_B/\beta} \qquad V_{CE} = V_{CC} - V_{R_C}$$
$$\qquad\qquad\qquad\qquad V_C = V_{CC} - V_{R_C}$$
$$I_B = \frac{I_C}{\beta} \qquad\qquad I_{C(SAT)} = \frac{V_{CC}}{R_C}$$
$$I_E = I_C + I_B$$
$$V_{R_B} = I_B \times R_B \qquad I_{B(SAT)} = \frac{I_{C(SAT)}}{\beta}$$
$$V_{R_C} = I_C \times R_C$$

17. Saturation is a condition wherein the BJT is conducting the maximum possible current and $V_{CE} \cong 0$ V.

18. Cutoff is a condition wherein the BJT does not conduct and $V_{CE} \cong V_{CC}$.

19. Important formulas for *emitter-feedback bias* are

$$I_C = \frac{V_{CC} - V_{BE}}{R_E + (R_B/\beta)} \qquad V_{CE} = V_{CC} - (V_{R_C} + V_{R_E})$$
$$\qquad\qquad\qquad\qquad V_E = V_{R_E}$$
$$I_B = \frac{I_C}{\beta} \qquad\qquad V_C = V_{CC} - V_{R_C}$$
$$\qquad\qquad\qquad\qquad V_B = V_{R_E} + V_{BE}$$
$$I_E = I_C + I_B$$
$$V_{R_B} = I_B \times R_B \qquad I_{C(SAT)} = \frac{V_{CC}}{R_C + R_E}$$
$$V_{R_C} = I_C \times R_C$$
$$V_{R_E} = I_E \times R_E \qquad I_{B(SAT)} = \frac{I_{C(SAT)}}{\beta}$$
$$V_{BE} = 0.7 \text{ V}$$

20. Important formulas for *collector–feedback bias* are

$$I_C = \frac{V_{CC} - V_{BE}}{R_C + (R_B/\beta)} \qquad V_{R_B} = I_B \times R_B$$
$$\qquad\qquad\qquad\qquad V_{R_C} = I_C \times R_C$$
$$I_B = \frac{I_C}{\beta} \qquad\qquad V_{CE} = V_{CC} - V_{R_C}$$
$$\qquad\qquad\qquad\qquad V_E = 0 \text{ V}$$
$$I_E = I_C + I_B \qquad V_B = V_{BE} = 0.7 \text{ V}$$
$$V_{BE} = 0.7 \text{ V} \qquad V_C = V_{CC} - V_{R_C}$$

21. The collector-feedback bias circuit is impossible to saturate.

22. The emitter-bias circuit requires two separate power supplies, allowing for a greater range of output voltages.

23. Important formulas for *emitter bias* are

$$V_{BE} = 0.7 \text{ V} \qquad\qquad I_C \cong I_E$$
$$V_{R_B} \cong 0 \text{ V} \qquad\qquad V_{R_C} = I_C \times R_C$$
$$V_{R_E} = V_{EE} - V_{BE} \qquad V_{BC} = V_{CC} - V_{R_C}$$
$$I_E = \frac{V_{R_E}}{R_E} \qquad\qquad V_{CE} = V_{BC} - (-V_{BE})$$

24. The voltage-divider bias is a very stable method of biasing since it is independent of beta.

25. Important formulas for *voltage-divider bias* are

$$I_{R1+R2} = \frac{V_{CC}}{R_1 + R_2}$$

$$V_{R1} = I_{R1} \times R_1$$

$$V_{R2} = I_{R2} \times R_2$$

$$V_{BE} = 0.7 \text{ V}$$

$$V_{R_E} = V_{R2} - V_{BE}$$

$$I_E = \frac{V_{R_E}}{R_E}$$

$$I_C \cong I_E$$

$$V_{Rc} = I_C \times R_C$$

$$V_{CE} = V_{CC} - (V_{Rc} + V_{R_E})$$

$$V_B = V_{R2}$$

$$V_E = V_{R_E}$$

$$V_C = V_{CC} - V_{Rc}$$

$$I_{C(SAT)} = \frac{V_{CC}}{R_C + R_E}$$

26. Increasing the voltage at the base (V_B) increases the voltage at the emitter (V_E) and decreases the voltage at the collector (V_C). Note that V_{CE} also decreases.

27. Decreasing I_B decreases both I_C and I_E.

28. The B–E and B–C junctions of a BJT are checked with an ohmmeter in the same fashion as a PN junction diode. There should be a higher reverse resistance than forward resistance in each case.

GLOSSARY

Beta: The ratio of collector current to base current.

Biasing: Designing a circuit for certain operating voltages.

BJT: Abbreviation for bipolar junction transistor, a PN junction device that relies on both majority and minority carriers for current flow.

Cutoff: The condition of a BJT that results in no current flow.

Majority Carrier: The carrier of a charge that constitutes more than one-half of the total charge carriers.

Minority Carrier: The carrier of a charge that constitutes less than one-half of the total charge carriers.

Saturation: The condition of a BJT that results in maximum current flow.

Thermal Stability: A condition that is unaffected by changes in temperature.

PROBLEMS

SECTION 4-1

4-1. Explain the meaning of bipolar devices.

4-2. Describe the doping levels of the various regions within the transistor.

4-3. What is the difference between an NPN and PNP transistor?

4-4. What are the names of the transistor legs?

4-5. Draw and label the schematic symbols for both the NPN and PNP transistor.

SECTION 4-2

4-6. Describe the proper method of biasing the PN junctions of the BJT.

4-7. Explain current flow in a BJT.

4-8. What is the reason for the B–E voltage drop?

4-9. Explain the importance of base current for the operation of a transistor.

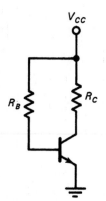

FIGURE 4-17   Base-bias circuit.

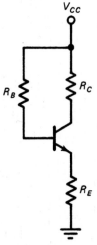

FIGURE 4-18   Emitter-feedback bias circuit.

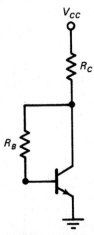

FIGURE 4-19   Collector-feedback bias circuit.

The following problems refer to the circuit of Fig. 4-17.

4-10.   Trace the current path.

4-11.   How does base bias offset changes in beta?

4-12. $V_{CC} = 12$ V, $R_C = 500\ \Omega$, $R_B = 75$ kΩ, and beta = 110. What are the values of I_B and V_C?

4-13. $V_{CC} = 15$ V, $R_C = 680\ \Omega$, $R_B = 120$ kΩ, and beta = 125. What are the values of I_B and V_C?

SECTION 4-4

4-14.   Define saturation and cutoff.

4-15.   How do saturation and cutoff occur?

4-16.   In Fig. 4-17, $V_{CC} = 12$ V, $R_C = 500\ \Omega$, and beta = 110. What are the values of $I_{C(SAT)}$ and $V_{C(SAT)}$?

4-17.   In Fig. 4-17, $V_{CC} = 15$ V, $R_C = 680\ \Omega$, and beta = 125. What are the values of $I_{C(SAT)}$ and $V_{C(SAT)}$?

SECTION 4-5

The following problems refer to the circuit of Fig. 4-18.

4-18.   Trace the current path.

4-19.   How does emitter-feedback bias offset changes in beta?

4-20. $V_{CC} = 10$ V, $R_C = 120\ \Omega$, $R_E = 100\ \Omega$, $R_B = 15$ kΩ, and beta = 125. What are the values of I_B and V_C?

4-21. $V_{CC} = 10$ V, $R_C = 120\ \Omega$, $R_E = 100\ \Omega$, and beta = 125. What are the values of $V_{C(SAT)}$ and $R_{B(SAT)}$?

4-22. $V_{CC} = 12$ V, $R_C = 330\ \Omega$, $R_E = 120\ \Omega$, $R_B = 47$ kΩ, and beta = 100. What are the values of I_B and V_C?

4-23. $V_{CC} = 12$ V, $R_C = 330\ \Omega$, $R_E = 120\ \Omega$, $R_B = 47$ kΩ, and beta = 100. What are the values of $V_{C(SAT)}$ and $R_{B(SAT)}$?

SECTION 4-6

The following problems refer to the circuit of Fig. 4-19.

4-24.   Trace the current paths.

4-25.   How does the collector-feedback bias offset changes in beta?

4-26. $V_{CC} = 15$ V, $R_C = 270\ \Omega$, $R_B = 33$ kΩ, and beta = 110. What are the values of I_B and V_C?

4-27. $V_{CC} = 10$ V, $R_C = 150\ \Omega$, $R_B = 27$ kΩ, and beta = 120. What are the values of I_B and V_C?

4-28. $V_{CC} = 10$ V, $R_C = 150\ \Omega$, $R_B = 27$ kΩ, and beta = 120. What are the values of I_B and V_C at cutoff?

SECTION 4-7

4-29.   Trace the current path through an emitter-bias circuit.

4-30.   What is the disadvantage of the emitter-bias circuit?

SECTION 4-8

The following problems refer to the circuit of Fig. 4-20.

4-31. Trace the current paths.

4-32. How does voltage-divider bias overcome changes in beta?

4-33. $V_{CC} = 12$ V, $R_1 = 100$ kΩ, $R_2 = 56$ kΩ, $R_E = 470$ Ω, and $R_C = 1$ kΩ. What are the values of V_B and V_C?

4-34. $V_{CC} = 12$ V, $R_1 = 100$ kΩ, $R_2 = 56$ kΩ, $R_E = 470$ Ω, and $R_C = 1$ kΩ. What is the value of $V_{C(SAT)}$?

4-35. $V_{CC} = 15$ V, $R_1 = 100$ kΩ, $R_2 = 68$ kΩ, $R_E = 1.2$ kΩ, and $R_C = 1.5$ kΩ. What are the values of V_B and V_C?

4-36. $V_{CC} = 15$ V, $R_1 = 100$ kΩ, $R_2 = 68$ kΩ, $R_E = 1.2$ kΩ, and $R_C = 1.5$ kΩ. What is the value of $V_{C(SAT)}$?

4-37. Design a voltage-divider bias circuit with $V_C = 10$ V and $I_C = 10$ mA using a 25-V power supply.

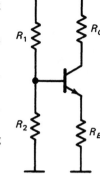

FIGURE 4-20 Voltage-divider bias circuit.

SECTION 4-9

4-38. What would be the value of V_C in the circuit of Fig. 4-17 if R_B were to open?

4-39. What would be the value of V_C in the circuit of Fig. 4-17 if the base–emitter leg were to short?

4-40. What would be the value of V_C in the circuit of Fig. 4-18 if R_E were to open?

4-41. What would be the value of V_C in the circuit of Fig. 4-18 if the base were shorted to V_{CC}?

4-42. What would be the value of V_C in the circuit of Fig. 4-19 if R_C were to open?

4-43. What would be the value of V_C in the circuit of Fig. 4-20 if R_1 were to open?

4-44. What would be the value of V_C in the circuit of Fig. 4-20 if a solder bridge shorted out R_E?

4-45. List all the possible causes for an increase and decrease in V_C in the circuit of Fig. 4-20.

ANSWERS TO REVIEW QUESTIONS

1. bipolar junction transistor	**2.** bipolar	**3.** transfer resistor
4. collector, emitter, base	**5.** emitter	**6.** base
7. NPN	**8.** emitter	**9.** two
10. $B–E$	**11.** 0.7 V	**12.** emitter
13. False	**14.** True	**15.** current
16. biased	**17.** forward	**18.** beta
19. dividing	**20.** True	**21.** saturation
22. saturation	**23.** 0	**24.** saturation
25. open	**26.** False	**27.** It includes R_E
28. False	**29.** True	**30.** True
31. False	**32.** maximum	**33.** V_{CC}
34. increase	**35.** False	**36.** Thermal
37. increases	**38.** cut off	**39.** 0.7 V
40. True	**41.** True	**42.** 0 V
43. $V_C – V_E$	**44.** No.	**45.** R_1 and R_2.

46. directly
47. inversely
48. 0 V
49. It equals V_{CC}
50. Zero.
51. Infinite.
52. Zero.
53. Usually V_{CC}
54. increases
55. decreases
56. increases
57. increases
58. increases
59. increases
60. False
61. True

Small-Signal Amplifiers

When you have completed this chapter, you should be able to:

- Correlate the relationship between dc biasing and ac amplification.
- Calculate the ac quantities of gain, input impedance, and output impedance.
- Analyze the characteristics of the common-emitter, common-collector, and common-base amplifiers.
- Prove the concepts of multistage amplifiers.
- Analyze the effects of various faults on the ac input and output signals.

Small-Signal Amplifier: An amplifier whose amplified power is usually one-half watt or less.

The previous chapter provided an introduction to the bipolar transistor along with guidelines for proper biasing. Once the transistor is correctly biased, it is ready to perform some useful function. This chapter deals with the application of a bipolar transistor as a small-signal amplifier.

A **small-signal amplifier** increases the current, voltage, or power of an ac signal. It is distinguished from the power amplifier in that the power rating is usually one-half watt or less. Because of the low power rating, the transistor case is usually a small plastic type.

The small-signal amplifier can be arranged in three different configurations: common emitter, common collector, and common base. Each of these configurations offers a different benefit in terms of current, voltage, and power amplification. Input and output impedance is another factor that changes with the configuration. Each of these configurations is explained in the following sections.

5-1   COMMON-EMITTER AMPLIFIER ————————————————————

The common-emitter configuration is the most widely used because it offers good voltage, current, and power gain. It offers low input impedance and high output impedance. The configuration is determined by the fact that the input is on the base and the output is taken off the collector; therefore, the emitter is common to both, as shown in Fig. 5-1.

From Fig. 5-1, we notice that voltage-divider biasing is used. The biasing sets up the transistor for proper operation by providing the necessary dc voltages. Once biased, the transistor performs the function of amplification. The ac signal at the base is amplified and appears at the collector.

Coupling Capacitor: A capacitor that couples, or connects, the ac signal from one stage to the next, yet isolates the dc.

The capacitors are called **coupling capacitors.** Their purpose is to connect, or couple, the ac signal while preventing, or blocking, dc current from entering or exiting the amplifier stage. If some stray dc voltage from the ac supply were allowed to enter the amplifier stage, it would change the dc biasing voltages on the resistors. This, in turn, would cause the operation of the amplifier to change and the result could be a distorted signal. Figure 5-1 is a single-stage amplifier, which means it is composed of one transistor and associated resistors.

The approach to this circuit is to analyze both dc and ac values. Such an analysis requires that the circuit of Fig. 5-1 be assigned some values. The coupling capacitor values are chosen so that a minimal amount of ac signal is lost across them, requiring their reactances to be relatively small.

DC Analysis   The dc analysis of this circuit is essentially the same as that of the voltage-divider bias circuit in the previous chapter. The first step is to deter-

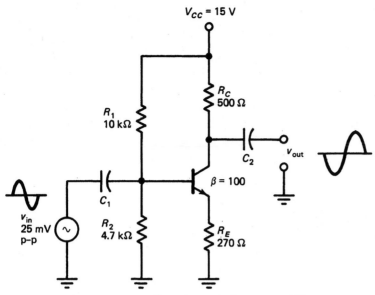

FIGURE 5-1 Common-emitter amplifier.

mine how the voltage divides across R_1 and R_2 in the circuit of Fig. 5-1. This can be accomplished through the use of the voltage-divider formula:

$$V_{R2} = \frac{R_2}{R_1 + R_2} V_{CC}$$

$$= \frac{4.7 \text{ k}\Omega}{14.7 \text{ k}\Omega} (15 \text{ V})$$

$$= 4.8 \text{ V}$$

This means that the balance of V_{CC} is dropped across V_{R1}:

$$V_{R1} = V_{CC} - V_{R2}$$

$$= 15 \text{ V} - 4.8 \text{ V}$$

$$= 10.2 \text{ V}$$

Finding the voltage across V_{R_E} is accomplished by subtracting the 0.7-V base-to-emitter voltage of a properly biased transistor from V_{R2}.

$$V_{R_E} = V_{R2} - V_{BE}$$

$$= 4.8 \text{ V} - 0.7 \text{ V}$$

$$= 4.1 \text{ V}$$

Now that the voltage across the emitter resistor is solved, we can solve for the dc emitter current using Ohm's law. It should be understood that the current flowing through the emitter resistor is also flowing into the emitter of the transistor.

$$I_E = \frac{V_{R_E}}{R_E}$$

$$= \frac{4.1 \text{ V}}{270 \ \Omega}$$

$$= 15.2 \text{ mA}$$

Since the collector current and emitter current are approximately the same, the same value is used for each.

$$I_C = I_E$$

$$= 15.2 \text{ mA}$$

If the current exiting the collector is 15.2 mA, then this same current must pass through R_C. Therefore, the voltage across this resistor can be calculated as follows:

$$V_{R_C} = I_C \times R_C$$

$$= 15.2 \text{ mA} \times 500 \ \Omega$$

$$= 7.6 \text{ V}$$

There are three places where voltage can drop on the output side of the transistor: V_{R_C}, V_{R_E}, and V_{CE}. Since V_{R_E} and V_{R_C} are known, the collector-to-emitter voltage equals the balance of V_{CC}.

$$V_{CE} = V_{CC} - (V_{R_C} + V_{R_E})$$

$$= 15 \text{ V} - (7.6 \text{ V} + 4.1 \text{ V})$$

$$= 3.3 \text{ V}$$

Quiescent Voltage: The dc operating-point voltage.

The last and most important voltage to be found in this circuit is V_C, the voltage at the collector with reference to ground. This is called the operating voltage, or **quiescent voltage,** in the common-emitter circuit. It is upon this voltage that the amplified ac signal rides, as we will see in the ac analysis.

There are several ways to solve this value, but the most direct and consistent method is to realize that the V_{R_C} portion of V_{CC} is not included in V_C. The following formula is the result of this observation.

$$V_C = V_{CC} - V_{R_C}$$

$$= 15 \text{ V} - 7.6 \text{ V}$$

$$= 7.4 \text{ V}$$

Figure 5-2 illustrates the paths of dc current in this circuit. Notice that no dc current flows through the coupling capacitors. This concludes the dc analysis of the common-emitter circuit.

DC Load Line   There is a wide range of possibilities for actively operating a transistor. These fall between the extremes of saturation and cutoff. One way of portraying the many possibilities is through the use of a **dc load line.**

DC Load Line: A graph showing all the possible operating conditions available to the transistor.

Let us begin by considering the two extremes. When a transistor enters cutoff, ideally, V_{CE} equals the source voltage. That would be 15 V in our example.

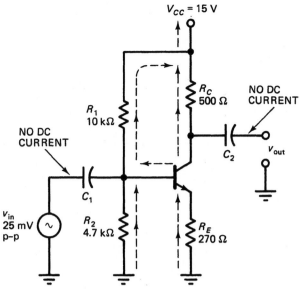

FIGURE 5-2 DC current flow.

When that same transistor is driven to saturation, I_C equals 19.5 mA. This is found by

$$I_{C(SAT)} = \frac{V_{CC}}{R_C + R_E}$$

$$= \frac{15\ \text{V}}{500\ \Omega + 270\ \Omega}$$

$$= 19.5\ \text{mA}$$

By connecting these two points on a graph by a straight line, all the possible conditions available to the transistor are represented. Figure 5-3 shows such a representation. The operating point (Q point) in our example includes a V_{CE} of 3.3 V. Normally, a Q point closer to the middle of the load line is desired to prevent distortion of the amplified ac signal. This is accomplished through changing the values of the biasing resistors.

Practice Problem: Draw a dc load line for circuit values of $V_{CC} = 12$ V, $R_E = 1$ kΩ, and $R_C = 2.7$ kΩ. *Answer:* A graph should be drawn with a line connecting the point of 12 V to the saturation current of 3.24 mA.

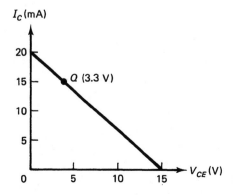

FIGURE 5-3 DC Load line.

r'_e: The ac emitter resistance.

AC Analysis   The common-emitter amplifier is a good all-purpose amplifier. It provides good voltage, current, and power gain. It has low input impedance and high output impedance. It also provides a 180-degree phase shift between input and output signals. Figure 5-1 illustrates this phase shift.

The ac analysis of this same circuit requires that we investigate a few values that have not been discussed thus far. The first of these is the ac emitter resistance, r'_e (pronounced r prime e). Notice that lowercase letters are used to indicate ac values and uppercase letters are used to indicate dc values. The formula for determining the approximate ac emitter resistance is

$$r'_e = \frac{25 \text{ mV}}{I_E}$$

The 25 mV is a constant value derived through the use of calculus. We accept this as the standard value without laboring through a maze of mathematics. The value I_E is the dc emitter current. Using the values of Fig. 5-1 given earlier and the dc analysis that followed, we can calculate r'_e as

$$r'_e = \frac{25 \text{ mV}}{15.2 \text{ mA}}$$

$$= 1.64 \ \Omega$$

Note that the resulting ac resistance is expressed in ohms. This value is rather small and has little affect in the following analysis. We are considering it now only because it eventually serves a very useful purpose, and the sooner we become familiar with it, the better.

Practice Problem: What is the ac emitter resistance if the dc emitter current is 1.5 mA?   *Answer:* $r'_e = 16.7 \ \Omega$.

AC Current Gain and Beta (h_{FE})   Another value to consider is the ac beta. Remember that beta represents current gain. There is a difference between dc beta (h_{FE}) and ac beta (h_{fe}). DC beta is the ratio of dc collector current to dc base current and ac beta is the ratio of ac collector current to ac base current. In our discussions, we consider both to be approximately equal in value. If beta represents current gain, then

$$A_i = \beta$$

$$= 100$$

Gain is symbolized by the letter A and ac current by lowercase i. The gain of the circuit currently under discussion is found in the values that were originally given.

AC Voltage Gain   The second type of gain to be considered is voltage gain. Voltage gain can be found in one of two ways, depending upon whether or not a bypass capacitor is used. Figure 5-1 does not contain a bypass capacitor, but Fig. 5-4 does. A bypass capacitor is connected across the emitter resistor to provide a path of very low resistance to ground for the ac signal. The use of a bypass

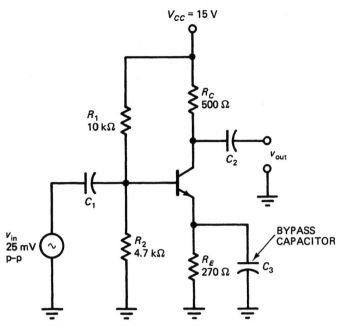

FIGURE 5-4 Common-emitter amplifier with a bypass capacitor.

capacitor increases a circuit's voltage gain. Consider the formula for finding voltage gain both with and without a bypass capacitor.

$$A_v = \frac{R_C}{R_E + r'_e} \quad \text{(without a bypass capacitor)}$$

$$A_v = \frac{R_C}{r'_e} \quad \text{(with a bypass capacitor)}$$

Since our ac analysis involves the circuit of Fig. 5-1, the formula without the bypass capacitor should be used. Inserting values into the formula gives us the following:

$$A_v = \frac{500 \ \Omega}{270 \ \Omega + 1.64 \ \Omega}$$

$$= 1.84$$

This value of voltage gain can now be used to determine the ac output signal, v_{out}.

$$v_{\text{out}} = v_{\text{in}} \times A_v$$

$$= 25 \ \text{mV p–p} \times 1.84$$

$$= 46 \ \text{mV p–p}$$

The voltage gain is rather low. This is because of the resistance that R_E offers to the ac signal. It would be ideal if we could reduce the resistance of R_E for the ac signal while maintaining that resistance for the dc biasing. The use of a bypass capacitor accomplishes this objective. The capacitor looks like an open to the dc current. This results in no effect on R_E and V_{R_E}. But the capacitor does have the property of reactance (capacitor's ac resistance). If a value of capacitor is chosen so that the capacitive reactance is low (ideally, 0 Ω), then the capacitor can act as a short for the ac signal. It is because R_E is shorted that it is removed from the formula by using a bypass capacitor. If a bypass capacitor were added to the circuit as in Fig. 5-4, the voltage gain would greatly increase. The new value of voltage gain is

$$A_v = \frac{R_C}{r_e'}$$

$$= \frac{500 \ \Omega}{1.64 \ \Omega}$$

$$= 305$$

As voltage gain increases, it stands to reason that the voltage level of the ac output signal increases. Using the new value of voltage gain, we find that

$$v_{\text{out}} = A_V \times v_{\text{in}}$$

$$= 305 \times 25 \ \text{mV p–p}$$

$$= 7.63 \ \text{V p–p}$$

The ac output signal without the bypass capacitor is 46 mV p–p and the value with the bypass capacitor is 7.63 V p–p. It is evident that the bypass capacitor does have a tremendous effect on the circuit.

Practice Problem: What is the ac voltage gain if $I_E = 2.5$ mA, $R_C = 1.5$ kΩ, and a bypass capacitor is used. *Answer:* $A_v = 150$.

AC Power Gain
The final gain that we will consider is power gain, A_p. Power gain can be found by applying Watt's law to the two previous gains. That is,

$$A_p = A_i \times A_v$$

By substituting the values found earlier for the circuit of Fig. 5-1, power gain would equal

$$A_p = 100 \times 1.84$$

$$= 184$$

Practice Problem: What is the ac power gain if $I_E = 2$ mA, $R_C = 1.2$ kΩ, beta = 100, and a bypass capacitor is used? *Answer:* $A_p = 9600$.

Input Impedance
The input impedance is the resistance that the ac signal sees when looking into the input of the circuit. The first step in solving the input impedance is to realize that the dc supply acts like a short to ground as far as the ac signal is concerned. The circuit of Fig. 5-1 would appear as the circuit of Fig. 5-5 to the ac source. From Fig. 5-5, you see that the ac source encounters a

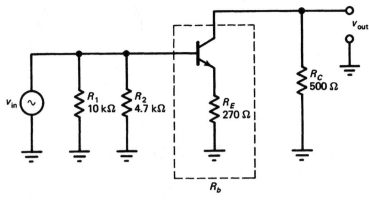

FIGURE 5-5 AC equivalent of the circuit of Fig. 5-1.

parallel circuit consisting of R_1, R_2, and the base resistance, R_b. Since the values of R_1 and R_2 are obvious, we focus our attention on R_b.

$$R_b = \beta(r'_e + R_E)$$
$$= 100(1.64 \ \Omega + 270 \ \Omega)$$
$$= 27.2 \ \text{k}\Omega$$

Note that if a bypass capacitor were used, R_E would be eliminated from the formula.

Solve the overall input impedance z_in for the circuit of Fig. 5-5 by using the formula for parallel circuits, which is

$$z_\text{in} = R_1 \parallel R_2 \parallel R_b \qquad \text{AKA } R_\text{in}$$

or

$$z_\text{in} = \frac{1}{1/R_1 + 1/R_2 + 1/R_b}$$
$$= \frac{1}{1/10 \ \text{k}\Omega + 1/4.7 \ \text{k}\Omega + 1/27.2 \ \text{k}\Omega}$$
$$= 2.86 \ \text{k}\Omega$$

Practice Problem: What is z_in of the circuit of Fig. 5-5 if a bypass capacitor is used? *Answer:* $z_\text{in} = 156 \ \Omega$.

Output Impedance The output impedance is much easier to solve. Using the circuit of Fig. 5-5 as a guide, we can see that the output is in parallel to R_C and the collector-to-ground resistance. Because the collector-to-ground resistance is extremely high, the overall resistance of the output parallel network is equal to the value of R_C.

$$z_\text{out} = R_C$$
$$= 500 \ \Omega$$

FIGURE 5-6 AC output of the circuit of Fig. 5-1.

Combined Analysis We now combine the information gained from both the dc and ac analyses. Each analysis was discussed separately, yet they combine to create the overall circuit operation. Figure 5-6 illustrates the results of the previous analyses. Recall that the dc voltage at the collector was found to equal 7.4 V. This is considered the operating voltage or quiescent voltage (Q point) of the circuit. In the ac analysis, it was found that the 25-mV input signal was amplified, resulting in a 46-mV p–p output signal. This ac signal "rides" on the dc signal.

There are limitations that are caused by a number of factors. The first is that a 15-V supply is used. This means that the broadest range available is 0 to 15 V. Also, due to the method of biasing, the voltage at the collector is approximately 5.26 V at saturation. This raises the lower limit of the signal swing from 0 to 5.26 V. Figure 5-7 shows the result of attempting to exceed this limitation. In this case, the 7.62-V p–p amplified signal rises 3.81 V above V_C and attempts to fall 3.81 V below V_C. A problem arises because the transistor saturates when the signal reaches 5.26 V. The result is a distorted signal that is clipped on its negative alternation. This problem can be corrected by either decreasing the input signal or adjusting the gain.

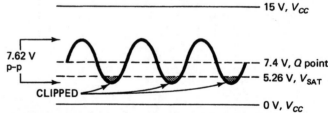

FIGURE 5-7 AC output of the circuit of Fig. 5-4.

EXAMPLE 5-1

The circuit of Fig. 5-4 is changed so that the values are $V_{CC} = 12$ V, $R_1 = 12$ kΩ, $R_2 = 5.1$ kΩ, $R_E = 1$ kΩ, $R_C = 1.5$ kΩ, $\beta = 125$, and $v_{in} = 10$ mV p–p. Determine (a) V_C, (b) v_{out}, (c) A_p, and (d) z_{in}.

Solution:
(a) The dc operating voltage at the collector of the common-emitter circuit is an important voltage. This is the voltage that the ac output signal rides upon. A dc analysis reveals the following:

$$V_{R2} = \frac{5.1 \text{ k}\Omega}{12 \text{ k}\Omega + 5.1 \text{ k}\Omega} (12 \text{ V})$$

$$= 3.58 \text{ V}$$

$$V_{R_E} = 3.58 \text{ V} - 0.7 \text{ V}$$
$$= 2.88 \text{ V}$$

$$I_E = \frac{2.88 \text{ V}}{1 \text{ k}\Omega}$$
$$= 2.88 \text{ mA}$$

$$V_{R_C} = 2.88 \text{ mA} \times 1.5 \text{ k}\Omega$$
$$= 4.32 \text{ V}$$

$$V_C = 12 \text{ V} - 4.32 \text{ V}$$
$$= 7.68 \text{ V}$$

(b) In order to determine the ac output voltage, it is first necessary to calculate the ac voltage gain. The voltage gain of Fig. 5-4 is high due to the presence of the bypass capacitor. The ac emitter resistance is found by using the emitter current calculated during the previous solution.

$$r_e' = \frac{25 \text{ mV}}{2.88 \text{ mA}}$$
$$= 8.68 \text{ }\Omega$$

$$A_v = \frac{1.5 \text{ k}\Omega}{8.68 \text{ }\Omega}$$
$$= 173$$

$$v_{\text{out}} = 10 \text{ mV p–p} \times 173$$
$$= 1.73 \text{ V p–p}$$

(c) The ac power gain is found by applying Watt's law. The voltage gain was found to be 173 in part (b) and the current gain is equal to beta. The power gain is, therefore,

$$A_p = 125 \times 173$$
$$= 21,625$$

(d) Input impedance for the common-emitter amplifier is generally low. The bypass capacitor eliminates R_E from the formula for base resistance.

$$R_b = 125(8.68 \text{ }\Omega + 0 \text{ }\Omega)$$
$$= 1085 \text{ }\Omega$$

The input impedance is found by considering the overall parallel resistance of R_1, R_2, and R_b.

$$z_{\text{in}} = \frac{1}{1/12 \text{ k}\Omega + 1/5.1 \text{ k}\Omega + 1/1085 \text{ }\Omega}$$
$$= 833 \text{ }\Omega \qquad \blacksquare$$

REVIEW QUESTIONS

1. What are three quantities that are amplified by a transistor?
2. What is the maximum wattage rating of a small-signal amplifier?
3. What type of case are small-signal BJTs packaged in?

4. The common-emitter amplifier offers _____ input impedance and _____ output impedance.

5. The common-emitter amplifier offers good voltage, current, and power gain. True or false?

6. In the common-emitter amplifier, the input is on the _____ and the output is off the _____.

7. The emitter current is approximately the same as the _____ current.

8. Which voltage in the common-emitter amplifier is considered the quiescent output voltage?

9. DC current does not flow through the coupling capacitors. True or false?

10. The region between cutoff and saturation is called the _____ region.

11. What is the name given to the graph that shows all the possible conditions available to a load?

12. What does r_e' represent?

13. What is the purpose of a bypass capacitor?

14. What is the formula for ac current gain in the common-emitter amplifier circuit?

15. What is the formula for ac voltage gain in the common-emitter amplifier circuit that contains a bypass capacitor?

16. What is the formula for ac power gain in the common-emitter amplifier circuit?

17. What is the formula for input impedance when a bypass capacitor is present?

18. What is the formula for output impedance when a bypass capacitor is present?

5-2 MULTISTAGE COMMON-EMITTER AMPLIFIER

If you understand the single-stage amplifier, then the multistage amplifier should pose no problem. A three-stage amplifier is shown in Fig. 5-8. Notice that each stage is isolated from the next stage by a coupling capacitor (C_1, C_2, C_4, and C_5). The coupling capacitors prevent the dc voltages of one stage from affecting the dc voltages of the next stage. Only the first stage contains a bypass capacitor (C_3).

The key to solving a circuit such as this is to approach each stage individually. It is similar to solving for three separate common-emitter amplifiers. The voltage divider bias is used for all three stages; therefore, our approach is the same as that used in Section 5-1.

First Stage Transistor Q_1 and the four associated resistors form the first stage. The dc values of this stage are solved by first realizing that the same source voltage of 10 V services all three stages. Therefore, the voltage at the base of Q_1 is

$$V_{B1} = \frac{6.8 \text{ k}\Omega}{16.8 \text{ k}\Omega} 10 \text{ V}$$

$$= 4.05 \text{ V}$$

and the voltage at the emitter is

$$V_{E1} = V_{B1} - 0.7 \text{ V}$$

$$= 3.35 \text{ V}$$

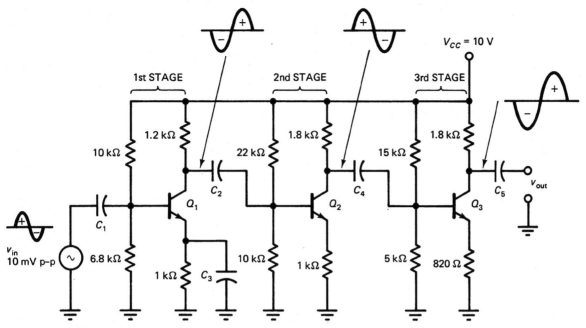

FIGURE 5-8 Multistage common-emitter amplifier.

so the collector current is

$$I_{C1} = I_{E1} = \frac{3.35 \text{ V}}{1 \text{ k}\Omega}$$

$$= 3.35 \text{ mA}$$

Therefore, the operating voltage at the collector of the first stage is

$$V_{C1} = V_{CC} - (I_{C1} \times R_C)$$

$$= 10 \text{ V} - (3.35 \text{ mA} \times 1.2 \text{ k}\Omega)$$

$$= 5.98 \text{ V}$$

The ac voltage gain of the first stage is greater than the other stages because of the bypass capacitor. The formulas are approximations because R_C is actually in parallel with the input impedance of the following stage. The formula is

$$A_v = \frac{R_C}{r_e'}$$

and r_e' is found by

$$r_e' = \frac{25 \text{ mV}}{I_E}$$

Therefore,

$$A_{v1} = \frac{1.2 \text{ k}\Omega}{7.46 \text{ }\Omega}$$

$$= 161$$

and the output voltage of the first stage is

$$v_{out} = v_{in} \times A_v$$

$$= 10 \text{ mV p-p} \times 161$$

$$= 1.61 \text{ V p-p}$$

The first stage amplifies the input signal 161 times and passes the amplified signal onto the second stage through coupling capacitor C_2. A small portion of that signal is lost across C_2; however, since it is so small, it is not considered in our example. This amplified signal is now the input signal for the second stage.

Second Stage The second stage is composed of Q_2 and its four connecting resistors. Once again, the dc analysis is performed without regard for the other stages. The voltage at the base of the Q_2 is found by

$$V_{B2} = \frac{10 \text{ k}\Omega}{32 \text{ k}\Omega} 10 \text{ V}$$

$$= 3.13 \text{ V}$$

and the voltage at the emitter is

$$V_{E2} = 3.13 \text{ V} - 0.7 \text{ V}$$

$$= 2.43 \text{ V}$$

so the collector current of the second stage is

$$I_{C2} = I_{E2} = \frac{2.43 \text{ V}}{1 \text{ k}\Omega}$$

$$= 2.43 \text{ mA}$$

Therefore, the voltage at the collector is

$$V_{C2} = 10 \text{ V} - (2.43 \text{ mA} \times 1.8 \text{ k}\Omega)$$

$$= 5.63 \text{ V}$$

The ac voltage gain of the second stage is considerably less than the first stage due to the absence of a bypass capacitor. In this case, the approximate voltage gain is found by

$$A_{v2} = \frac{R_C}{R_E + r_e'}$$

$$= \frac{1.8 \text{ k}\Omega}{1 \text{ k}\Omega + 10.3 \text{ }\Omega}$$

$$= 1.78$$

and the output voltage of the second stage is

$$v_{out} = 1.61 \text{ V p-p} \times 1.78$$

$$= 2.87 \text{ V p-p}$$

The original ac input signal of 10 mV is growing larger and larger as it travels through the stages. It is close to 300 times larger in amplitude.

Third Stage   The final stage is composed of Q_3 and the connecting resistors. Using the same steps as before, we find that

$$V_{B3} = 2.5 \text{ V}$$

$$V_{E3} = 1.8 \text{ V}$$

$$I_{C3} = 2.2 \text{ mA}$$

$$V_{C3} = 6.04 \text{ V}$$

$$A_{v3} = 2.17$$

and the final output signal is

$$v_{\text{out}} = 2.87 \text{ V p–p} \times 2.17$$

$$= 6.23 \text{ V p–p}$$

The overall gain afforded by this multistage amplifier may be found in one of two ways. A ratio of output voltage to input voltage yields

$$\text{Total } A_v = \frac{v_{\text{out}}}{v_{\text{in}}}$$

$$= \frac{6.23 \text{ V}}{10 \text{ mV}}$$

$$= 623$$

or we can multiply all of the gains:

$$\text{Total } A_v = A_{v1} \times A_{v2} \times A_{v3}$$

$$= 161 \times 1.78 \times 2.17$$

$$= 622$$

(The slight discrepancy is due to rounding.)

It should be understood that if there is too much gain, distortion will occur. Using different-value resistors varies the gain and biasing voltages.

EXAMPLE 5-2

What would be the value of the output signal if the bypass capacitor, C_3, were removed from the circuit of Fig. 5-8?

Solution: Removing the bypass capacitor from the first stage greatly reduces the gain of that stage. It was previously found to have a value of 161 with the bypass capacitor. The gain of the first stage without the bypass capacitor is

$$A_{v1} = \frac{1.2 \text{ k}\Omega}{1 \text{ k}\Omega}$$

$$= 1.2$$

Therefore, the overall gain of the amplifier is

$$\text{Total } A_v = 1.2 \times 1.78 \times 2.17$$
$$= 4.64$$

The input signal is amplified by the value of total voltage gain. The value of the output signal is

$$v_{\text{out}} = 4.64 \times 10 \text{ mV p–p}$$
$$= 46.4 \text{ mV p–p} \qquad \blacksquare$$

REVIEW QUESTIONS

19. Which capacitor in the circuit of Fig. 5-8 is the bypass capacitor?
20. What is the total A_v of a three-stage amplifier whose individual stage gains are 2, 50, and 5?

5-3 COMMON-COLLECTOR AMPLIFIER

Common-Collector Amplifier: This amplifier provides good voltage, current, and power gain. It has low input impedance and high output impedance. The input is on the base and the output is off the collector.

The **common-collector amplifier** is distinguished from other configurations because the input is on the base and the output is taken off the emitter. It has good current and power gain, but unity (gain of 1) voltage gain. It has high input impedance and low output impedance. Figure 5-9 illustrates the common-collector configuration.

DC Analysis The circuit of Fig. 5-9 uses a voltage-divider network at the input. This voltage divider is approached in the same as any other. The voltage at the base can be found through the use of the voltage-divider formula:

$$V_B = V_{R2} = \frac{R_2}{R_1 + R_2} V_{CC}$$

$$= \frac{100 \text{ k}\Omega}{100 \text{ k}\Omega + 100 \text{ k}\Omega} (12 \text{ V})$$

$$= 6 \text{ V}$$

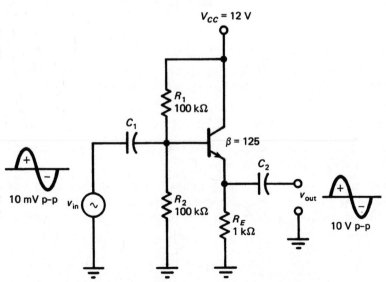

FIGURE 5-9 Common-collector amplifier.

The voltage at the emitter is found by subtracting the base–emitter junction from the base voltage. Since the output is taken off the emitter, this is the operating point of the common-collector amplifier.

$$V_E = V_B - 0.7 \text{ V}$$

$$= 5.3 \text{ V}$$

This allows the emitter current to be found:

$$I_E = \frac{V_E}{R_E}$$

$$= \frac{5.3 \text{ V}}{1 \text{ k}\Omega}$$

$$= 5.3 \text{ mA}$$

AC Analysis   The common-collector amplifier is sometimes referred to as an emitter follower, because the voltage at the emitter follows the input voltage both in phase and amplitude. Notice the in-phase condition of both the input and output waveforms in Fig. 5-9.

The common collector is used in situations where current and power gains are required with unity voltage gain. It has a high input impedance and a very low output impedance.

AC Voltage Gain   Many times the voltage gain is stated as unity, or 1. The voltage gain is found by

$$A_v = \frac{R_E}{R_E + r'_e}$$

$$= \frac{1 \text{ k}\Omega}{1 \text{ k}\Omega + 4.72 \text{ }\Omega}$$

$$= 0.995$$

As we can see, the resulting ac voltage gain is very close to unity for this circuit.

AC Current Gain   The current gain is dependent upon beta and the values of R_1 and R_2. If the resistors of the voltage divider are large enough, then the current gain is approximately equal to the value of beta. For our purposes, we assume that beta and current gain are equal; therefore,

$$A_i = \beta$$

$$= 125$$

AC Power Gain   Power gain is the product of voltage gain and current gain. If voltage gain equals unity, then it makes sense that power gain is approximately equal to current gain. In a formula, it is stated as

$$A_p = A_i \times A_v$$

$$= 125 \times 1$$

$$= 125$$

AC Input Impedance The ac input impedance of the common-collector amplifier is dependent upon three factors. These are the values of R_1 and R_2 of the voltage divider and the base impedance of the transistor. Figure 5-10 shows the ac equivalent of the common-collector amplifier circuit. The first step is to determine the value of the base impedance, R_b.

$$R_b = \beta(r'_e + R_E)$$
$$= 125(4.72 \ \Omega + 1 \ k\Omega)$$
$$= 126 \ k\Omega$$

From Fig. 5-10, we see that R_1, R_2, and R_b form a parallel network. It is from this arrangement that we arrive at the following formula for input impedance to the circuit.

$$z_{in} = R_1 \parallel R_2 \parallel R_b$$
$$= 100 \ k\Omega \parallel 100 \ k\Omega \parallel 126 \ k\Omega$$
$$= 35.8 \ k\Omega$$

Practice Problem: What would be the input impedance of the circuit if R_1 and R_2 were changed to 10 kΩ resistors in the circuit of Fig. 5-10? *Answer:* $z_{in} = 4.81 \ k\Omega$.

AC Output Impedance Finding the ac output impedance requires a knowledge of the internal resistance of the ac source supply (R_s). The logic behind the formula is that if current increases by the value of beta, the resistance must decrease by that value of beta. This resistance and the value of R_E form the approximate output impedance.

$$z_{out} = \frac{R_s}{\beta} \parallel R_E$$

If the ac source supply had an internal resistance of 5 kΩ, then the output impedance could be approximated as

$$z_{out} \cong \frac{5 \ k\Omega}{125} \parallel 1 \ k\Omega$$
$$= 38.5 \ \Omega$$

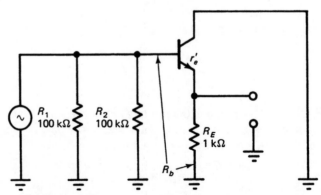

FIGURE 5-10 AC equivalent of the circuit of Fig. 5-9.

EXAMPLE 5-3

The circuit of Fig. 5-9 is changed so that the values are $V_{CC} = 10$ V, $R_1 = 470$ kΩ, $R_2 = 470$ kΩ, $R_E = 2$ kΩ, $\beta = 100$, and $v_{in} = 250$ mV p–p. Determine (a) V_E, (b) v_{out}, (c) A_p, and (d) z_{in}.

Solution:

(a) The voltage at the emitter is 0.7 V less than the voltage at the base. A voltage divider is created by R_1 and R_2. The voltage across R_2 is equal to the voltage at the base.

$$V_B = \frac{470 \text{ k}\Omega}{470 \text{ k}\Omega + 470 \text{ k}\Omega} (10 \text{ V})$$

$$= 5 \text{ V}$$

and

$$V_E = 5 \text{ V} - 0.7 \text{ V}$$

$$= 4.3 \text{ V}$$

(b) A characteristic of the common-collector amplifier is that it has a voltage gain of approximately 1, or unity. The output voltage is, therefore, approximately equal to the input voltage.

$$v_{out} = v_{in}$$

$$= 250 \text{ mV p–p}$$

(c) The power gain is equal to the voltage gain multiplied by the current gain. We have already established the voltage gain as 1. The current gain is approximately equal to the beta of the transistor. A beta of 100 was stated among the opening values of this example; therefore,

$$A_p = 1 \times 100$$

$$= 100$$

(d) The initial step in finding the ac input impedance is to find the value of base impedance. In finding the value of base impedance, we must first find the value of ac emitter resistance.

$$r'_e = \frac{25 \text{ mV}}{V_E/R_E}$$

$$= \frac{25 \text{ mV}}{4.3 \text{ V}/2 \text{ k}\Omega}$$

$$= 11.6 \ \Omega$$

$$R_b = 100 \times (11.6 \ \Omega + 2 \text{ k}\Omega)$$

$$= 201 \text{ k}\Omega$$

The ac model of Fig. 5-10 illustrates that R_1, R_2, and R_b are in parallel. Input impedance is, therefore, equal to the parallel sum of these resistances:

$$z_{in} = 470 \text{ k}\Omega \parallel 470 \text{ k}\Omega \parallel 201 \text{ k}\Omega$$

$$= 108 \text{ k}\Omega \qquad \blacksquare$$

REVIEW QUESTIONS

21. The common-collector amplifier offers _____ input impedance and _____ output impedance.

22. The common-collector amplifier offers good voltage, current, and power gain. True or false?

23. In the common-collector amplifier, the input is on the _____ and the output is off the _____ .

24. Which voltage in the common-collector amplifier is considered the quiescent output voltage?

5-4 MULTISTAGE COMMON-COLLECTOR AMPLIFIER

Darlington Pair: A name commonly given to a two-stage common-collector amplifier.

Darlington amplifier in a single package.

The multistage common-collector amplifier that we are about to discuss is also known as the **Darlington pair** amplifier. It is so widely used that the transistor combination comes prepackaged as a single component. It is used in applications where high current gain is required but no voltage gain. It also provides an extremely high input impedance.

DC Analysis The dc analysis begins with the voltage-divider bias at the input of the amplifier in Fig. 5-11. The voltage at the base of Q_1 is found through the familiar voltage-divider formula.

$$V_B = V_{R2} = \frac{R_2}{R_1 + R_2} V_{CC}$$

$$= 6 \text{ V}$$

From Fig. 5-11, we notice that there are two base–emitter junctions between the base of Q_1 and the emitter of Q_2. This means that if a total of 1.4 V is dropped in the process of forward biasing the transistors, the balance of V_B will be found across R_E.

$$V_E = V_B - 1.4 \text{ V}$$

$$= 4.6 \text{ V}$$

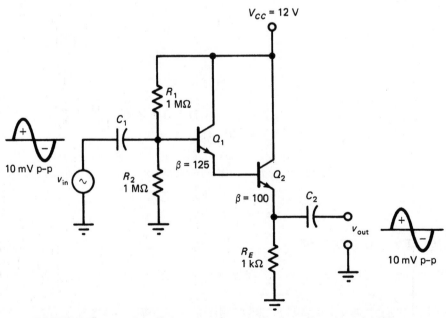

FIGURE 5-11 Darlington pair common-collector amplifier.

This produces an emitter current of

$$I_E = \frac{V_E}{R_E}$$

$$= 4.6 \text{ mA}$$

AC Analysis   The total gain of any multistage amplifier is found by multiplying the gains of the stages. This is true for current, voltage, and power gains. The circuit of Fig. 5-11 is essentially composed of two common-collector amplifiers.

AC Voltage Gain   The overall ac voltage gain for the circuit of Fig. 5-11 remains approximately 1. This is because overall voltage gain is found by multiplying the unity gain of the first stage by the second stage, which results in a total voltage gain of unity.

$$A_v \cong 1$$

AC Current Gain   The current gain of a single-stage common-collector amplifier is approximately equal to beta. The total current gain of the circuit of Fig. 5-11 is the product of the gains of Q_1 and Q_2.

$$A_i \cong \beta_1 \times \beta_2$$

$$\cong 125 \times 100$$

$$\cong 12,500$$

AC Power Gain   The total power gain can be found by multiplying the power gains of the individual stages or by multiplying the total voltage gain by the total current gain. In either case, the total power gain is approximately equal to the total current gain.

$$A_p \cong A_v \times A_i$$

$$\cong 12,500$$

AC Input Impedance   The high input impedance of the Darlington amplifier is a very useful characteristic. The ac input impedance of Q_1 can be approximated as the product of the total current gain of the circuit and R_E.

$$R_{b1} \cong A_i \times R_E$$

Because the base impedance of the first transistor is extremely high, it may be approximated that the total input impedance of the Darlington amplifier is equal to the parallel value of R_1 and R_2.

$$z_{in} = R_1 \parallel R_2$$

Practice Problem: What is the input impedance of the circuit of Fig. 5-11?   *Answer:* $z_{in} = 500 \text{ k}\Omega$.

SEC. 5-4 / MULTISTAGE COMMON-COLLECTOR AMPLIFIER

113

AC Output Impedance   The ac output impedance of the Darlington amplifier is very low. We end this analysis with the general statement that the output
impedance is usually less than 50 Ω.

$$z_{out} < 50 \ \Omega$$

EXAMPLE 5-4

The circuit values of Fig. 5-11 are changed so that $V_{CC} = 10$ V, $R_1 = 470$ kΩ,
$R_2 = 470$ kΩ, $R_E = 2$ kΩ, $\beta_1 = 75$, $\beta_2 = 150$, and $v_{in} = 1$ V p–p. Determine the
values of (a) V_{E2}, (b) A_p, and (c) z_{in}.

Solution:
(a) The voltage at the emitter of Q_2 is 1.4 V less than the voltage at the base
    of Q_1. The base voltage of Q_1 is equal to the voltage across R_2.

$$V_{B1} = \frac{470 \text{ k}\Omega}{470 \text{ k}\Omega + 470 \text{ k}\Omega} \ (10 \text{ V})$$
$$= 5 \text{ V}$$

and

$$V_{E2} = 5 \text{ V} - 1.4 \text{ V}$$
$$= 3.6 \text{ V}$$

(b) The power gain is determined by multiplying the values of the current
    and voltage gains. A characteristic of the common-collector amplifier is
    that it has a voltage gain of 1 and the current gain is equal to beta. When
    two stages are cascaded, as in Fig. 5-11, the total current gain is the
    product of the individual stage gains.

$$A_v = 1$$

$$A_i = 75 \times 150$$
$$= 11{,}250$$

$$A_p = 1 \times 11{,}250$$
$$= 11{,}250$$

(c) Finally, the input impedance for a Darlington pair amplifier is approxi-
    mately equal to the parallel resistance of R_1 and R_2.

$$z_{in} = 470 \text{ k}\Omega \parallel 470 \text{ k}\Omega$$
$$= 235 \text{ k}\Omega \qquad \blacksquare$$

REVIEW QUESTIONS

25. Is there a phase shift between input and output for the common-collector
     amplifier?
26. What is another name for a two-stage common-collector amplifier?
27. What is the approximate voltage gain for all common-collector amplifiers?
28. What is the formula for the z_{in} of the circuit of Fig. 5-11?
29. The overall power gain of a multistage common-collector amplifier is ap-
     proximately equal to the overall current gain. True or false?

5-5 COMMON-BASE AMPLIFIER

The **common-base amplifier** can be recognized because the input is on the emitter and the output is off the collector. This makes the base common to both input and output. The common-base amplifier has good voltage and power gain, but no current gain. The input impedance is low, whereas the output impedance is high.

Common-Base Amplifier: This amplifier provides good voltage and power gain, but unity current gain. It has low input impedance and high output impedance. The input is on the emitter and the output is off the collector.

DC Analysis   The common-base amplifier in Fig. 5-12 uses a voltage-divider bias. It looks different because the transistor is lying on its side to facilitate the input and output signal in a smoother fashion. The voltage divider is made up of R_1 and R_2. Therefore, the voltage at the base with reference to ground is equal to V_{R2}.

$$V_B = \frac{R_2}{R_1 + R_2} V_{CC}$$

$$= 2.7 \text{ V}$$

The base–emitter junction must still be forward biased; therefore, the voltage at the emitter is 0.7 V less than the base.

$$V_E = V_B - 0.7 \text{ V}$$

$$= 2 \text{ V}$$

This emitter voltage means that the emitter current must be

$$I_E = \frac{V_E}{R_E}$$

$$= 2 \text{ mA}$$

It is still legitimate to approximate the collector current as being equal to the emitter current.

$$I_C = I_E = 2 \text{ mA}$$

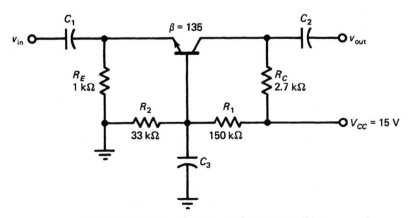

FIGURE 5-12   Common-base amplifier.

The operating point of this circuit is equal to the voltage at the collector. The collector voltage is found in the same way as in earlier examples.

$$V_C = V_{CC} - (I_C \times R_C)$$
$$= 15 \text{ V} - (2 \text{ mA} \times 2.7 \text{ k}\Omega)$$
$$= 9.6 \text{ V}$$

AC Analysis The common base is least popular of the three configurations that we have studied. The low input impedance is a detriment in most applications. However, let us consider the various ac aspects of this amplifier.

AC Voltage Gain The voltage gain of the common base is similar to that of the common-emitter amplifier. The formula for finding the ac emitter resistance is also the same. Therefore,

$$A_v = \frac{R_C}{r'_e}$$
$$= \frac{2.7 \text{ k}\Omega}{12.5 \text{ }\Omega}$$
$$= 216$$

Practice Problem: What is the ac voltage gain if $I_E = 3.5$ mA and $R_C = 1.8$ kΩ? *Answer:* $A_v = 252$.

AC Current Gain The collector and emitter currents are approximately equal. This means that there must be unity current gain from emitter to collector. We conclude that the current gain is approximately 1.

$$A_i = 1$$

AC Power Gain By using the procedures of the last two configurations, the power gain is the product of the voltage and current gain. Since the current gain is unity, the power gain equals the voltage gain.

$$A_p = A_i \times A_v$$
$$= 1 \times 216$$
$$= 216$$

AC Input Impedance The input signal is applied to the emitter and encounters a resistance equal to that of R_E in parallel with r'_e. Because the ac emitter resistance is so small, we may say

$$z_{\text{in}} \cong r'_e$$
$$\cong 12.5 \text{ }\Omega$$

AC Output Impedance The output signal is taken off the collector. Looking at the output of the circuit, we see R_C in parallel with the ac collector resistance (r'_c). Because the resistance of the reverse-biased collector is so large, we may say

$$z_{\text{out}} \cong R_C$$
$$\cong 2.7 \text{ k}\Omega$$

EXAMPLE 5-5

The circuit of Fig. 5-12 is changed so that the values are $V_{CC} = 12$ V, $R_1 = 100$ kΩ, $R_2 = 22$ kΩ, $R_E = 1$ kΩ, $R_C = 1.5$ kΩ, $\beta = 125$, and $v_{in} = 10$ mV p–p. Determine (a) V_C, (b) v_{out}, (c) A_p, and (d) z_{in}.

Solution:

(a) Finding the voltage at the collector begins with finding the voltage at the base. A voltage divider is formed by R_1 and R_2; therefore,

$$V_B = \frac{22 \text{ k}\Omega}{100 \text{ k}\Omega + 22 \text{ k}\Omega} (12 \text{ V})$$
$$= 2.16 \text{ V}$$

$$V_E = 2.16 - 0.7 \text{ V}$$
$$= 1.46 \text{ V}$$

$$I_E = \frac{1.46 \text{ V}}{1 \text{ kV}}$$
$$= 1.46 \text{ mA}$$

$$I_c = I_E = 1.46 \text{ mA}$$

$$V_{R_C} = 1.46 \text{ mA} \times 1.5 \text{ k}\Omega$$
$$= 2.2 \text{ V}$$

$$V_C = 12 \text{ V} - 2.2 \text{ V}$$
$$= 9.8 \text{ V}$$

(b) Calculating the value of the output voltage requires that we first find the voltage gain. The voltage gain is determined by the collector resistor and ac emitter resistances.

$$r_e' = \frac{25 \text{ mV}}{1.46 \text{ mA}}$$
$$= 17.1 \text{ }\Omega$$

$$A_v = \frac{1.5 \text{ k}\Omega}{17.1 \text{ }\Omega}$$
$$= 87.7$$

$$v_{out} = 10 \text{ mV p–p} \times 87.7$$
$$= 877 \text{ mV p–p}$$

(c) A characteristic of the common-base amplifier is that it has a current gain of approximately 1. The power gain is, therefore, equal to the value of the voltage gain.

$$A_p = 1 \times 87.7$$
$$= 87.7$$

(d) The input impedance of the common-base amplifier is very small. It is equal to the value of r_e' found earlier.

$$z_{in} = 17.1 \text{ }\Omega \qquad \blacksquare$$

TABLE 5-1 Amplifier Characteristics

	COMMON EMITTER	COMMON COLLECTOR	COMMON BASE
Input	Base	Base	Emitter
Output	Collector	Emitter	Collector
Voltage gain	Good	Unity	Good
Current gain	Good	Good	Unity
Power gain	Good	Good	Good
Input impedance	Low	High	Low
Output impedance	High	Low	High
180° Phase shift	Yes	No	No

REVIEW QUESTIONS

30. The common-base amplifier offers _____ input impedance and _____ output impedance.

31. The common-base amplifier offers good voltage, current, and power gain. True or false?

32. In the common-base amplifier, the input is on the _____ and the output is off the _____.

33. Which voltage in the common-base amplifier is considered the quiescent output voltage?

Table 5-1 compares the characteristics of the three amplifier configurations.

5-6 TROUBLESHOOTING

Burning-In: A term describing a test in which a device is left powered on for a determined length of time. The burn-in period is used to test for faults due to heat.

Open Circuit: A circuit that has infinite resistance and allows no current flow.

Cold Solder Joint: A condition that occurs when insufficient heat is applied to a joint while soldering.

Prototype Board: A board that is used to build temporary circuits. It eliminates the need for soldering components.

Short Circuit: A circuit that has zero resistance and allows maximum current flow.

Solder Bridge: A condition that occurs when solder runs from one point on a printed-circuit board to another, thus connecting the two points.

There are two environments in which we may troubleshoot. The first is where the circuit was known to be working at some time or other and suddenly developed a fault. In this scenario, the design of the circuit does not come into question. This circuit may exist in a consumer product that has worked dependably for months or years. It is likely that a faulty component or loosened solder joint is the problem.

The second environment in which we may troubleshoot is in the developmental stages of a circuit. In this situation, the problem can range from a bad component to bad design, or from a cold solder joint to an oversight in assembly. The circuit may have never worked, as is often the case in lab experiments, or it may have failed during the **burn-in period.**

We discuss troubleshooting from the perspective of both environments. We must realize, however, that when an **open** is given as a possible cause, it can reveal itself in one of several ways. An open can be caused by a faulty component, a **cold solder joint,** or a component that is misplaced on a **prototype board.** Likewise, a **short** may mean a physically shorted component, a **solder bridge,** or a component that is misplaced on a prototype board.

A variety of amplifier problems are considered. Although the common-emitter amplifier is used in the following examples, the theory is applicable to all configurations. The purpose of the circuit of Fig. 5-13 is to amplify the input signal. The input signal is rather small in amplitude. The output signal should be larger, yet undistorted.

No Input Signal (Fig. 5-13) In order for a BJT to perform its task, it needs a signal at its base. If a signal is not found at point *A*, the signal source

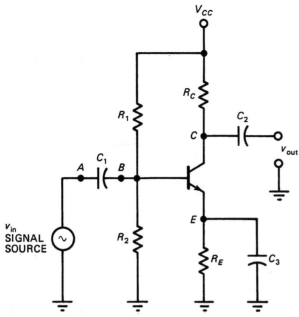

FIGURE 5-13 Troubleshooting a common-emitter amplifier.

should be examined or replaced. If there is a signal at point A and it disappears at point B, capacitor C_1 should be removed and checked for an open. Another possibility for the absence of a signal at point B is a short across R_1 or R_2 due to a solder bridge or misplaced component.

No Output Signal (Fig. 5-13)
V_{CC} may not be present due to a faulty dc supply; check the dc voltage. The power supply is usually the first place to start troubleshooting procedures. Then check the dc voltages on each leg to see if the Q point is correct.

Assume that there is a signal at point B; otherwise, see the previous section. Q_1 may have an open internally; it should be removed from the circuit and checked with an ohmmeter.

An open may exist at point C or E due to a cold solder joint or a misplaced component. Resistor R_1, R_C, or R_E may be open; the resistor may look burned or cracked. Any of these faults prevents both dc and ac current flow at the output.

Point C may be shorted to ground through a solder bridge. A short to ground holds that point at a potential of zero volts.

If coupling capacitor C_2 opened, no ac signal could reach the output of the circuit. A resistance or capacitance check of C_2 out of circuit reveals if the component is faulty.

Positive Peak of the Output Signal Is Clipped (Fig. 5-13)
The amplifier in this case is working; however, it is causing distortion in the output signal. Clipping of the positive peak can mean one of two things; either the input is too large or the quiescent voltage at the collector is high and the transistor is driven into cutoff. A quick comparison of the input signal against the circuit specifications reveals if the signal source must be decreased.

The resistance of R_1 may be too large or that of R_2 too small. Either of these problems lowers the dc voltage at the base and therefore increases the quiescent collector voltage.

The resistance of R_E may be too large or that of R_C too small. These changes in resistance also cause V_C to increase, by decreasing the voltage across R_C.

Negative Peak of the Output Signal Is Clipped (Fig. 5-13)

Distortion of the negative alternation of the output signal is caused by a low V_C. This, in turn, allows the transistor to be driven into saturation.

A lower R_1 or higher R_2 causes an increase in the voltage at the base. The increase in V_B causes a decrease in V_C. A glance back at Fig. 5-7 reveals the danger of a quiescent collector voltage that is too low.

If R_C were to increase or R_E were to decrease, the same effect would be experienced at V_C. A lower R_E increases emitter and, therefore, collector current. An increase in I_C causes an increase in V_{R_C}, which, in turn, causes a decrease in V_C. A larger R_C also drops more voltage, lowering V_C.

A leaky or shorted bypass capacitor (C_3) decreases the capacitor's resistance to dc current, which is normally infinite. The bypass capacitor affects R_E since it is in parallel with it. A decrease in the overall resistance at point E with reference to ground reduces V_C. The capacitor should be removed and checked with an ohmmeter or capacitance meter. A leak or short in C_1 can also have the same effect.

Finally, an input signal that is too large, riding on a low quiescent collector voltage, can clip the negative peak. Compare the level of the signal source to the circuit specification and adjust it if necessary.

Both Peaks of the Output Signal Are Clipped (Fig. 5-13)

The clipping of both peaks indicates that either the input signal is too large or the circuit is designed for too much gain. The easiest solution is to reduce the amplitude of the signal source. If adjusting the signal source is not possible, then a reconsideration of r'_e or the value of R_C is in order. Voltage gain can be lowered by either increasing r'_e or decreasing R_C, or removing C_3.

Output Signal Is Smaller Than Expected (Fig. 5-13)

This can be the result of either a design flaw or an accidental change in a circuit value due to a fault. Voltage gain is determined by

$$r'_e = \frac{25 \text{ mV}}{I_E}$$

$$A_v = \frac{R_C}{r'_e} \quad \text{(with a bypass capacitor)}$$

By examining the influences behind the values in these formulas, we can find the probable cause of the problem. A decrease in R_C for example, lowers the voltage gain.

A decrease in dc emitter current increases r'_e, which, in turn, decreases the voltage gain. The emitter current decreases if R_1 increases or R_2 decreases. A change in these resistances lowers the voltage at the base and, therefore, lowers the voltage at the emitter. If emitter voltage is lowered, emitter current must decrease.

Another way the emitter current is decreased is through an increase in R_E. A damaged resistor may show an increase in resistance.

The most obvious factor in affecting the voltage gain is the bypass capacitor. If C_3 were to open, the circuit would operate as if there were no bypass capacitor. This faulty component can be detected by removing it and testing it with either an ohmmeter or capacitance meter. A cold solder joint on the capacitor would produce the same effect. Also, a reduction in the values of C_1 or C_2 lowers the gain.

Finally, a smaller signal at the output may be due to a smaller signal at the input. Check the signal source for proper amplitude and adjust if necessary.

REVIEW QUESTIONS

34. What effect would an open at point E have on the output signal of the circuit of Fig. 5-13?
35. What effect would a short across R_C have on the output signal of the circuit of Fig. 5-13?
36. What effect would an open in C_1 have on the output signal of the circuit of Fig. 5-13?
37. What effect would an open in C_3 have on the output signal of the circuit of Fig. 5-13?
38. Name one fault that would cause the output signal to be clipped on both peaks.
39. Name two faults that would cause the output signal to be clipped on the negative peak.
40. A good capacitor should measure very little resistance in both directions. True or false?

SUMMARY

1. A small-signal amplifier is used to increase the current, voltage, or power of an applied ac signal.
2. The dc load line illustrates the operating range of a transistor between cutoff and saturation.
3. The Q point refers to the dc operating voltage at the output of an amplifier.
4. AC emitter resistance (r'_e) is found by

$$r'_e = \frac{25 \text{ mV}}{I_E}$$

5. Base resistance (R_b) is found by

$$R_b = B(r'_e + R_E)$$

6. A transistor has two current-gain specifications: dc beta (h_{FE}) and ac beta (h_{fe}).
7. The ac beta or current gain in this text is considered equivalent to the dc beta with the exception of the common-base amplifier. Therefore,

$$A_i = \beta$$

8. A bypass capacitor connected across the emitter resistor provides a low-resistance path to ground for an ac signal.
9. A bypass capacitor increases the voltage gain of the amplifier.

10. The ac output voltage rides on a dc Q point.

11. The common-emitter (CE) amplifier offers good voltage, current, and power gain as well as low input impedance and high output impedance.

12. The input of a CE amplifier is on the base and the output is taken off the collector.

13. A phase shift of 180° is produced by the CE amplifier.

14. Important formulas for the CE amplifier are

$$A_v = \frac{R_C}{R_E + r_e'} \quad \text{(without a bypass capacitor)}$$

$$A_v = \frac{R_C}{r_e'} \quad \text{(with a bypass capacitor)}$$

$$v_{\text{out}} = v_{\text{in}} \times A_v$$

$$A_i = \beta$$

$$A_p = A_i \times A_p$$

$$z_{\text{in}} = R_1 \parallel R_2 \parallel R_b$$

$$z_{\text{out}} = R_C$$

15. Coupling capacitors are used to isolate dc voltages.

16. The overall gain of a multistage amplifier is equivalent to the product of the gains of the individual stages.

17. The common-collector (CC) amplifier offers good current and power gain as well as high input impedance and low output impedance.

18. The CC amplifier offers unity voltage gain.

19. The input of a CC amplifier is on the base and the output is taken off the emitter.

20. The CC amplifier does not provide a phase shift.

21. Important formulas for the CC amplifier are

$$A_v = \frac{R_E}{R_E + r_e'} \qquad A_p = A_i \times A_p$$

$$\qquad\qquad\qquad z_{in} = R_1 \parallel R_2 \parallel R_b$$

$$v_{\text{out}} = v_{\text{in}} \times A_v$$

$$A_i = \beta \qquad\qquad z_{\text{out}} = \frac{R_s}{\beta} \parallel R_E$$

22. A Darlington pair amplifier is composed of two CC amplifiers.

23. The Darlington pair provides extremely high base impedance, high current gain, but unity voltage gain.

24. The input impedance of a Darlington pair circuit using voltage-divider bias is found by

$$z_{\text{in}} = R_1 \parallel R_2$$

25. The output impedance of a Darlington pair circuit can be approximated as being less than 50 Ω.

26. The common-base (CB) amplifier offers good voltage and power gain as well as low input impedance and high output impedance.

27. The CB amplifier offers unity current gain.

28. The input of a CB amplifier is on the emitter and the output is taken off the collector.

29. The CB amplifier does not provide a phase shift.

30. Important formulas for the CB amplifier are

$$A_v = \frac{R_C}{r_e'} \qquad A_p = A_i \times A_p$$

$$z_{in} = r_e'$$

$$v_{out} = v_{in} \times A_v \qquad z_{out} = R_C$$

$$A_i = 1$$

31. No output signal in an amplifier can be due to the lack of an input signal, improper dc biasing voltages, a faulty BJT, or an open coupling capacitor.

32. Clipping distortion can be due to too large an input signal, higher than normal dc biasing voltages, a faulty BJT, or a shorted bypass capacitor.

33. A smaller than normal output signal can be due to an improper circuit gain design, too small an input signal, a faulty BJT, or an open bypass capacitor.

GLOSSARY

Amplifier: An amplifier boosts, or increases, the voltage, current, or power of a signal.

Burning-In: A term describing a test in which a device is left powered on for a determined length of time. The burn-in period is used to test for faults due to heat.

Cold Solder Joint: A condition that occurs when insufficient heat is applied to a joint while soldering.

Common-Base Amplifier: This amplifier provides good voltage and power gain, but unity current gain. It has low input impedance and high output impedance. The input is on the emitter and the output is off the collector.

Common-Collector Amplifier: This amplifier provides good current and power gain, but unity voltage gain. It has high input impedance and low output impedance. The input is on the base and the output is off the emitter.

Common-Emitter Amplifier: This amplifier provides good voltage, current, and power gain. It has low input impedance and high output impedance. This input is on the base and the output is off the collector.

Coupling Capacitor: A capacitor that couples, or connects, the ac signal from one stage to the next, yet isolates the dc.

DC Load Line: A graph showing all the possible operating conditions available to the transistor.

Darlington Pair: A name commonly given to a two-stage common-collector amplifier.

Open Circuit: A circuit that has infinite resistance and allows no current flow.

Prototype Board: A board that is used to build temporary circuits. It eliminates the need for soldering components.

Quiescent Voltage: The dc operating-point voltage.

r_e': The ac emitter resistance.

Short Circuit: A circuit that has zero resistance and allows maximum current flow.

Small-Signal Amplifier: An amplifier whose amplified signal power is usually one-half watt or less.

Solder-Bridge: A condition that occurs when solder runs from one point on a printed-circuit board to another, thus connecting the two points.

PROBLEMS

SECTION 5-1

The following problems refer to the circuit of Fig. 5-14.

5-1. What is the value of V_B?

5-2. Find the value of V_C.

5-3. What is the value of A_v?

5-4. What is the value of A_p?

5-5. Determine the value of v_{out}.

5-6. What is the value of the input impedance?

5-7. What is the value of the output impedance?

5-8. What is the value of $V_{C(SAT)}$?

5-9. Is the output signal distorted?

5-10. What are the characteristics of the common-emitter amplifier?

SECTION 5-2

The following problems refer to the circuit of Fig. 5-15.

5-11. Find the value of V_C of Q_1.

5-12. What is the value of A_v of Q_1?

5-13. Determine the peak-to-peak value of v_c of Q_1.

5-14. What is the value of V_C of Q_2?

5-15. Calculate the value of A_v of Q_2.

5-16. Determine the peak-to-peak value of v_c of Q_2.

5-17. Find the value of V_C of Q_3.

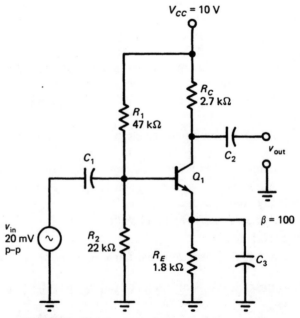

FIGURE 5-14 Common-emitter amplifier.

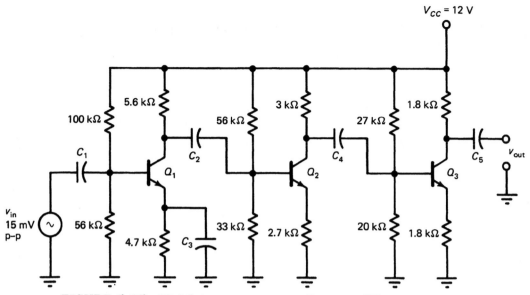

FIGURE 5-15 Multistage common-emitter amplifier.

5-18. Determine the peak-to-peak value of v_c of Q_3.

5-19. What is the overall A_v?

SECTION 5-3

The following problems refer to circuit of Fig. 5-16.

5-20. What is the value of V_B?

5-21. What is the value of V_E?

5-22. Calculate the value of A_v.

5-23. Calculate the value of A_i.

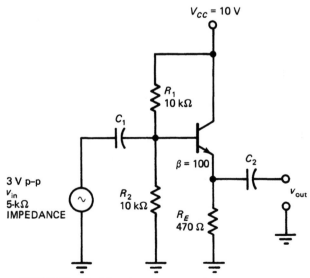

FIGURE 5-16 Common-collector ampli-fier.

5-24. Find the value of v_{out}.

5-25. Determine the value of the input impedance.

5-26. Determine the value of the output impedance. (The power supply has a 5-kΩ impedance.)

5-27. What are the characteristics of the common-collector amplifier?

SECTION 5-4

The following problems refer to circuit of Fig. 5-17.

5-28. What is the value of V_B of Q_1?

5-29. What is the value of V_E of Q_1?

5-30. Calculate the value of the overall A_v.

5-31. Calculate the value of the overall A_i.

5-32. Determine the value of the input impedance.

5-33. Determine the value of the output impedance.

5-34. What is the value of V_E of Q_2?

SECTION 5-5

The following problems refer to circuit of Fig. 5-18.

5-35. What is the value of V_C?

5-36. Calculate the value of A_v.

5-37. Calculate the value of A_i.

5-38. Determine the value of the input impedance.

5-39. Determine the value of the output impedance.

5-40. What are the characteristics of the common-base amplifier?

SECTION 5-6

5-41. What is the difference between a bypass capacitor and a coupling capacitor?

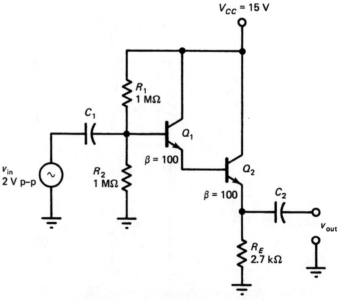

FIGURE 5-17 Darlington pair.

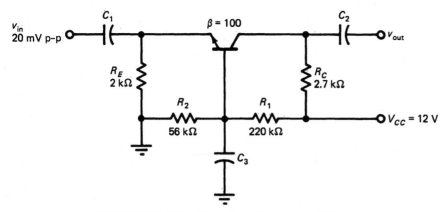

FIGURE 5-18 Common-base amplifier.

5-42. List the possible causes for a clipped signal in the circuit of Fig. 5-14.

5-43. List the possible causes for no output signal in the circuit of Fig. 5-16.

5-44. What problems might affect the voltage gain of a circuit?

ANSWERS TO REVIEW QUESTIONS

1. voltage, current, power	2. one-half watt	3. small plastic case
4. low, high	5. True	6. base, collector
7. collector	8. V_C	9. True
10. active	11. dc load line	12. ac emitter resistance
13. To increase gain	14. $A_i = \text{beta}$	15. $\dfrac{R_C}{r'_e}$
16. $A_i \times A_v$	17. $R_1 \parallel R_2 \parallel R_b$	18. $Z_{out} = R_C$
19. C_3	20. 500	21. high, low
22. False	23. base, emitter	24. V_E
25. No	26. Darlington pair	27. Unity
28. $R_1 \parallel R_2$	29. True	30. low, high
31. False	32. emitter, collector	33. Collector voltage
34. Decreases to zero	35. Decreases to zero	36. Decreases to zero
37. Reduces to amplitude	38. The input is too large	39. Short across C_3, R_E, or R_2. Open R_C.
40. False		

6

Power Amplifiers

When you have completed this chapter, you should be able to:

- Recognize the difference between small-signal and power amplifiers.
- Analyze the operational differences between the various classes of amplifier.
- Become familiar with the advantages and disadvantages offered by each class of amplifier.
- Discern the types of distortion along with their causes.
- Troubleshoot the power stages of an amplifier circuit.

INTRODUCTION

The discussion in the previous chapter was concerned with small-signal amplifiers. The **small-signal amplifier** is used in the earlier stages of an amplifier circuit to provide current gain and voltage gain. The final stage of an amplifier circuit is most often a **power amplifier.** This is because the final stage is usually connected to a small impedance load and, therefore, needs the extra boost of power that a power amplifier can provide.

Power amplifiers are categorized according to class. Each amplifier receives a full 360° input signal, but provides output signals of varying angles. The class is determined by the angle of the output. There are class A, AB, B, C, D, E, F, and S amplifiers. We focus on three of the more popular classes: A, B, and C.

When speaking of the **angle** of the waveform, we mean the percentage of the input signal that is seen at the output. This percentage is expressed in degrees. For example, an input signal consists of 360°, or one complete **cycle.** If the output signal consists of only the positive **alternations,** then it has an angle of 180°, because only half of the input signal is reproduced at the output.

Notice within the discussion of each amplifier that there is a trade-off between the size of the output angle and the percentage of efficiency. The larger the output angle, the more closely it resembles the input signal. However, the larger the output angle, the greater the amount of power that is wasted in operating the amplifier. Keep these thoughts in mind as you read the following sections.

Small-Signal Amplifier: An amplifier that is mainly concerned with voltage or current gain.

Power Amplifier: An amplifier that provides a larger amount of power than a small-signal amplifier.

Angle: Used to describe the percentage of one cycle of a sine wave in degrees.
Cycle: A complete sine wave composed of a positive and negative alternation. It equals 360°.
Alternation: A half cycle, which equals 180°.

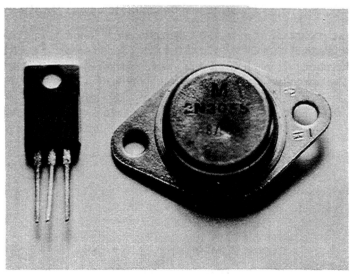

Power transistors.

Class A Amplifier: An amplifier that conducts for 360° of the input. It offers a maximum efficiency of 25–50%, depending on the type of coupling used.

The **class A amplifier** conducts during the full 360° input signal. This means that a complete 360° signal is seen at its output. Since the transistor is conducting all of the time, a large amount of power is dissipated. The circuit of Fig. 6-1 shows both the input and output waveforms.

A common-emitter configuration is used in Fig. 6-1, but either of the other configurations could have been used. An *RC* coupling network connects the amplifier stage to the load resistance. The *RC* coupling network consists of C_2 and R_C. The purpose of *RC* coupling is to pass the ac signal while isolating the dc.

Attention will be focused on power-related topics since the topic of this chapter is power amplifiers. Let us begin by refreshing our memory with power gain.

AC Loaded Power Gain   The formula used in the previous chapter is used here. The only difference now is that the load resistance is being emphasized. Watt's law states that $P = V \times I$. Therefore, before discussing power gain, let us examine the voltage and current gains.

The ac voltage gain is found by

$$A_v = \frac{r_c}{r'_e}$$

where r_c = ac resistance seen by collector

r'_e = ac emitter resistance.

The ac emitter resistance, which was studied in the last chapter, is found by

$$r'_e = \frac{25 \text{ mV}}{I_E}$$

$$= \frac{25 \text{ mV}}{542 \text{ } \mu\text{A}}$$

$$= 46.2 \text{ } \Omega$$

Shunted: Connected in parallel.

The ac resistance as viewed by the collector is found by calculating R_L in parallel (or **shunted**) with R_C. The ac signal treats the dc supply as a direct short to ground; therefore, R_C is tied to ground and is parallel with R_L in the circuit of Fig. 6-1. The formula can be written as

$$r_c = R_L \parallel R_C$$

$$= 10 \text{ k}\Omega \parallel 15 \text{ k}\Omega$$

$$= 6 \text{ k}\Omega$$

Finally, returning to the voltage gain,

$$A_v = \frac{6 \text{ k}\Omega}{46.2 \text{ } \Omega}$$

$$= -130$$

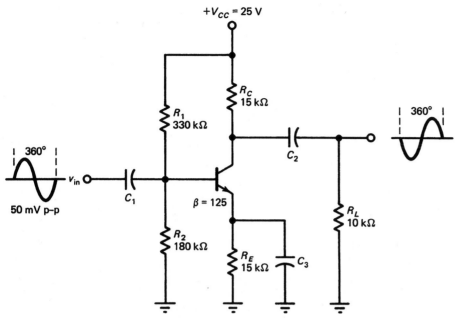

FIGURE 6-1   *RC* coupled class A amplifier.

The common-emitter amplifier provides good ac voltage gain. The minus sign indicates the phase shift. A power amp is used to provide power gain and not necessarily to increase the amplitude of the voltage.

The next step in pursuing power gain is to determine the ac current gain. Use the same approximation as used earlier: current gain equals beta. This value is taken from Fig. 6-1.

$$A_i = \beta$$
$$= 125$$

The power gain of this circuit can now be determined from the product of the voltage and current gains.

$$A_p = A_v \times A_i$$
$$= 130 \times 125$$
$$= 16,250 \qquad \blacksquare$$

EXAMPLE 6-1 _____

Determine the power gain of the circuit of Fig. 6-1 if the load resistance (R_L) were changed to 100 Ω.

Solution: Lowering the load resistance reduces the power gain. This occurs because a lower load resistance reduces the ac voltage gain and the power gain is the product of the ac voltage and current gains.

$$r_c' = \frac{25 \text{ mV}}{542 \ \mu\text{A}}$$
$$= 46.2 \ \Omega$$

$$r_c = 100 \ \Omega \ \| \ 15 \ \text{k}\Omega$$
$$= 99.3 \ \Omega$$

$$A_v = \frac{99.3 \ \Omega}{46.2 \ \Omega}$$
$$= 2.15$$

$$A_i = \beta = 125$$

$$A_p = 2.15 \times 125$$
$$= 269 \qquad \blacksquare$$

Practice Problem: Different size load resistances affect the gains in the circuit of Fig. 6-1. What is the power gain for a load resistance of 1 kΩ? *Answer:* $A_p = 2538$.

Power Dissipated by the Transistor
One physical difference between a small-signal BJT and a large-signal, or power, BJT is the casing. The small-signal BJT is usually housed in a small plastic case. It does not need to dissipate much power. The power BJT is usually housed in a larger case with a metal back plate. The size of the BJT and the metal allow it to dissipate much more heat.

Heat Sink: A piece of metal attached to a component to help dissipate heat.

Heat sinks are also used to help dissipate power in the form of heat. They are fin-shaped pieces of metal that are attached to the transistor. The reason for the fins is that the larger the surface area, the more heat that can be dissipated. This results in the heat not given a chance to build up inside the component. Heat is a common cause of damage to components.

The power dissipated by the transistor in Fig. 6-1 is determined by using the quiescent, or operating, current and voltage of the transistor. These values are found by doing a dc analysis of the common-emitter circuit.

$$P_Q = V_{CEQ} \times I_{CQ}$$

where P_Q = quiescent power of BJT

V_{CEQ} = quiescent collector-emitter voltage

I_{CQ} = quiescent collector current

$$P_Q = 8.75 \ \text{V} \times 542 \ \mu\text{A}$$
$$= 4.74 \ \text{mW}$$

Power Transistor: A transistor designed to handle a power greater than one-half watt.

This is a relatively small amount of power for any transistor. **Power transistors** can handle one-half watt or more. There is not much chance of damaging the transistor in this circuit.

Practice Problem: How much power would the transistor dissipate if R_C and R_E were changed to 1-kΩ resistors? *Answer:* 71.1 mW.

Power Dissipated by the Load
Power is very useful. A load may require a certain amount of power to do its job. When you perform a strenuous task, your body dissipates heat. The load also dissipates heat in the performance of its job.

There are two ways in which to find the **root-mean-square (rms) power** dissipated by the load. The reason for two formulas is that one is easier if measurements are made with an oscilloscope and the other is more convenient for use with a digital voltmeter.

The first step is to determine the output voltage. This is accomplished by using the voltage gain value found earlier.

$$v_{out} = v_{in} \times A_v$$
$$= 50 \text{ mV p–p} \times 130$$
$$= 6.5 \text{ V p–p}$$

This is the value of voltage that would be read on an oscilloscope. A convenient formula to use in this situation is

$$P_L = \frac{v_{L(p\text{–}p)}^2}{8R_L}$$
$$= \frac{(6.5 \text{ V})^2}{8 \times 10 \text{ k}\Omega}$$
$$= 528 \text{ }\mu\text{W}$$

If a digital voltmeter were used, the values would be measured in rms. Let us convert these values to rms in order to demonstrate the use of the next formula for finding the power of the load.

$$v_{L(rms)} = \frac{v_{L(p\text{–}p)}}{2} 0.707$$
$$= 2.3 \text{ V}$$

The alternate formula for finding the power of the load is

$$P_L = \frac{v_{L(rms)}^2}{R_L}$$
$$= \frac{5.28 \text{ V}}{10 \text{ k}\Omega}$$
$$= 528 \text{ }\mu\text{W}$$

The result is the same; the difference is the starting point. If you are simply calculating the circuit or measuring it with an oscilloscope, then the first formula is the best choice. If you are using a digital voltmeter, then the last formula would save time converting values to peak-to-peak.

Practice Problem: How much power would the load resistance use if the input signal were increased to 75 mV p–p? *Answer:* 1.19 mW.

Power Efficiency The dc supply provides the power for the amplifier circuit and the load. If the load uses most of the power offered by the dc supply, then the circuit is said to be efficient. Ideally, 100% of the dc supply's power should be used by the load. The maximum efficiency that can be expected from an *RC* coupled class A amplifier is 25%. Much of the power furnished by the supply is not used by the load; it is wasted.

Root-Mean-Square Power: The root-mean-square (rms) value is an expression of the heating effects of an ac signal.

In order to determine an amplifier's efficiency, we only need to ask two questions. "How much power is the dc supply providing?" And, "how much power is the load using?"

Begin by solving for the amount of power that the dc supply is providing for the output. In the circuit of Fig. 6-1, the power furnished by the dc supply is determined by

$$P_{dc} = I_{CC} \times V_{CC}$$

where P_{dc} = power furnished by the dc supply

I_{CC} = total dc supply current

V_{CC} = dc supply voltage

The total dc supply current is the sum of the quiescent collector current and the voltage-divider current.

$$I_{CC} = I_{CQ} + I_{VD}$$
$$= 542 \ \mu A + 49 \ \mu A$$
$$= 591 \ \mu A$$

so

$$P_{dc} = 591 \ \mu A \times 25 \ V$$
$$= 14.8 \ mW$$

Next, recall the power used by the load. This value is the same as the rms power of the load found earlier. This value is 528 μW.

Armed with these values, we can determine the efficiency of the amplifier in Fig. 6-1. This is based on the following formula:

% Efficiency: The percentage of the dc supply power used by the ac load.

$$\% \ \textbf{efficiency} = \frac{P_L}{P_{dc}} \times 100$$
$$= \frac{528 \ \mu W}{14.8 \ mW} \ 100$$
$$= 3.57$$

Considering that maximum efficiency is 25%, this circuit is operating far below its potential. What can be done to increase the efficiency of this circuit? This question can be answered in one of two ways. Either change component values or increase the input signal.

Figure 6-2 illustrates the condition of the circuit in Fig. 6-1. The uppermost and lowermost limits of circuit operation are determined by the dc supply voltage. The operating point, or quiescent voltage, in the circuit is 16.9 V. A 6.5-V p–p output signal is currently riding on V_{CQ}. An important point to notice is that the lower voltage limit is not 0 V, but 12.5 V. This is because $V_{CQ} = 12.5$ when the transistor is driven into saturation.

Since a transistor is conducting at maximum when saturated, V_{CQ} cannot be lowered any further. Therefore, the maximum voltage of the output signal is 8.8 V p–p. This is technically referred to as the ac output compliance. The ac output

134

$V_{CC} = 25\ V$

$V_L = 6.5\ V\ p\text{-}p$ $V_{CQ} = 16.9\ V$

$V_{SAT} = 12.5\ V$

$V_{CC} = 0\ V$

FIGURE 6-2   Operating range.

compliance is the maximum unclipped peak–peak voltage that can occur at the output.

If we use this new output voltage to determine its effect on the efficiency of the circuit, we find that

$$P_L = \frac{(8.8\ V)^2}{8 \times 10\ k\Omega}$$

$$= 968\ \mu W$$

$$\%\ \text{efficiency} = \frac{968\ \mu W}{14.8\ mW} \times 100$$

$$= 6.54$$

The efficiency almost doubled, yet it is far from the 25% maximum that we discussed earlier. But we do begin to see that a larger output signal makes greater use of the dc power and results in a more efficient circuit. If a larger signal is combined with a change of components so that V_{CQ} allows a greater signal swing, then efficiency will continue to increase. All of these factors go into the proper design of an amplifier.

EXAMPLE 6-2

The load resistor in the circuit of Fig. 6-1 is changed to 100 Ω. How does this affect (a) the power dissipated by the transistor, (b) the power dissipated by the load, and (c) the power efficiency?

Solution:
(a) Changing the load resistance does not affect the power dissipated by the transistor. The values found before still hold true.

$$P_Q = 4.74\ mW$$

(b) The voltage gain is affected by the change in load resistance. The first step is to establish the new voltage gain. AC emitter resistance remains the same; however, the ac collector resistance changes.

$$r_c = 100\ \Omega \parallel 15\ k\Omega$$

$$= 99.3\ \Omega$$

Voltage gain is found by dividing the ac collector resistance by the ac emitter resistance.

$$A_v = \frac{99.3\ \Omega}{46.2\ \Omega}$$
$$= 2.15$$

$$v_{\text{out}} = 2.15 \times 50\ \text{mV p–p}$$
$$= 108\ \text{mV p–p}$$

$$P_L = \frac{(108\ \text{mV})^2}{8 \times 100\ \Omega}$$
$$= 14.6\ \mu\text{W}$$

(c) Since the power of the load has changed, so will the efficiency of the circuit. Changing the load resistance has no effect on the dc values of the amplifier. The power generated by the dc supply remains at 14.8 mW. The new power dissipated by the load was found to be 14.6 μW; therefore, the efficiency decreases to

$$\% \text{ efficiency} = \frac{14.6\ \mu\text{W}}{14.8\ \text{mW}} \times 100$$
$$= 0.099$$ ∎

Transformer Coupling: The use of a transformer to connect the stages of an amplifier.

Impedance: The overall opposition of a circuit to ac.

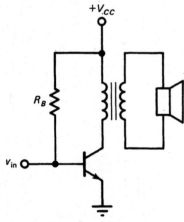

FIGURE 6-3 Transformer coupling.

Transformer Coupling Transformer coupling is illustrated in Fig. 6-3. It offers several benefits, including impedance matching, which allows us to connect the output of an amplifier that has high **impedance** to a load that has low impedance. Transformer coupling also isolates the load from the dc supply.

The efficiency of a class A amplifier can be increased to almost 50% by the use of transformer coupling. This is due to the magnetic self-induction of the transformer. The voltage produced by the changing magnetic field is added to the dc supply, providing almost twice the value of V_{CC} to the load. This, in turn, increases the power for the load and the overall efficiency of the amplifier.

High cost is what deters most from using the transformer. It has been mostly replaced by transistor circuits in audio amplifiers. RF applications still use transformer coupling.

REVIEW QUESTIONS

1. The power amplifier is usually one of the last stages of an amplifier circuit. True or false?
2. Small-signal amplifiers are usually used to provide voltage or current gain. True or false?
3. The class of an amplifier is determined by the _____ of the output.
4. One complete cycle of a sine wave is equivalent to _____ degrees.
5. The class A amplifier has an output angle of _____ degrees.
6. The input of the common-emitter configuration is on the _____ and the output is taken off of the _____ .
7. Watt's law states that $P = E \times$ _____ .
8. Current gain is approximately equal to _____ .

9. As load resistance decreases, power gain _____.

10. The metal finlike devices used to dissipate heat are called _____ _____.

11. The transistor in a class A amplifier is continuously conducting. True or false?

12. A BJT dissipates power in the form of _____.

13. As the voltage of the input signal increases, the power used by the load _____.

14. The _____ _____ provides the power for the amplifier circuit and load.

15. The maximum efficiency of an *RC* coupled class A amplifier is _____.

16. What is the formula for % efficiency?

17. What term describes the maximum unclipped peak–peak voltage that can occur at the output?

18. Transformer coupling is useful for matching _____.

19. Transformer coupling can increase class A efficiency to a maximum of _____.

20. One deterrent of the use of transformer coupling is _____.

6-2 CLASS B AMPLIFIERS

Class B amplifiers provide a 180° output for a 360° input. This results in only half of the input wave reproduced at the output. Because the transistor is on for only half the time, less dc power is generated. The outcome is greater efficiency. The problem, however, is that the output signal is very distorted.

  The **push–pull** arrangement using class B amplifiers takes advantage of higher efficiency and eliminates the **distortion.** Figure 6-4 shows a typical push–pull configuration.

Class B Amplifier: An amplifier that conducts for 180° of the input. It offers a maximum efficiency of 78.5%.

Push–Pull: An arrangement in which two components or subcircuits conduct on an alternate basis. It is typically composed of class B amplifiers.

Distortion: The unwanted change of shape of a signal.

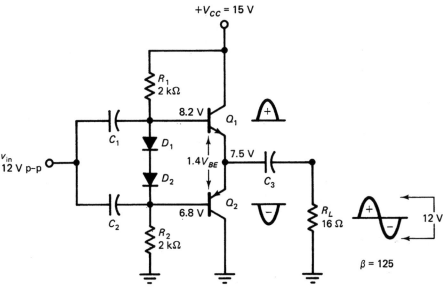

FIGURE 6-4   Class B push–pull configuration.

Push–Pull Operation   Beginning at the input side of the circuit, the input signal is coupled through C_1 and C_2 to the base of both transistors. The coupling capacitors isolate the ac source from the dc voltages within the voltage divider.

The voltage divider consists of two resistors and two diodes. The two diodes are connected in parallel to the two base–emitter junctions of the transistors. This is done to keep the transistors slightly turned on, thus eliminating distortion. The diodes drop 1.4 V, leaving the remainder of V_{CC} to split between R_1 and R_2. The voltages at the base of each transistor are shown in Fig. 6-4.

The transistors consist of an NPN and a PNP that are matched. Matching the specifications of the transistor is necessary to avoid distortion of the output signal. The base of Q_1 is P-type and the base of Q_2 is N-type. Remember that the base–emitter junction of a transistor must be forward biased for it to conduct. Therefore, the base of Q_1 needs a more positive voltage and the base of Q_2 needs a more negative voltage.

As the input ac signal goes through its positive alternation, it turns on Q_1. As Q_1 begins to conduct, its V_{CE} decreases, but the voltage at its emitter increases. These transistors are configured as common-collector amplifiers. The emitter signal follows the base signal.

As the input ac signal goes through its negative alternation, it turns on Q_2. The voltage at the emitter of Q_2 gradually decreases and increases as the input signal swings through its negative alternation.

The action of Q_1 and Q_2 conducting alternately continues with every cycle of the input signal. Q_1 passes the positive alternations and Q_2 passes the negative alternations.

The dc voltage at the emitters is 7.5 V. This can be determined by subtracting V_{BE} from the base voltage of Q_1 or adding V_{BE} to the base voltage of Q_2. The ac input voltage of 12 V p–p causes the dc emitter voltage to swing 6 V in either direction. This results from the common-collector voltage gain of 1. Figure 6-5 shows the condition at the emitters of the push–pull arrangement.

The remaining coupling capacitor, C_3, connects the load resistance R_L to the emitters. As the signal at the emitters fluctuates, the signal across the load resistance fluctuates. The positive and negative alternations from each transistor are rejoined to create a complete 360° signal once again.

Power Dissipated by the Load   The formula for finding the power dissipated by the load in this circuit is the same as the one used with the class A amplifier. The circuit of Fig. 6-4 shows an input signal of 12 V p–p. Assuming no losses, the signal at the load is also 12 V p–p.

$$P_L = \frac{v_{L(p-p)}^2}{8R_L}$$

$$= 1.125 \text{ W}$$

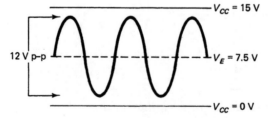

FIGURE 6-5 Operating range of a push–pull amplifier.

Power Efficiency   Theoretically, the efficiency of a class B push–pull amplifier can run as high as 78.5%. The increased efficiency is one of the advantages of this circuit. As explained earlier, the efficiency of an amplifier is determined by the percentage of dc supply power that it uses.

The current drain on the dc supply is determined by the sum of two currents. One is the constant current flowing through the voltage-divider circuit and the other is the average of the peak load current.

$$I_{VD} = \frac{V_{CC} - 1.4}{R_1 + R_2}$$

$$= 3.4 \text{ mA}$$

The value 1.4 V comes from the addition of the two diode voltage drops.

From the topic of half-wave rectifiers discussed in Chapter 2, recall the formula for arriving at the average output voltage. The reason for returning to this topic is because the transistors of Fig. 6-4 only conduct on one alternation. They act like half-wave rectifiers. Finding the average collector current allows us to determine the total dc supply current, I_{CC}.

$$I_{C(\text{ave})} = \frac{v_{L(\text{pk.})} \times 0.318}{R_L}$$

$$= \frac{6 \text{ V pk.} \times 0.318}{16 \ \Omega}$$

$$= 119 \text{ mA}$$

Therefore,

$$I_{CC} = I_{VD} + I_{C(\text{ave})}$$

$$= 122 \text{ mA}$$

The formula for finding the power generated by the dc supply is also the same.

$$P_{\text{dc}} = I_{CC} \times V_{CC}$$

$$= 122 \text{ mA} \times 15 \text{ V}$$

$$= 1.83 \text{ W}$$

Using this value and the power of the load resistance found earlier, we can determine the efficiency of this amplifier.

$$\% \text{ efficiency} = \frac{P_L}{P_{\text{dc}}} \times 100$$

$$= \frac{1.125 \text{ W}}{1.83 \text{ W}} \times 100$$

$$= 61.5$$

This is a very good efficiency rating. The efficiency can be raised by increasing the size of the input signal. This takes greater advantage of the power provided by the dc supply. On the other hand, the dc supply voltage could be

---

decreased. This would decrease the amount of generated power, thereby increasing efficiency.

EXAMPLE 6-3

The load resistance in the circuit of Fig. 6-4 is changed to 8Ω. How does this affect (a) the power dissipated by the load, and (b) the efficiency of the amplifier?

Solution:
(a) Lowering the load resistance increases the amount of power used by the load.

$$P_L = \frac{(12 \text{ V})^2}{8 \times 8 \text{ }\Omega}$$

$$= 2.25 \text{ W}$$

(b) Lowering the load resistance also increases the amount of power supplied by the dc supply. This occurs because of the increase in the average collector current.

$$I_{C(ave)} = \frac{(6 \text{ V pk.} \times 0.318)}{8 \text{ }\Omega}$$

$$= 239 \text{ mA}$$

The voltage-divider current remains at 3.4 mA. Therefore, the total supply current is

$$I_{CC} = 239 \text{ mA} + 3.4 \text{ mA}$$

$$= 242 \text{ mA}$$

and

$$P_{dc} = 242 \text{ mA} \times 15 \text{ V}$$

$$= 3.63 \text{ W}$$

Finally, we learn that the efficiency increases slightly.

$$\% \text{ efficiency} = \frac{2.25 \text{ W}}{3.63 \text{ W}} \times 100$$

$$= 62 \qquad \blacksquare$$

Practice Problem: What is the % efficiency if the input signal of the circuit of Fig. 6-4 were increased to 15 V p–p? *Answer:* 76.9%.

REVIEW QUESTIONS

21. Class B amplifiers provide a _____ degree output.
22. The transistor in a class B amplifier is on all the time. True or false?
23. The class B amplifier is more efficient than the class A. True or false?
24. Coupling capacitors pass _____, but block _____ .
25. When no input signal is applied, neither transistor in the push–pull arrangement conducts. True or false?
26. The emitter voltage of a PNP transistor is 0.7 V more (positive or negative) than the voltage at the base.
27. The output of the push–pull arrangement is _____ degrees.

28. What is the maximum efficiency that can be obtained with a class B amplifier?

29. How is the average voltage of an alternation determined?

30. The larger the input signal, the greater the efficiency. True or false?

6-3 CLASS C AMPLIFIERS

The **class C amplifier** produces less than 180° output for a 360° input signal. This amplifier is useful for narrowband RF applications and is discussed in this context.

The circuit of Fig. 6-6 is a tuned class C amplifier. Notice that the base–emitter junction is not biased by the dc supply. Instead, it is connected to the signal source via an *RC* network. Only the positive alternations at the base can turn on the transistor.

Capacitor C_1 provides dc clamping for the circuit. The capacitor charges to 2.3 V on the very first positive alternation. On each subsequent positive alternation, the 3-V peak input signal must overcome this 2.3-V series opposing voltage. The result is that only 0.7 V of the positive peaks is passed onto the base. These short positive pulses on the base cause the transistor to conduct in a pulselike fashion.

Capacitor C_1 does not discharge on the negative alternation of the input because the *RC* time constant provided by R_1 is too long. The next positive pulse occurs before the capacitor has a chance to discharge to any degree. Each positive pulse recharges the capacitor through the base–emitter junction rather than through the resistor.

When a pulse at the base causes Q_1 to conduct, the brief surge of collector current charges C_2. Components C_2 and L_1 form a parallel resonant **tank circuit.** It is so named because the circuit serves as a storage tank for energy. After the pulse has occurred, C_2 discharges and, in turn, causes a magnetic field to develop around the coil. When the magnetic field reaches its maximum, it begins to collapse and the induced voltage charges C_2 with the opposite polarity. This exchange of current between C_2 and L_1 continues: first in a clockwise fashion and

Class C Amplifier: An amplifier that conducts for less than 180° of the input. It offers a maximum efficiency of 100%.

Tank Circuit: An *LC* circuit designed to oscillate at its resonant frequency.

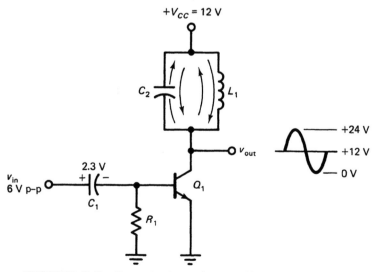

FIGURE 6-6 Tuned class C amplifier using signal bias.

Oscillation: The action of current flowing back and forth alternately.

then counterclockwise. The action of current flowing back and forth between the capacitor and inductor in the tank circuit is called **oscillation.** The transistor is pulsed on periodically by the signal at the base, which causes short surges of collector current to replace any losses within the tank circuit.

The action in the tank circuit causes a sinusoidal ac output. The strength and frequency of the output is determined by both the input frequency and the value of the components in the tank circuit. In order for the tank circuit to work effectively, it should be resonant with the input signal. **Resonant frequency** can be determined by the formula

Resonance: A condition where the capacitive reactance equals the inductive reactance at one frequency for an LC circuit. $f_r = 1/2 \pi \sqrt{LC}$.

$$f_r = \frac{1}{2\pi \sqrt{LC}}$$

For example, if $L = 0.1$ mH and $C = 0.01$ μF, then $f_r = 159$ kHz.

An output signal of the greatest strength is created when the input frequency matches the resonant frequency of the tank circuit. As the input frequency shifts either higher or lower than the resonant frequency, the output signal becomes weaker and weaker. The tuned class C amplifier circuit responds favorably to only a narrow band of input frequencies, in a fashion similar to a **band-pass filter.** This is how stations are selected or tuned for **radio** or TV reception.

Band-pass Filter: A circuit that passes a certain band or range of frequencies and filters, or blocks all others.

Radio Frequencies: Radio frequencies (RF) range from 10 kHz to 3000 GHz.

The operating point of the output of the circuit in Fig. 6-6 is equal to V_{CC}. This is because the transistor is cut off for most of the time. When the transistor is turned on, C_2 can charge to 12 V. The transistor returns to its cut-off condition and the quiescent collector voltage returns to the value of V_{CC}. As the tank circuit oscillates, the polarity of its 12 V is changing back and forth. It either aids or opposes the voltage of the dc supply. When it aids the dc supply voltage, the net result at the collector is twice the value of V_{CC}, or 24 V in our example. When it opposes the dc supply voltage, it has the effect of subtracting itself from the value of V_{CC} and this results in 0 V.

EXAMPLE 6-4

The $+V_{CC}$ of the circuit of Fig. 6-6 is changed to 10 V and the v_{in} to 4 V p–p. A 0.1-mH coil and 0.0022-μF capacitor are utilized in the tank circuit. (a) During what portion of the input signal does Q_1 turn on, and (b) what is the frequency and voltage level of the output signal?

Solution:

(a) A 4-V p–p signal is composed of two alternations, each measuring 2 V pk. Only the positive alternations have the ability to turn on the transistor since it has a P-type base. However, the positive alternations must overcome the voltage of the capacitor, which is charged to 1.3 V. Therefore, only the top 0.7 V peak of each positive alternation turns on Q_1.

(b) The frequency of the output is determined by the resonant frequency of the tank circuit.

$$f_r = \frac{1}{2\pi \sqrt{(0.1 \text{ mH})(0.0022 \text{ } \mu F)}}$$
$$= 339 \text{ kHz}$$

Assuming the input frequency matches the resonant frequency of the tank, a signal of maximum strength appears at the output. Under ideal conditions, this is 20 V p–p. ∎

TABLE 6-1 Classes of Amplifiers

	Class A	Class B	Class C
Input angle	360	360	360
Output angle	360	180	< 180
Maximum efficiency	25%	78.5%	80– 100%

Types of Coupling

The output in Fig. 6-6 uses **direct coupling**. The dc biasing voltage of 12 V is passed on to the load along with the signal. This is the third type of coupling used in this chapter. The first two were capacitive and transformer coupling. Direct coupling works over a broader range of frequencies, but does not isolate the dc. Capacitive coupling isolates the dc, but is restricted to higher frequencies because of the **capacitive reactance.** Transformer coupling also isolates the dc, but is the most expensive of the three.

Direct Coupling: Connecting stages of an amplifier directly by a piece of wire.

Capacitive Reactance: A capacitor's opposition to ac. It is expressed as $X_C = 1/2\pi fC$.

Power Efficiency

The efficiency of this circuit is between 80 and 100%. DC current only flows for short periods of time. Most of the power generated by the dc supply is utilized by the load resistance.

Table 6-1 compares the angles and efficiency of three classes of amplifiers.

REVIEW QUESTIONS

31. The class C amplifier conducts for less than _____ degrees.
32. What is the function of capacitor C_1 in the circuit of Fig. 6-6?
33. What is the name of the circuit formed by C_2 and L_1 in the circuit of Fig. 6-6?
34. The output is greatest when the input frequency equals the resonant frequency. True or false?
35. The larger the inductance or capacitance, the higher the resonant frequency. True or false?
36. The transistor in a class C amplifier is cut off most of the time. True or false?
37. The action of current flowing back and forth between a capacitor and inductor in a tank circuit is called _____ .
38. What is one advantage of direct coupling?
39. What is one advantage of capacitive coupling?
40. What is one disadvantage of transformer coupling?

6-4 TYPES OF DISTORTION

Nonlinear Distortion

Linear means that the output is proportional to the input. **Nonlinear** means that the output differs from the input. The purpose of an amplifier is to increase the strength of a signal without changing its shape. Nonlinear distortion causes the output to have a different size or shape than that of the input. Figure 6-7 shows a signal with an abnormally large positive alternation and a condensed negative alternation. This is one example of nonlinear distortion.

Linear: A condition where the output is proportional to the input.
Nonlinear: A condition where the output is not proportional to the input.

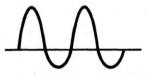

FIGURE 6-7 Nonlinear distortion.

FIGURE 6-8 Harmonic distortion.

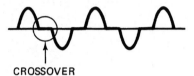

CROSSOVER

FIGURE 6-9 Crossover distortion.

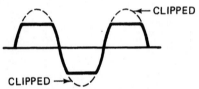

CLIPPED

CLIPPED →

FIGURE 6-10 Clipping distortion.

Harmonic Distortion Harmonic distortion can be caused by nonlinearity. Harmonics are integral multiples of a frequency. If the input frequency were 5 kHz, then the harmonics of that frequency would be 10, 15, 20, 25 kHz, etc. This type of distortion affects class A amplifiers to a greater degree.

When a pure sine wave passes through a nonlinear device such as a transistor, harmonic frequencies are produced. A 10-kHz signal passing through a large-signal amplifier may produce harmonic frequencies of 20, 30, 40 kHz, etc. These signals combine with the original input frequency and the output looks much like Fig. 6-8. This type of distortion affects class A amplifiers more than class B. Class B amplifiers are able to cancel out the even harmonics because of their push–pull configuration.

Crossover Distortion Crossover distortion is found in class B amplifiers. It is caused by the delay in time between one transistor cutting off and the other transistor turning on. It takes 0.7 V across the $B–E$ junction to turn on a transistor. The use of the diodes in the circuit of Fig. 6-4 helps to eliminate this type of distortion by providing almost enough voltage to turn on the two $B–E$ junctions. Figure 6-9 illustrates severe crossover distortion.

Clipping Distortion Clipping of only one peak may be due to improper setting of the Q point. The positive or negative peaks of the output may be clipped if the biasing voltage at the operating point is either too high or too low. Too large an input signal causes the clipping of both peaks, as shown in Fig. 6-10. All classes of amplifiers are susceptible to this type of distortion.

EXAMPLE 6-5 _____

The positive alternation of the output signal for a common-emitter amplifier is clipped. What is the cause of this distortion?

Solution: Clipping on the positive peak most likely indicates that the quiescent voltage at the collector is too high. The transistor, therefore, is being driven into cutoff. ■

REVIEW QUESTIONS _____

41. What is the term that means that the output is proportional to the input?
42. What type of distortion is unique to class B push–pull amplifiers?
43. A transistor is a (linear or nonlinear) device.
44. What are the three harmonic frequencies just above an input frequency of 2 kHz?
45. Clipping distortion on the positive peak of the output means that the operating voltage is too (high or low).

6-5 TROUBLESHOOTING _____

All of the problems encountered with small-signal amplifiers in Chapter 5 apply to the power amplifiers of this chapter. We now pick up with some of the new components used in the circuits of this chapter.

Bad Coil A coil is essentially a piece of wire wrapped around some type of core. The core may be anything from paper to iron. The wire itself is insulated so that the windings do not short. Copper wire has very little resistance. If the resistance of a good coil were measured with an ohmmeter, it would register rather low.

The windings of a coil may partially short, in which case a lower than normal reading is observed. A total short, of course, measures zero ohms. Although shorts do occur in coils, an open is a more likely problem.

An open in the windings of a coil measures infinite resistance. A coil that is damaged is usually unrepairable and must be replaced with a new coil.

A coil reacts differently to dc and ac. The coil has a very low resistance to dc. Very little dc voltage is dropped across a coil. AC, on the other hand, encounters an impedance based on the frequency. A coil has a property called inductive reactance, which can be considered as the coil's ac impedance. Higher frequencies create a larger **inductive reactance** as the following formula shows.

$$X_L = 2\pi f L$$

where X_L = inductive reactance in ohms (Ω)

π = 3.1415926

f = frequency in hertz (Hz)

L = value of the coil in henries (H)

Inductive Reactance: The opposition to ac offered by a coil. It is expressed as $X_L = 2\pi f L$.

For example, a 10-mH coil at 1 MHz has an X_L of 62.8 kΩ.

It is the ability to react differently to different frequencies that enables the coil to be used effectively in tuning circuits like the one in the class C amplifier. The ac measured across a coil varies according to the frequency.

An open coil does not pass any current. A shorted or partially shorted coil measures a lower than normal ac or dc voltage.

Bad Transformer A **transformer** is basically two coils that pass an ac signal to each other through the principle of **magnetic induction.** The transformer consists of a primary, which is the input side, and a secondary, which is the output side. An ac signal applied to the primary causes a magnetic field to expand and collapse over and over again. This magnetic field cuts through the secondary windings and induces a voltage, which, in turn, starts current flowing through the secondary.

Transformer: A device composed of a primary coil and a secondary coil that passes ac signals through the process of magnetic induction. It does not pass dc voltages.

Magnetic Induction: The action of the magnetic field of one coil causing current to flow in another coil. It is usually associated with transformers.

No load current can flow if the primary or secondary coil is open and there would be no output signal. A transformer coil is checked with an ohmmeter in the same fashion as the ordinary coil discussed earlier.

A short in the primary or secondary coil prevents a magnetic field from developing. This also prevents an output signal from developing.

A bad transformer may give off a distinct odor. Large transformers are used in high-voltage applications; dangerous voltages may be encountered while testing with the power on! RF transformers, such as those used in the amplifier discussion, are about the size of a thumbnail. Although they are not as dangerous as power transformers, they should be treated with caution.

REVIEW QUESTIONS

46. A coil measures _____ resistance.

47. DC views the coil as having _____ resistance.

48. A low-frequency ac causes the coil to have a _____ inductive reactance.

49. An open coil measures _____ resistance.

50. A transformer works on the principle of _____ _____ .
51. The input side of a transformer is called the _____ .
52. A transformer passes ac but not dc. True or false?
53. What is the voltage level of the signal across a load that is transformer coupled with an open in the primary?
54. An open secondary measures _____ resistance.
55. A shorted primary measures _____ resistance.
56. What is the reason both transistors should be replaced in a push–pull arrangement although only one may be bad?

SUMMARY

1. The final stage of a multistage amplifier is most often a power amplifier.
2. The class A amplifier provides 360° output for a 360° input signal.
3. *RC* coupling passes the ac signal while isolating the dc.
4. The ac resistance as viewed by the collector in a common-emitter class A amplifier is equivalent to R_C in parallel with R_L.

$$r_c = R_L \parallel R_C$$

5. The power dissipated by a transistor is determined by

$$P_Q = V_{CEQ} \times I_{CQ}$$

6. The power dissipated by a load is found by

$$P_L = \frac{v_{L(p-p)}^2}{8R_L}$$

or

$$P_L = \frac{v_{L(rms)}^2}{8R_L}$$

7. The maximum efficiency of an *RC* coupled class A amplifier is 25%.
8. The power furnished by the dc supply is found by

$$P_{dc} = I_{CC} \times V_{CC}$$

9. The efficiency with which the load utilizes the power offered by the dc power supply is found by

$$\% \text{ efficiency} = \frac{P_L}{P_{dc}} \times 100$$

10. The ac output compliance is the maximum unclipped peak–peak voltage that can occur at the output of an amplifier.
11. Transformer coupling isolates the load from the dc supply and offers impedance matching.
12. Transformer coupling can increase the maximum efficiency of a class A amplifier to almost 50%.
13. Class B amplifiers provide 180° output for a 360° input signal.
14. Each transistor in the push–pull arrangement passes one-half, or 180°, of the input signal to the output for a combined total output angle of 360°.

15. The push–pull amplifier consists of a matched pair of PNP and NPN transistors that conduct alternately within the circuit.

16. The purpose of the diodes in the voltage divider of the push–pull amplifier is to keep the transistors slightly turned on, thus limiting crossover distortion.

17. The maximum efficiency of the class B push–pull amplifier is 78.5%.

18. The class C amplifier produces less than 180° output for a 360° input signal.

19. Only the positive peaks of the input signal turn on the class C amplifier due to the clamping capacitor.

20. The tank circuit of the class C amplifier should be resonant with the input signal.

21. Nonlinear distortion causes the output to have a different size or shape than the input.

22. Harmonic distortion occurs when harmonic frequencies of the original signal interfere with the original signal.

23. Crossover distortion in push–pull class B amplifiers is caused by the delay between one transistor cutting off and the other transistor turning on.

24. Clipping distortion occurs as a result of an improper Q point or the input signal is too large.

25. Coils have a very low resistance.

GLOSSARY

Alternation: A half cycle, which equals 180°.

Angle: Used to describe the percentage of one cycle of a sine wave in degrees.

Band-pass Filter: A circuit that passes a certain band or range of frequencies and filters, or blocks, all others.

Capacitive Reactance: A capacitor's opposition to ac. It is expressed as $X_C = 1/2\pi fC$.

Class A Amplifier: An amplifier that conducts for 360° of the input. It offers a maximum efficiency of 25–50%, depending on the type of coupling used.

Class B Amplifier: An amplifier that conducts for 180° of the input. It offers a maximum efficiency of 78.5%.

Class C Amplifier: An amplifier that conducts for less than 180° of the input. It offers a maximum efficiency of 100%.

Cycle: A complete sine wave composed of a positive and negative alternation. It equals 360°.

Direct Coupling: Connecting stages of an amplifier directly by a piece of wire.

Distortion: The unwanted change of shape of a signal.

% Efficiency: The percentage of the dc supply power used by the ac load.

Heat Sink: A piece of metal attached to a component to help dissipate heat.

Impedance: The overall opposition of a circuit to ac.

Inductive Reactance: The opposition to ac offered by a coil. It is expressed as $X_L = 2\pi fL$.

Linear: A condition where the output is proportional to the input.

Magnetic Induction: The action of the magnetic field of one coil causing current to flow in another coil. It is usually associated with transformers.

Nonlinear: A condition where the output is not proportional to the input.

Oscillation: The action of current flowing back and forth alternately.

Power Amplifier: An amplifier that provides a larger amount of power than a small-signal amplifier.

Power Transistor: A transistor designed to handle a power greater than one-half watt.

Push–Pull: An arrangement in which two components or subcircuits conduct on an alternate basis. It is typically composed of class B amplifiers.

Radio Frequencies: Radio frequencies (RF) range from 10 kHz to 3000 GHz.

Resonance: A condition where the capacitive reactance equals the inductive reactance at one frequency for an *LC* circuit. $f_r = 1/2\pi\sqrt{LC}$.

Root-Mean-Square Power: The root-mean-square (rms) value is an expression of the heating effect of an ac signal.

Shunted: To connect in parallel.

Small-Signal Amplifier: An amplifier that is mainly concerned with voltage or current gain.

Tank Circuit: An *LC* circuit designed to oscillate at its resonant frequency.

Transformer: A device composed of a primary coil and a secondary coil that passes ac signals through the process of magnetic induction. It does not pass dc voltages.

Transformer Coupling: The use of a transformer to connect the stages of an amplifier.

PROBLEMS

SECTION 6-1

The following problems refer to the circuit of Fig. 6-11.

6-1. What is the value of dc current through the voltage divider?

6-2. What is the value of dc voltage at the base of Q_1?

6-3. What is the value of dc voltage at the emitter of Q_1?

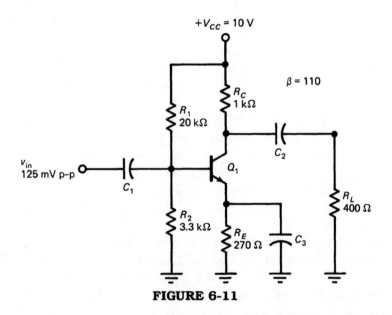

FIGURE 6-11

6-4. What is the value of dc emitter current?

6-5. Find the value of V_{CEQ}.

6-6. Find the value of V_{CQ}.

6-7. Find the value of power dissipated by the transistor?

6-8. What is the total power generated by the dc supply?

6-9. What is the ac resistance seen by the collector?

6-10. Find the value of ac voltage gain.

6-11. Determine the value of ac current gain.

6-12. Calculate the total power gain.

6-13. What is the peak-to-peak voltage across the load resistance?

6-14. Determine the power dissipated by the load resistance.

6-15. Calculate the % efficiency.

6-16. Draw the output waveform, indicating the operating point and the upper and lower limits as per Fig. 6-2.

6-17. Determine the class of amplifier and record your reasons.

6-18. Identify the name of each of the capacitors and explain its purpose.

6-19. Explain the effects of different size load resistances.

6-20. What configuration of amplifier is being used (CE, CC, or CB)?

SECTION 6-2

The following problems refer to the circuit of Fig. 6-12.

6-21. What value of current is flowing through the voltage divider?

6-22. Calculate the voltage at the base of Q_1.

6-23. Calculate the voltage at the base of Q_2.

6-24. Determine the quiescent voltage at the emitters.

6-25. What is the average current flowing through each transistor?

6-26. Which transistor is a PNP transistor?

6-27. Calculate the power generated by the dc supply.

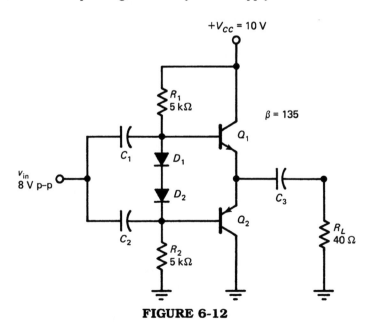

FIGURE 6-12

6-28. Find the voltage gain.

6-29. Find the current gain.

6-30. Find the power gain.

6-31. Determine the peak-to-peak voltage across the load.

6-32. Calculate the power used by the load resistance.

6-33. What is the % efficiency of this circuit?

6-34. What is the name given to the transistor arrangement in this circuit?

6-35. What is the angle of the output?

6-36. Determine which alternation turns on each transistor.

6-37. Prove whether increasing the input signal increases or decreases the efficiency of the circuit.

SECTION 6-3

The following problems refer to the circuit of Fig. 6-13.

6-38. Calculate the dc clamping voltage across C_1.

6-39. What is the quiescent voltage at the collector?

6-40. Determine the resonant frequency of the tank circuit.

6-41. What is the peak-to-peak output voltage?

6-42. What would be the best input frequency for this circuit?

6-43. What type of coupling is used?

6-44. Describe the oscillation of the tank circuit.

6-45. Explain what class of amplifier this is.

SECTION 6-4

6-46. Define the terms linear, nonlinear, and distortion.

6-47. List the three classes of amplifiers and the type(s) of distortion associated with each.

6-48. Draw, label, and define each type of distortion.

6-49. Give an example of three harmonics for the resonant frequency of the circuit of Fig. 6-13.

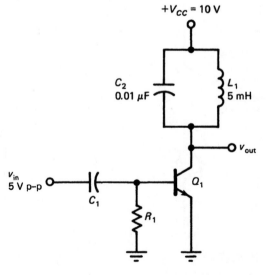

FIGURE 6-13

6-50. Make a list of the following devices along with the proper resistance measurements for each: diode, transistor, capacitor, coil, and transformer.

6-51. If R_1 were to open in the circuit of Fig. 6-11, what would happen to the output signal?

6-52. What effect does a short across R_C have on the output of the circuit of Fig. 6-11?

6-53. If C_1 were to open in the circuit of Fig. 6-12, what would happen to the output signal?

6-54. What effect does an open in the $B-E$ junction of Q_2 have on the output of the circuit of Fig. 6-12?

6-55. If L_1 were to short in the circuit of Fig. 6-13, what would happen to the output signal?

6-56. What effect does an open in the $B-E$ junction of Q_2 have on the output of the circuit of Fig. 6-13?

ANSWERS TO REVIEW QUESTIONS

1. True	**2.** True	**3.** angle
4. 360°	**5.** 360°	**6.** base, collector
7. I	**8.** beta	**9.** decreases
10. heat sinks	**11.** True	**12.** heat
13. increases	**14.** dc supply	**15.** 25%
16. $\frac{P_L}{P_{dc}} \times 100$	**17.** AC output compliance	**18.** impedances
19. 50%	**20.** cost	**21.** 180°
22. False	**23.** True	**24.** ac, dc
25. True	**26.** positive	**27.** 360°
28. 78.5%	**29.** Peak voltage times 0.318	**30.** True
31. 180°	**32.** DC clamping	**33.** Tank circuit
34. True	**35.** False	**36.** True
37. oscillation	**38.** A broad frequency range	**39.** It isolates dc
40. High cost	**41.** Linear	**42.** Crossover
43. nonlinear	**44.** 4 kHz, 6 kHz, and 8 kHz	**45.** high
46. low	**47.** low	**48.** low
49. infinite	**50.** magnetic inductance	**51.** primary
52. True	**53.** 0 V	**54.** infinite
55. Zero	**56.** They should be a balanced pair	

Regulated Power Supplies

When you have completed this chapter, you should be able to:

- Analyze the concept of regulation.
- Become familiar with different methods of achieving regulation.
- Examine different types of regulator devices, from zener diodes to ICs.
- Undertake sensible steps in troubleshooting regulator circuits.

INTRODUCTION

In the last two chapters, the transistor was used in ac applications. Its purpose was to amplify an ac signal. The present chapter makes use of a transistor in a purely dc application. This chapter explains those characteristics of a transistor that apply to the design and operation of regulator circuits.

The word "regulate" means adjusting something to function accurately. A **regulator** is used to provide a constant, or regulated, dc voltage.

The whole process begins with 120 VAC coming from the wall socket. This ac voltage is usually stepped down by a transformer. It is then rectified through the use of diodes. At this point, it is a pulsing dc voltage. This dc voltage is smoothed, or filtered, by a capacitor. The final stage that makes this dc voltage usable is the regulator circuit.

Regulator: A circuit that maintains a constant output voltage.

7-1 ZENER REGULATOR

A zener diode provides the most basic form of regulation. A zener diode is operated in reverse bias. It maintains a steady zener voltage over a wide range of load conditions. Figure 7-1 shows the zener regulator. The input is connected to a filtered rectifier circuit and the output to a load resistance.

The current-limiting resistor R_1 limits the total current of the regulator circuit to a constant 85.1 mA. This is determined by the fact that if 10 V enters the regulator and the zener diode is rated to drop 6 V, then the remaining 4 V must be dropped across R_1.

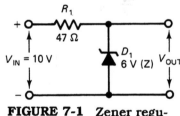

FIGURE 7-1 Zener regulator.

$$V_{R1} = V_{IN} - V_{ZD}$$
$$= 10 \text{ V} - 6 \text{ V}$$
$$= 4 \text{ V}$$

so

$$I_T = I_{R1} = \frac{V_{R1}}{R_1}$$
$$= \frac{4 \text{ V}}{47 \text{ }\Omega}$$
$$= 85.1 \text{ mA}$$

The power dissipated by the current-limiting resistor under normal operating conditions is

$$P_{R1} = I_{R1} \times V_{R1}$$

$$= 85.1 \text{ mA} \times 4 \text{ V}$$

$$= 340 \text{ mW}$$

Therefore, a 0.5-W resistor is required.

Effects of Varying Load A regulated power supply should be able to provide a constant output voltage over a broad range of load conditions. We begin by analyzing the circuit of Fig. 7-1 when no load is connected to the output.

A no-load condition in Fig. 7-1 results in a simple series circuit. The zener diode passes the total current. This causes a maximum amount of power to be dissipated by the diode.

$$P_{D1} = 85.1 \text{ mA} \times 6 \text{ V}$$

$$= 511 \text{ mW}$$

If a load resistance of 10 kΩ were connected to the output, an alternate path for current would be provided. The current through the zener diode decreases by this amount. Since the load is in parallel with the zener voltage, it too drops 6 V.

$$I_{R_L} = \frac{V_{D1}}{R_L}$$

$$= \frac{6 \text{ V}}{10 \text{ k}\Omega}$$

$$= 600 \text{ }\mu\text{A}$$

The current that is now flowing through the zener diode is found by

$$I_{D1} = I_T - I_{R_L}$$

$$= 85.1 \text{ mA} - 600 \text{ }\mu\text{A}$$

$$= 84.5 \text{ mA}$$

The slight decrease in the zener current causes a slight decrease in the power that the zener diode has to dissipate. The new value of power is 507 mW.

What if a 100-Ω load were connected across the output? How would this affect the regulator circuit? The answer is twofold. It causes a reduction in the current and power of the zener diode. Let us take a look at the drastic effect of this load condition.

$$I_{R_L} = \frac{6 \text{ V}}{100 \text{ }\Omega}$$

$$= 60 \text{ mA}$$

$$I_{D1} = I_T - I_{R_L}$$

$$= 85.1 \text{ mA} - 60 \text{ mA}$$

$$= 25.1 \text{ mA}$$

$$P_{D1} = 25.1 \text{ mA} \times 6 \text{ V}$$
$$= 151 \text{ mW}$$

Under no-load conditions, the power dissipated by the zener diode was 511 mW and the zener current was 84.5 mA. When a load of 100 Ω is connected, the power dissipated by the zener diode is reduced to 151 mW and the current to 25.1 mA. There is a minimum amount of current required for a zener diode to conduct and provide its zener voltage. This is a limitation of the zener diode.

The final point in this discussion is the current drain caused by the load resistance. As the load resistance decreases, more current is drawn. All power supplies have limits on the amount of current that can be supplied at a given voltage. In this case, the maximum current is limited by R_1.

EXAMPLE 7-1 _____

The circuit of Fig. 7-1 is modified so that $V_{IN} = 15$ V and a load resistance of 2.7 kΩ is connected across the output. Determine (a) V_{OUT} and (b) P_{D1}.

Solution:
(a) The voltage at the output is regulated by the zener diode to 6 V. Changes in the input voltage should not affect the output voltage.
(b) In order to determine the power dissipated by the zener diode, the value of current through the zener diode is needed. Current through the zener diode is the difference between the total current and that through the load resistance, R_L.

$$I_T = \frac{15 \text{ V} - 6 \text{ V}}{47 \ \Omega}$$
$$= 191 \text{ mA}$$

and

$$I_{R_L} = \frac{6 \text{ V}}{2.7 \text{ k}\Omega}$$
$$= 2.22 \text{ mA}$$

Therefore,

$$I_{D1} = 191 \text{ mA} - 2.22 \text{ mA}$$
$$= 189 \text{ mA}$$

so

$$P_{D1} = 189 \text{ mA} \times 6 \text{ V}$$
$$= 1.13 \text{ W} \qquad \blacksquare$$

Practice Problem: In the circuit of Fig. 7-1, what value of load current is drawn by a 1-kΩ load resistor and how much power is dissipated by the zener diode? *Answer:* $I_{R_L} = 6$ mA and $P_{D1} = 475$ mW.

Design Considerations The simple example of Fig. 7-1 illustrates some regulator design considerations. These include the current and power limitations of components as well as the operating conditions for the regulator itself. The circuit of Fig. 7-1 was designed to provide a regulated 6-V output to a minimum load resistance of approximately 100 Ω.

TABLE 7-1

Zener Diodes by Ascending Zener Voltage					
Device No.	V_z (V) Nom	Z_z (Ω) @ I_z Max	mA	P_D (mW) (25°C)	Package
1N5226B	3.3	28	20	500	DO-35
1N746A	3.3	28	20	500	DO-35
1N4728A	3.3	10	7.6	1000	DO-41
1N5227B	3.6	24	20	500	DO-35
1N747A	3.6	24	20	500	DO-35
1N4729A	3.6	10	69	1000	DO-41
1N5228B	3.9	23	20	500	DO-35
1N748A	3.9	23	20	500	DO-35
1N4730A	3.9	9.0	64	1000	DO-41
1N5229B	4.3	22	20	500	DO-35
1N749A	4.3	22	20	500	DO-35
1N4731A	4.3	9.0	58	1000	DO-41
1N5230B	4.7	19	20	500	DO-35
1N750A	4.7	19	20	500	DO-35
1N4732A	4.7	8.0	53	1000	DO-41
1N5231B	5.1	17	20	500	DO-35
1N751A	5.1	17	20	500	DO-35
1N4733A	5.1	7.0	49	1000	DO-41
1N5232B	5.6	11	20	500	DO-35
1N752A	5.6	11	20	500	DO-35
1N4734A	5.6	5.0	45	1000	DO-41
1N5233B	6.0	7.0	20	500	DO-35
1N5234B	6.2	7.0	20	500	DO-35
1N753A	6.2	7.0	20	500	DO-35
1N4735A	6.2	2.0	41	1000	DO-41
1N5235B	6.8	5.0	20	500	DO-35
1N754A	6.8	5.0	20	500	DO-35
1N957B	6.8	4.5	18.5	500	DO-35
1N4736A	6.8	3.5	37	1000	DO-41
1N5236B	7.5	6.0	20	500	DO-35
1N755A	7.5	6.0	20	500	DO-35
1N958B	7.5	5.5	16.5	500	DO-35
1N4737A	7.5	4.0	34	1000	DO-41
1N5237B	8.2	8.0	20	500	DO-35
1N756A	8.2	8.0	20	500	DO-35
1N959B	8.2	6.5	15	500	DO-35
1N4738A	8.2	4.5	34	1000	DO-41
1N5238B	8.7	8.0	20	500	DO-35
1N5239B	9.1	10	20	500	DO-35
1N757A	9.1	10	20	500	DO-35
1N960B	9.1	7.5	14	500	DO-35
1N4739A	9.1	5.0	8	1000	DO-41
1N5240B	10	17	20	500	DO-35
1N758A	10	17	20	500	DO-35
1N961B	10	8.5	12.5	500	DO-35
1N4740A	10	7.0	25	1000	DO-41
1N5241B	11	22	20	500	DO-35
1N962B	11	9.5	11.5	500	DO-35
1N4741A	11	8.0	23	1000	DO-41
1N5242B	12	30	20	500	DO-35
1N759A	12	30	20	500	DO-35
1N963B	12	11.5	10.5	500	DO-35
1N4742A	12	9.0	21	1000	DO-41
1N5243B	13	13	9.5	500	DO-35
1N964B	13	13	9.5	500	DO-35

All zener diodes need a certain recommended reverse current in order to maintain their zener voltage. A value of 25 mA was obtained from a manufacturer's spec sheet for the zener diode in this circuit. The other important information gained from the spec sheet was the 6-V operating voltage and the zener power rating of 1 W. Table 7-1 shows a sample spec sheet.

The minimum load resistance of 100 Ω is the smallest value that still permits 25 mA to flow through the diode when a load is attached.

All of the information is combined in a worst-case scenario to determine a value for the current-limiting resistor. The voltage across R_1 under normal operating conditions should be 4 V. The current flowing through the current-limiting resistor under a maximum load condition should be

156

$$I_{R1} = \frac{V_{D1}}{R_{L(MIN)}} + I_{D1(MIN)}$$

$$= \frac{6 \text{ V}}{100 \text{ }\Omega} + 25 \text{ mA}$$

$$= 85 \text{ mA}$$

Ohm's law can be used to determine the resistor size.

$$R_1 = \frac{V_{R1}}{I_{R1}}$$

$$= \frac{4 \text{ V}}{85 \text{ mA}}$$

$$= 47.1 \text{ }\Omega$$

A 47-Ω resistor suffices. The wattage size is determined by calculating the amount of power that the resistor has to dissipate and then doubling it. A 0.5-W resistor is a safe solution in this example since it only dissipates 340 mW. The rule of thumb is to select a wattage size roughly twice the value of the maximum power that it has to dissipate.

$$P_{R1} = \frac{V_{R1}^2}{R_1}$$

$$= 340 \text{ mW}$$

The maximum power that the zener diode has to dissipate is 511 mW, as determined from our discussion earlier. This zener diode is rated at 1 W, which is roughly twice the value of power it will actually dissipate. Therefore, the circuit of Fig. 7-1 should operate safely for any value of load from 100 to infinite ohms.

Practice Problem: In the circuit of Fig. 7-1, what size current-limiting resistor is necessary for a minimum load condition of 1 kΩ, assuming a recommended reverse current of 30 mA for the zener diode? *Answer:* 111 Ω.

REVIEW QUESTIONS

1. The definition of _____ is to adjust something to function accurately.
2. A regulator is used to provide a regulated _____ voltage.
3. A zener diode provides a zener voltage when _____ biased.
4. The voltage applied to the input of a regulator circuit is _____ dc voltage.
5. The _____ _____ resistor restricts current to a predetermined level.
6. What is the maximum current that can flow in the circuit of Fig. 7-1 while maintaining a 6 V output if R_1 were replaced with a 100-Ω resistor?
7. What happens to the voltage across a zener diode as current through it increases?
8. A lower load resistance draws (more or less) current.

9. A regulator circuit with no load connected draws a (minimum or maximum) amount of current.

10. A regulator whose output is shorted draws a (minimum or maximum) amount of current.

11. How much current does a 2-kΩ load draw in the circuit of Fig. 7-1?

7-2 EMITTER-FOLLOWER REGULATOR

Emitter Follower: A regulator in which the output from the emitter follows the value of the zener diode at the base.

Pass Transistor: A transistor through which total load current passes and is controlled in a regulator circuit.

The **emitter follower** is so named because the voltage at the emitter follows the base voltage. Another name for the emitter follower is the common collector. A characteristic of the common collector is current gain. The transistor of Fig. 7-2 increases the current capabilities of the regulator circuit.

The transistor of Fig. 7-2 is called the **pass transistor** because all of the load current must pass through it. The output voltage is equal to the voltage at the emitter and the emitter is always 0.7 V less than the base for NPN transistors. In the circuit of Fig. 7-2, the output voltage is equal to 5.3 V.

$$V_{OUT} = V_{D1} - V_{BE}$$
$$= 6 \text{ V} - 0.7 \text{ V}$$
$$= 5.3 \text{ V}$$

The zener regulator uses R_1 to limit current. The emitter-follower regulator uses Q_1 to limit current. The collector–emitter leg of Q_1 can be viewed as a variable resistor. When the load is drawing a small current, the C–E resistance is large and when the load is drawing a large current, the C–E resistance is small. The collector–emitter voltage is equal to the difference between the input and output voltage.

$$V_{CE} = V_{IN} - V_{OUT}$$
$$= 10 \text{ V} - 5.3 \text{ V}$$
$$= 4.7 \text{ V}$$

V_{CE} remains constant, but the current through the transistor is dependent on the load. In Fig. 7-3, the load resistance has a value of 100 Ω. The current through

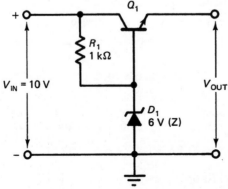

FIGURE 7-2 Emitter-follower regulator.

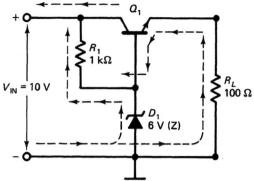

FIGURE 7-3 Emitter-follower current flow.

the load is the same current that enters the emitter of the transistor. Therefore, the current flowing through the transistor can be calculated as

$$I_E = I_{R_L} = \frac{V_{OUT}}{R_L}$$

$$= \frac{5.3 \text{ V}}{100 \text{ }\Omega}$$

$$= 53 \text{ mA}$$

The pass transistor is perhaps the most critical device in the circuit. It may have to handle a large amount of power under certain load conditions. It is usually heat sinked in case of such an occurrence. The power that the pass transistor must dissipate is equal to

$$P_{Q1} = I_E \times V_{CE}$$

$$= 53 \text{ mA} \times 4.7 \text{ V}$$

$$= 249 \text{ mW}$$

The total current in a zener regulator was a constant determined by the value of the current-limiting resistor. The total current in the emitter follower is determined to a large degree by the value of the load, because the pass transistor changes its resistance to compensate for changes of the load resistance.

R_1 is used to limit the current flow through the zener diode. View R_1 and D_1 as being in parallel to the input voltage. Subtract the zener voltage from the input voltage; the remaining voltage is dropped across R_1. In Fig. 7-3, the voltage across R_1 is 4 V.

EXAMPLE 7-2

The values in the circuit of Fig. 7-3 are changed so that $V_{IN} = 24$ V, $R_1 = 1$ kΩ, $R_L = 680$ Ω, and $D_1 = 9$ V (Z). Determine (a) V_{OUT} and (b) P_{Q1}.

Solution:
(a) Remember that for an NPN transistor to conduct, the emitter must be 0.7 V more negative than the base. The base of the transistor is connected to a 9-V zener diode. Output voltage in this circuit is, therefore, 0.7 V less than this value.

$$V_{OUT} = 9 \text{ V} - 0.7 \text{ V}$$
$$= 8.3 \text{ V}$$

(b) The power of the pass transistor is determined by multiplying its current by the collector-to-emitter voltage. Current into the pass transistor is determined by the size of the load.

$$I_E = \frac{8.3 \text{ V}}{680 \text{ }\Omega}$$
$$= 12.2 \text{ mA}$$

and

$$V_{CE} = 24 \text{ V} - 8.3 \text{ V}$$
$$= 15.7 \text{ V}$$

Therefore,

$$P_{Q1} = 12.2 \text{ mA} \times 15.7 \text{ V}$$
$$= 192 \text{ mW} \quad \blacksquare$$

Practice Problem: What is the value of P_{Q1} if the load in the circuit of Fig. 7-3 were changed to 270 Ω? *Answer:* 92.3 mW.

REVIEW QUESTIONS

12. What is another name for the emitter follower?
13. The emitter follower provides a large voltage gain. True or false?
14. Which component does the emitter follower use to limit output current flow?
15. The total current in the emitter follower is determined to a large degree by the (diode, transistor, or load).
16. What would be the output voltage of the circuit of Fig. 7-3 if the load resistance were changed to 270 Ω?
17. What would be the load current of the circuit of Fig. 7-3 if the load resistance were changed to 330 Ω?
18. What would be the output voltage of the circuit of Fig. 7-3 if the voltage rating of the zener diode were 9.1 V?
19. What is the name of the device that is attached to a transistor to help dissipate power?
20. How much power would the pass transistor of the circuit of Fig. 7-3 dissipate if the load resistance were 56 Ω?

7-3 VARIABLE-FEEDBACK REGULATOR

Variable Output Voltage   The regulators up to this point provided a nonadjustable output voltage. This is fine if the power supply only needs to yield a single value of voltage. The feedback regulator of Fig. 7-4 allows us to vary the output between 0.7 and 15.7 V.

Potentiometer R_3 is connected in parallel with a 15-V zener diode. The voltage across the entire potentiometer is 15 V. Capacitor C_1 ensures that the voltage across D_1 and R_3 does not suddenly change. The voltage at the wiper of R_3 can vary from 0 to 15 V, depending on its position.

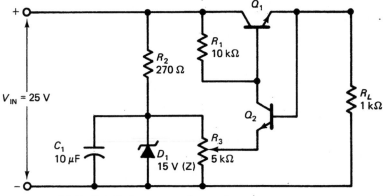

FIGURE 7-4 Variable-feedback regulator.

Q_2 is a **feedback transistor** and the voltage at its base is 0.7 V more positive than the voltage at its emitter. Since the voltage at the emitter is equal to the voltage at the wiper, adjusting the potentiometer changes the voltage at the base of Q_2. The base of Q_2 is tied to the output and is responsible for providing the value of output voltage.

Feedback Transistor: Also called the error amplifier, it monitors the output voltage, using feedback to make any necessary adjustments.

If the wiper of R_3 were adjusted half way, then the voltage at the wiper would be half of the zener voltage, or 7.5 V. The voltage at the output is found by adding the base–emitter voltage of Q_2 to the voltage of the wiper.

$$V_{OUT} = V_{R3(wiper)} + V_{BE2}$$

$$= 7.5 \text{ V} + 0.7 \text{ V}$$

$$= 8.2 \text{ V}$$

Transistor Q_1 is still considered a pass transistor. The total load current must pass through it. In the circuit of Fig. 7-4, the load current can be calculated as

$$I_{R_L} = \frac{V_{OUT}}{R_L}$$

$$= \frac{8.2 \text{ V}}{1 \text{ k}\Omega}$$

$$= 8.2 \text{ mA}$$

The voltage across the pass transistor is found in the same way as earlier.

$$V_{CE1} = V_{IN} - V_{OUT}$$

$$= 25 \text{ V} - 8.2 \text{ V}$$

$$= 16.8 \text{ V}$$

The current passing through Q_1 is equal to the current through the load. The small base current of Q_2 is ignored in this discussion. Using the previous values, calculate the power of the pass transistor.

$$P_{Q1} = I_E \times V_{CE1}$$

$$= 8.2 \text{ mA} \times 16.8 \text{ V}$$

$$= 138 \text{ mW}$$

The transistors are prevented from saturating by the presence of R_1. The remaining resistor, R_2, is used to limit the current flow through the zener diode. It drops the remaining 10 V of the input voltage.

Feedback: A condition where output information is returned back to the input.

Practice Problem: What is the power dissipated by the pass transistor if the wiper of R_3 is turned all the way up in the circuit of Fig. 7-4?   *Answer:*   146 mW.

The Concept of Feedback

When we ask someone how well we are performing a particular task, the response that we receive can be called "**feedback.**" We can adjust our performance based on this response.

The variable-feedback regulator works on this same principle. Q_2 is a feedback transistor and its job is to help provide and sample the output voltage. It aids in offsetting any change in the output voltage.

The voltage at the base of Q_2 is directly related to the output voltage. Consequently, the base voltage would change if the output voltage were to increase or decrease. The base and collector of Q_2 are 180° out of phase. If its base voltage were to increase due to an increase of the output voltage, then its collector voltage would decrease. The collector of Q_2 controls the base of Q_1. As the voltage at the base of Q_1 decreases, the collector–emitter resistance of the pass transistor increases. This increase in resistance lowers the load current, which, in turn, lowers the output voltage. This offsets the attempted increase in output voltage. Here is the feedback action in eight steps.

1. The output voltage attempts to increase.
2. The base voltage of Q_2 also increases.
3. The collector voltage of Q_2 decreases due to phase shift.
4. This causes the base voltage of Q_1 to decrease.
5. The collector–emitter resistance of Q_1 increases.
6. The emitter current of Q_1 decreases.
7. The load current equals the emitter current.
8. The decreased load current lowers the output voltage.

The opposite of these steps provides the action for an attempted decrease in the output voltage. The ability to compensate for changes in the output is the mark of a good regulator.

EXAMPLE 7-3

The circuit of Fig. 7-4 is modified so that $D_1 = 10\ V_z$ and $R_L = 330\ \Omega$. Determine (a) the output voltage when the wiper of R_3 is all the way down and (b) P_{Q1} when the wiper of R_3 is all the way up.

Solution:

(a) Turning the wiper all the way down places it at ground potential (0 V). The output voltage continues to be 0.7 V higher than the voltage at the emitter of Q_2.

$$V_{OUT} = 0\ V + 0.7\ V$$
$$= 0.7\ V$$

(b) When the wiper of R_3 is turned all the way up, its voltage is equal to that of the zener diode. The output voltage in this case is the 10 V (Z) plus the 0.7-V base-to-emitter voltage of Q_2.

$$V_{OUT} = 10\ V + 0.7\ V$$
$$= 10.7\ V$$

The output voltage across the load resistance of 330 Ω determines the emitter current of the pass transistor.

$$I_E = \frac{10.7\ V}{330\ \Omega}$$
$$= 32.4\ mA$$

and

$$V_{CE1} = 25\ V - 10.7\ V$$
$$= 14.3\ V$$

so

$$P_{Q1} = 32.4\ mA \times 14.3\ V$$
$$= 463\ mW \qquad \blacksquare$$

REVIEW QUESTIONS

21. Which transistor in the circuit of Fig. 7-4 is considered the pass transistor?
22. What is the output voltage in the circuit of Fig. 7-4 if the wiper is adjusted all the way down?
23. What is the current through the load in the circuit of Fig. 7-4 with the wiper adjusted all the way down?
24. What is the collector–emitter voltage in the circuit of Fig. 7-4 with the wiper adjusted all the way down?
25. What is the power dissipated by Q_1 in the circuit of Fig. 7-4 with the wiper adjusted all the way down?
26. As the voltage at the output of the circuit of Fig. 7-4 attempts to decrease, V_{C2} _____ .
27. As the voltage at the output of the circuit of Fig. 7-4 attempts to decrease, V_{CE1} _____ .
28. As the voltage at the output of the circuit of Fig. 7-4 attempts to decrease, V_{E2} _____ .
29. Shorting the output of the circuit of Fig. 7-4 (increases, decreases, or does not affect) the power dissipated by the pass transistor.
30. The input voltage of the circuit of Fig. 7-4 is _____ dc.

7-4 DARLINGTON FEEDBACK REGULATOR

This regulator is named to emphasize the use of a Darlington pair in place of the single series pass transistor. The Darlington pair provides a high current gain. A small base current can produce a relatively large emitter current at the output. This is useful for driving small load resistances.

Darlington Feedback Regulator: A regulator that uses the Darlington pair as the pass transistor in order to enhance the current capabilities.

Output Voltage Figure 7-5 shows an unregulated input voltage of 20 V. This voltage is split across R_1 and D_1. The diode drops its zener voltage of 5 V and the remaining 15 V is found across R_1.

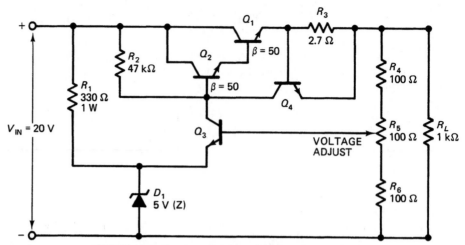

FIGURE 7-5 Darlington feedback regulator.

The base of Q_3 is 0.7 V more positive than the zener voltage at its emitter. This means that 5.7 V can be found at the wiper of R_5. A voltage divider is formed by resistors R_4, R_5, and R_6. The total resistance of this voltage divider is 300 Ω. If the wiper of R_5 is adjusted half way, then the resistance from the wiper to ground is 150 Ω. The output voltage is determined by using these values.

$$V_{OUT} = \frac{R_{VD}}{R_w} V_{B3}$$

where R_{VD} = total resistance of voltage divider

   R_w = resistance from wiper to ground

   V_{B3} = voltage at the base of Q_3

So,

$$V_{OUT} = \frac{300\ \Omega}{150\ \Omega} (5.7\ V)$$

$$= 11.4\ V$$

This regulator is designed to provide an output voltage in the area of 12 V. Potentiometer R_5 is used to provide any necessary voltage adjustment.

EXAMPLE 7-4 _____

Determine the output voltage in the circuit of Fig. 7-5 if the wiper of R_5 were adjusted all the way down.

Solution: It was mentioned before that voltage at the base of Q_3 is 5.7 V due to the value of the zener diode and the base-to-emitter junction of Q_3. Adjusting the wiper all the way down places this voltage across R_6. The result is a maximum output voltage.

$$V_{OUT} = \frac{300\ \Omega}{100\ \Omega} 5.7\ V$$

$$= 17.1\ V$$ ∎

CHAP. 7 / REGULATED POWER SUPPLIES

Practice Problem: What is the value of the output voltage if the wiper of R_5 is turned all the way up in the circuit of Fig. 7-5? *Answer:* $V_{OUT} = 8.55$ V.

Power of the Darlington Pair

Now that an output voltage of 11.4 V has been established, determine the current through the load resistance and the voltage divider.

$$I_L = \frac{11.4 \text{ V}}{1 \text{ k}\Omega}$$

$$= 11.4 \text{ mA}$$

and

$$I_{VD} = \frac{11.4 \text{ V}}{300 \text{ }\Omega}$$

$$= 38 \text{ mA}$$

These two currents join and eventually flow through R_3 on their way to the emitter of Q_1 in the circuit of Fig. 7-5. The current entering the Darlington pair is equal to the sum of these two currents.

$$I_{E1} = I_L + I_{VD}$$

$$= 49.4 \text{ mA}$$

This current must first pass through R_3. As a result, a small voltage is developed across this resistor.

$$V_{R3} = I_{E1} \times R_3$$

$$= 133 \text{ mV}$$

The voltage across the collector–emitter region of Q_1 can now be found. The formula is almost the same as that used earlier. The only difference is the small voltage developed across R_3.

$$V_{CE1} = V_{IN} - (V_{OUT} + V_{R3})$$

$$= 20 \text{ V} - (11.4 \text{ V} + 133 \text{ mV})$$

$$= 8.47 \text{ V}$$

and the power that Q_1 dissipates is equal to

$$P_{Q1} = I_{E1} \times V_{CE1}$$

$$= 418 \text{ mW}$$

The values for Q_2 are dependent upon the values of Q_1. The voltage at the emitter of Q_2 is 0.7 V higher than the voltage at the emitter of Q_1. This is due to the base–emitter junction of Q_1. The emitter current of Q_2 is equal to the emitter current of Q_1 divided by the beta of Q_1.

$$V_{E2} = V_{E1} + V_{BE1}$$

$$= 11.5 \text{ V} + 0.7 \text{ V}$$

$$= 12.2 \text{ V}$$

and

$$I_{E2} = \frac{I_{E1}}{\beta_1}$$

$$= \frac{49.4 \text{ mA}}{50}$$

$$= 988 \text{ } \mu\text{A}$$

Notice that the voltage at the emitter of Q_1 is 11.5 V. This is the result of adding the small voltage across R_3 to the output voltage. The collector–emitter voltage of Q_2 is the difference between the emitter voltage and the input voltage.

$$V_{CE2} = V_{IN} - V_{E2}$$

$$= 20 \text{ V} - 12.2 \text{ V}$$

$$= 7.8 \text{ V}$$

The power dissipated by the remaining half of the Darlington pair is found by

$$P_{Q2} = I_{E2} \times V_{CE2}$$

$$= 988 \text{ } \mu\text{A} \times 7.8 \text{ V}$$

$$= 7.71 \text{ mW}$$

EXAMPLE 7-5

Determine the power dissipated by Q_1 and Q_2 in the circuit of Fig. 7-5 if the wiper of R_5 is adjusted all the way down.

Solution: It was determined in Example 7-4 that the output voltage is 17.1 V when the wiper is turned all the way down. The next step is to find the currents flowing through the voltage divider and R_L.

$$I_{VD} = \frac{17.1 \text{ V}}{300 \text{ } \Omega}$$

$$= 57 \text{ mA}$$

and

$$I_L = \frac{17.1 \text{ V}}{1 \text{ k}\Omega}$$

$$= 17.1 \text{ mA}$$

so

$$I_{E1} = 57 \text{ mA} + 17.1 \text{ mA}$$

$$= 74.1 \text{ mA}$$

In order for the current to reach the emitter of Q_1, it must pass through R_3. A small voltage drop results. This small voltage, along with the output voltage, is taken into consideration in finding the voltage across Q_1.

$$V_{R3} = 74.1 \text{ mA} \times 2.7 \text{ } \Omega$$

$$= 200 \text{ mV}$$

so

$$V_{CE1} = 20 \text{ V} - (17.1 \text{ V} + 200 \text{ mV})$$
$$= 2.7 \text{ V}$$

The power of Q_1 can now be found.

$$P_{Q1} = 74.1 \text{ mA} \times 2.7 \text{ V}$$
$$= 200 \text{ mW}$$

The voltage at the emitter of Q_2 is 0.7 V higher than the voltage at the emitter of Q_1. We just found that the voltage at the emitter of Q_1 is equal to 17.3 V ($V_{\text{OUT}} + V_{R3}$). The voltage at the emitter of Q_2 is, therefore, 18 V. We proceed to find the voltage across Q_2.

$$V_{CE2} = 20 \text{ V} - 18 \text{ V}$$
$$= 2 \text{ V}$$

The current entering the emitter of Q_2 is equal to the base current of Q_1. The emitter current of Q_1 and its beta are known. It is an easy matter to determine its base current.

$$I_{E2} = I_{B1} = \frac{74.1 \text{ mA}}{50}$$
$$= 1.48 \text{ mA}$$

Finally,

$$P_{Q2} = 1.48 \text{ mA} \times 2 \text{ V}$$
$$= 2.96 \text{ mW} \qquad \blacksquare$$

Practice Problem: How much power will Q_1 and Q_2 dissipate if the wiper of R_5 is turned all the way up in the circuit of Fig. 7-5? *Answer:* $P_{Q1} = 421$ mW and $P_{Q2} = 7.9$ mW.

Current Limiting
The current-limiting section of the regulator is made up of R_3 and Q_4. The resistor is used to monitor the current that flows on to the Darlington pair pass transistor. For this reason, it is called a current-sensing resistor. The purpose of this section of the circuit is to limit the amount of current entering the pass transistor.

The current-limiting transistor, Q_4, is turned off most of the time. It turns on when its base–emitter junction receives the proper biasing voltage of 0.7 V. This occurs only when a large enough current is produced by the load.

Figure 7-6 shows the condition of the circuit when the potentiometer is adjusted for maximum output. Under this condition, the voltage across R_3 is only 0.2 V. This is not enough to turn on Q_4. The current-limiting section is off during the normal operation of the circuit.

If the load resistance were to decrease, load current would increase. This also causes an increase in the current through R_3. The current-limiting section eventually turns on in order to protect the pass transistor from the unusually high current. The maximum value of current that the pass transistor receives before the current-limiting section turns on can be calculated as

$$I_{E1(\text{MAX})} = \frac{0.7 \text{ V}}{2.7 \text{ } \Omega}$$
$$= 259 \text{ mA}$$

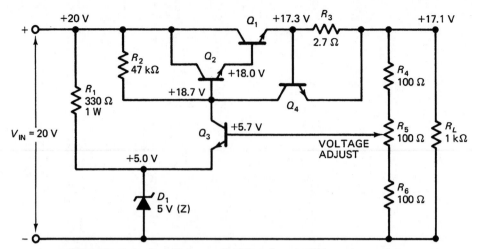

FIGURE 7-6 Darlington feedback regulator at maximum output.

When the combination of load current and voltage-divider current reaches 259 mA, Q_4 turns on. The collector current of Q_4 flows through R_2, thus increasing V_{R2}. Consequently, the voltage at the base of Q_2 decreases, resulting in the reduction of the output voltage.

The current-limiting transistor is like a safety valve. It restricts the current of the Darlington pass transistor to a safe level. This level is determined by the specifications of the pass transistor. The value of R_3 is chosen in conjunction with this value of current. If the maximum current through the pass transistor is 210 mA, then the value of R_3 would be

$$R_3 = \frac{0.7 \text{ V}}{I_{MAX}}$$

$$= \frac{0.7 \text{ V}}{210 \text{ mA}}$$

$$= 3.3 \ \Omega$$

Using this new value for the current-sensing resistor causes the current-limiting transistor to turn on when the current through R_3 reaches 210 mA.

Practice Problem: What size current-sensing resistor is needed to limit the current flow of the Darlington pass transistor to 100 mA in the circuit of Fig. 7-6? *Answer:* $R_3 = 7 \ \Omega$.

REVIEW QUESTIONS

31. Name a benefit of the Darlington pair with regard to the load.
32. Maximum voltage is seen at the output of the circuit of Fig. 7-5 when the wiper of R_5 is turned all the way (down or up).
33. What is the output voltage of the circuit of Fig. 7-5 if the wiper is adjusted 75% of the way up?
34. Will Q_4 turn on if the wiper is adjusted 75% of the way up?

35. What is the power dissipated by Q_1 when the wiper is adjusted 75% of the way up in the circuit of Fig. 7-5?

36. Which resistor is the current-sensing resistor in the circuit of Fig. 7-5?

37. What voltage is required across R_3 in the circuit of Fig. 7-5 in order to turn on the current-limiting transistor?

38. The maximum amount of current entering the pass transistor can be reduced by (increasing or decreasing) R_3 in the circuit of Fig. 7-5.

39. What value of R_3 is necessary in the circuit of Fig. 7-5 to limit the maximum current of the pass transistor to 150 mA?

7-5 SWITCHING REGULATOR

All of the regulators up to this point are considered to be **linear regulators.** This is because the pass transistor is operating somewhere between cutoff and saturation. Therefore, it is always on and dissipating power. The efficiency of linear regulators is usually 50% or less. Efficiency is determined by the ratio of input power to output power.

Linear Regulator: Regulators that operate between cutoff and saturation.

The switching regulator has an efficiency of 90% or more. This is because the pass transistor is operated like a switch. It is either saturated (closed) or cut off (open). Another reason for its high efficiency is the use of an LC filter as a storage tank for energy. The LC filter maintains the output voltage during the time that the pass transistor is cut off.

Figure 7-7 is a simplified example of a switching regulator. The block diagrams may contain components such as operational amplifiers, transformers, or other ICs. These components are arranged to produce the resulting subcircuits and their waveforms.

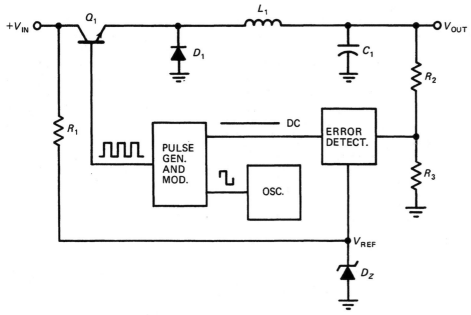

FIGURE 7-7 Switching regulator.

Transistor Switching Action   An **unregulated** positive dc **voltage** is supplied to the collector of Q_1 in the circuit of Fig. 7-8(A). A series of pulses are sent to the base of Q_1. The transistor saturates on each of the positive pulses. This is because an NPN transistor needs a positive voltage at its base to turn on. A saturated transistor can be thought of as a closed switch.

Each time the transistor switch is closed, a path for current is provided from ground to $+V_{IN}$. Figure 7-8(A) illustrates the path of current. Notice that as current flows through the coil, it builds a magnetic field. The capacitor also charges to the value of the output voltage. The diode is reverse biased at this point and does not conduct.

Eventually, the positive pulse at the base returns to a zero state and the transistor enters cutoff. The transistor acts like an open switch, as depicted in the circuit of Fig. 7-8(B). This removes the path of current to V_{IN}. The coil opposes this change in current, as its magnetic field collapses by self-inducing a voltage. Its polarity changes to that of Fig. 7-8(B). The coil keeps current flowing by returning the energy stored in its magnetic field to the circuit. The diode now provides a path for the current to reach the load at the output. The capacitor aids in filtering the output voltage. The transistor soon receives another positive pulse at its base and the whole process of opening and closing the transistor switch is repeated.

Output Voltage   The output voltage is determined by two factors: the value of V_{IN} and the pulse width at the base of the pass transistor. We begin by examin-

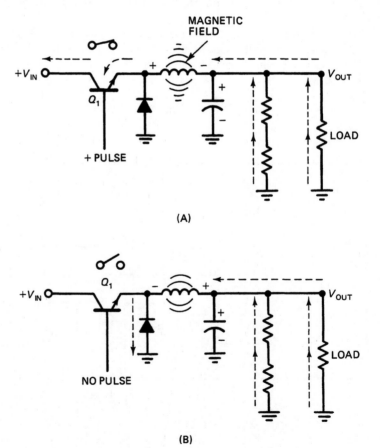

FIGURE 7-8   Switching action.

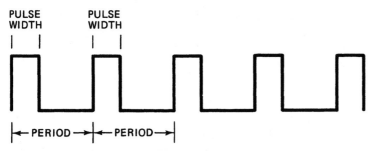

FIGURE 7-9 Determining the duty cycle.

ing the pulse width. Figure 7-9 shows a series of pulses. The period and pulse width are marked off.

The period is equal to the time of one cycle of the waveform. The pulse width is the length of time within a cycle that the waveform is pulsed high. The ratio of these two values is the duty cycle.

$$D = \frac{W}{P}$$

where D = duty cycle

W = width of the pulse in seconds

P = period of the cycle in seconds

The duty cycle of the waveform in Fig. 7-10 is

$$D = \frac{15\ \mu s}{40\ \mu s}$$

$$= 0.375$$

The duty cycle can be viewed as the percentage of time within a cycle that the pulse exists. In this example, the pulse exists for 37.5% of the time. The transistor switch is closed 37.5% of the time and opened for the remaining 62.5%. This means that the output voltage does not get the chance to reach the full value of V_{IN}. In the circuit of Fig. 7-10, the value of V_{IN} is 25 V. The value of V_{OUT} can be determined by

$$V_{OUT} = V_{IN} \times D$$

$$= 25\ V \times 0.375$$

$$= 9.38\ V$$

The importance of pulse width can be understood from this formula. The longer the pulse width, the longer the transistor acts like a closed switch. And the longer the transistor acts like a closed switch, the better the chances of the *LC* circuit charging to, and maintaining, a larger percentage of V_{IN} at the output.

Practice Problem: What is the value of V_{OUT} if the pulse width of the circuit of Fig. 7-10 were increased to 35 μs? *Answer:* V_{OUT} = 21.9 V.

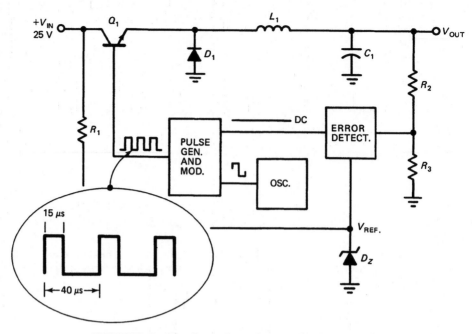

FIGURE 7-10 Sample pulse at the base of Q_1.

Pulse Generation and Error Detection

Pulse generation begins at the oscillator. An **oscillator** is a circuit that turns dc into ac. The frequency of the oscillator is generally 20 kHz or more. This waveform is then shaped into a triangle wave by a circuit within the pulse generator and modulator block. These circuits can be made up of operational amplifiers or special-purpose ICs.

A voltage divider is formed by R_2 and R_3 in the circuit of Fig. 7-10. The voltage divider is in parallel with the output. Any changes in output voltage are also experienced by the voltage divider. The error amplifier samples the voltage across R_3 and compares it to a reference voltage provided by the zener diode. The dc output of the error amplifier is adjusted for any changes in the output voltage.

The pulse generator uses the triangle wave that it creates and the dc level from the error amplifier to produce the rectangular wave at its output. Figure 7-11(A) shows how the dc level and triangle wave are compared to produce the normal rectangular output. If the output voltage were to increase, the dc level of the error detector would also increase. Figure 7-11(B) shows how increasing the dc level decreases the pulse width of the rectangular output. Remember that decreasing the pulse width decreases the output voltage. This offsets the attempted increase in the output voltage. The end result is a regulated dc output voltage.

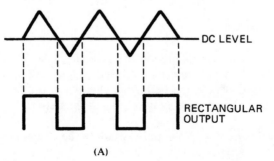

FIGURE 7-11 Using the dc level to adjust switching time.

CHAP. 7 / REGULATED POWER SUPPLIES

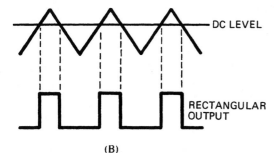

DC LEVEL

RECTANGULAR
OUTPUT

(B)   **FIGURE 7-11** (*cont.*)

EXAMPLE 7-6

What would be the output voltage of the circuit of Fig. 7-10 if the period of the waveform at the base of Q_1 were 150 μs and the pulse width were 35 μs?

Solution: The longer the pulse width, the higher the output voltage. In this example, the output voltage would be determined by the duty cycle. The duty cycle is

$$D = \frac{35 \ \mu s}{150 \ \mu s}$$
$$= 0.233$$

so

$$V_{OUT} = 25 \text{ V} \times 0.233$$
$$= 5.83 \text{ V} \qquad \blacksquare$$

REVIEW QUESTIONS

40. _____ regulators operate somewhere between cutoff and saturation.

41. The _____ regulator is very efficient because the pass transistor is not conducting all of the time.

42. A saturated transistor is similar to a _____ switch.

43. A positive pulse at the base of Q_1 in the circuit of Fig. 7-7 causes the transistor to (saturate or cut off).

44. What is the duty cycle of a waveform whose period is 28 μs and pulse width is 12 μs?

45. What is the output voltage if V_{IN} is 150 V, the period is 25 μs, and the pulse width is 18 μs?

46. Increasing the pulse width at the base of the pass transistor _____ the output voltage.

47. If the error detector of Fig. 7-10 detects a higher than normal output voltage, it eventually causes the pulse width at the pass transistor to _____ .

7-6   IC REGULATORS

Regulators are an important part of most power supplies, especially those that power digital and microprocessor circuits. Any circuit in widespread use is usually packaged in a single integrated circuit. This was true of the bridge rectifier and is also true of the regulator.

78M05 5-Volt regulator and LM317 variable regulator.

7805 Regulator   The 7805 IC is an example of 5-V fixed-voltage regulator. The input voltage can range anywhere from about 6 to 35 V. The output is a regulated 5 V dc.

Figure 7-12(A) illustrates the physical packaging of the 7805. It looks very much like a power transistor. Mounting it on a heat sink is usually recommended.

Figure 7-12(B) depicts the regulator operating circuit. The application of the 7805 is straightforward. An unregulated voltage of more than 5 V at the input produces a regulated 5-V dc output. The capacitors are not an absolute necessity, though they do aid in filtering any **noise** or **transients.**

Noise: An unwanted signal or signals.
Transient: A sudden surge or spike of high voltage.

7812 Regulator   The 7812 is a 12-V fixed-voltage regulator. Regulator ICs come in a variety of values. The 7812 is used here to illustrate how a fixed-voltage regulator can provide an adjustable output voltage through the use of external resistors. The output voltage can range from 12 V on up.

The voltage of the regulator is seen across R_1 in the circuit of Fig. 7-13. A modified voltage-divider formula aids in determining the output voltage. The small current (I_Q) required to operate the regulator is included at the end of the formula. The output voltage can be determined by

$$V_{\text{OUT}} = V_R \left(\frac{R_2}{R_1} + 1 \right) + (I_Q \times R_2)$$

where V_R = IC regulator voltage

I_Q = quiescent current of IC

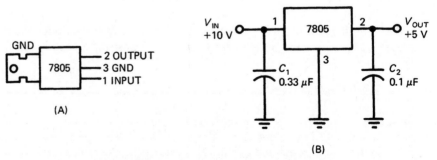

(A)

(B)

FIGURE 7-12   Five-volt IC regulator.

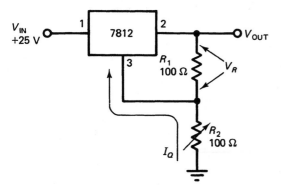

FIGURE 7-13 Twelve-volt IC regulator with adjustable output.

The maximum I_Q of a particular regulator is shown on the specification sheet. Most fall between 5 and 10 mA. Changes in load current do not change this value by very much. We use 8 mA in our example. If the rheostat, R_2, is adjusted to 50 Ω, then the output voltage would be

$$V_{OUT} = 12\ V\left(\frac{50\ \Omega}{100\ \Omega} + 1\right) + (8\ mA \times 50\ \Omega)$$

$$= 18.4\ V$$

Practice Problem: What is the output voltage if the rheostat is adjusted to 75 Ω in the circuit of Fig. 7-13? (Assume an I_Q of 8 mA.) *Answer:* $V_{OUT} = 21.6$ mA.

LM317 Regulator   Although a fixed regulator can be configured for an adjustable output voltage as shown before, the range is rather limited. The LM317 is a variable IC regulator. It offers a range of 1.25 to 37 V. The quiescent current of the LM317 is small, 50 μA, when compared to the 8 mA of a fixed regulator. The physical packaging is the same.

Figure 7-14 shows the LM317 connected in a circuit. The potentiometer is connected as a rheostat. This is done because of the greater availability of potentiometers. Assuming that R_2 is adjusted to 5 kΩ, the output voltage is

$$V_{OUT} = 1.25\left(\frac{R_2}{R_1} + 1\right)$$

$$= 1.25\left(\frac{5\ k\Omega}{390\ \Omega} + 1\right)$$

$$= 17.3\ V$$

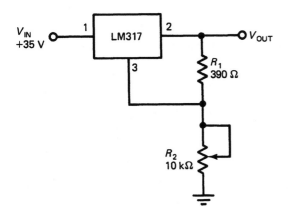

FIGURE 7-14 Variable IC regulator, 1.25–37 V.

EXAMPLE 7-7

Determine the output voltage in the circuit of Fig. 7-14 if R_2 is adjusted to 3 kΩ.

Solution:

$$V_{OUT} = 1.25 \left(\frac{3 \text{ k}\Omega}{390 \text{ }\Omega} + 1 \right)$$

$$= 10.9 \text{ V} \qquad \blacksquare$$

Practice Problem: What is the output voltage of the circuit of Fig. 7-14 if R_2 is adjusted to 2.5 kΩ? *Answer:* $V_{OUT} = 9.26$ V.

Other IC Regulators The regulators discussed so far have been positive linear voltage regulators. Regulator ICs do come in negative values. The 7905, for example, is a -5-V fixed regulator. The 7912 is a -12-V fixed regulator.

There are also regulators that provide a dual output. The LM325 provides both a positive and negative 15-V output. The input voltages can range from -30 to $+30$ V.

Finally, Fairchild manufactures what it calls the Universal Switch Regulator Subsystem. The uA78S40 is a switching regulator in a single IC. The output varies from 1.3 to 40 V.

REVIEW QUESTIONS

48. The 7805 is an example of a _____ .

49. Any regulator that is not considered a variable regulator is called a _____ regulator.

50. The minimum output voltage of the circuit of Fig. 7-13 is _____ .

51. R_2 in the circuit of Fig. 7-14 is a _____ connected as a _____ .

52. The 7912 is an example of a _____ regulator IC.

7-7 TROUBLESHOOTING

Troubleshooting usually begins in the power-supply section. This section consists of a filtered rectifier followed by a regulator. Rectifiers were discussed in Chapter 2, so this discussion focuses on regulator circuits. Let us begin by examining the Darlington feedback regulator of Fig. 7-15.

Effect of an Open Zener When the zener diode in the circuit of Fig. 7-15 opens, the 5-V bias on the emitter is lost. The voltage at the emitter is approximately equal to V_{IN}.

The Darlington pair is able to conduct because the emitter of Q_1 is at a lower positive potential than the base of Q_2. The base of Q_2 receives its biasing voltage through R_2.

The current through the current-sensing resistor, R_3, is not enough to turn on Q_4. The current-limiting transistor, therefore, remains in a cut-off state.

The circuit is conducting without the ability of Q_3 to monitor the output voltage. The result is a runaway condition. V_{OUT} almost equals V_{IN}. A small voltage is dropped across the collector–emitter region of Q_1. Varying the potentiometer has almost no effect on the output voltage.

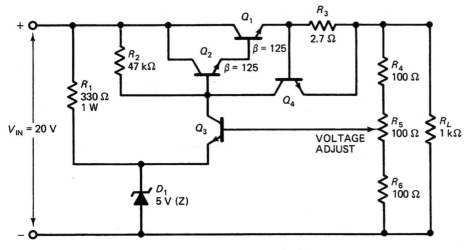

FIGURE 7-15 Darlington feedback regulator.

Effect of a Shorted Zener Diode   Shorting out the zener diode places the emitter of Q_3 at ground potential. This causes a reference voltage of 0.7 V at the base of Q_3. The range of the output under this condition is roughly 1 to 2.1 V.

Effect of an Open Q_{BE1}   If the base–emitter junction of Q_1 opens, no current can flow through either of the Darlington pair transistors. No load current can flow; therefore, the output voltage is zero volts.

Effect of a Shorted Q_{BE1}   Shorting the base–emitter junction of Q_1 actually causes the operational design of the circuit to change. The circuit that began as the Darlington feedback regulator of Fig. 7-15 ends up as the emitter-feedback circuit of Fig. 7-16.

The nonconducting transistors have been eliminated from Fig. 7-16. The base–emitter junction of Q_1 is damaged and Q_1, therefore, is taken out of the picture. The current-limiting transistor (Q_4) does not receive 0.7 V across its base–emitter junction and remains in a cut-off condition. Finally, Q_3 does not receive the proper base–emitter biasing voltage and also remains in a cut-off condition.

The emitter-bias circuit was discussed in Chapter 4. Let us begin by solving the approximate collector current. Since collector current and emitter current are viewed as approximately equal, the value of collector current can be used to solve for the output. The formula requires beta and a value of 125 has been assumed in the circuit of Fig. 7-16. The emitter resistor is equal to the sum of R_3 and R_{OUT}.

$$R_{OUT} = R_L \parallel (R_4 + R_5 + R_6)$$

$$R_E = R_{OUT} + R_3$$

$$I_c = \frac{V_{IN} - V_{BE2}}{R_E + (R_2/\beta)}$$

$$= \frac{20 \text{ V} - 0.7 \text{ V}}{233 \ \Omega + (47 \text{ k}\Omega/125)}$$

$$= 31.7 \text{ mA}$$

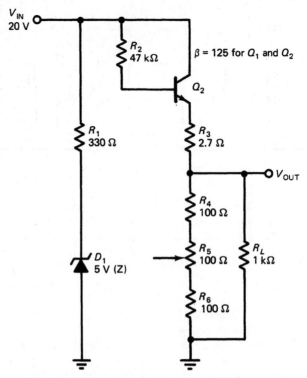

FIGURE 7-16 Darlington feedback regulator after a fault.

The emitter current flows through the equivalent output resistance R_{OUT}. The resulting voltage is

$$V_{OUT} = I_E \times R_{OUT}$$

$$= 31.7 \text{ mA} \times 231 \text{ } \Omega$$

$$= 7.32 \text{ V}$$

Adjusting the potentiometer has no effect on the output voltage. A combination of symptoms such as these would point toward the pass transistor.

The pass transistor is one of the more susceptible components in the regulator circuit. It must operate under a wide range of load conditions and is often dissipating large amounts of power. Power generates heat and heat destroys components. The pass transistor is a prime suspect in faulty regulator circuits.

Effect of an Open Q_{BE3} An open in the base–emitter junction of Q_3 in the circuit of Fig. 7-15 puts that transistor out of operation. The resulting circuit looks much like that of Fig. 7-16 with one exception: both transistors of the Darlington pair are conducting. This results in a much larger emitter current. We can use the emitter-feedback formula to prove this point. Notice the multiplication of the two betas.

$$I_C = \frac{V_{IN} - (V_{BE1} + V_{BE2})}{R_E + R_2/(\beta \times \beta)}$$

$$= \frac{20 \text{ V} - 1.4 \text{ V}}{233 \text{ } \Omega + 47 \text{ k}\Omega/(125 \times 125)}$$

$$= 78.8 \text{ mA}$$

178

The higher collector current means that a higher emitter current exists. A larger output voltage results.

$$V_{\text{OUT}} = I_E \times R_{\text{OUT}}$$
$$= 31.7 \text{ mA} \times 231 \text{ }\Omega$$
$$= 7.32 \text{ V}$$

Adjusting the potentiometer does not vary this voltage. This is another key symptom in recognizing this fault.

Effect of a Shorted Q_{BE3} Q_3 is an important transistor for monitoring the output voltage. A shorted base–emitter junction prevents it from operating in the circuit. The emitter feedback with the Darlington pair is formed once again. The only difference is that now the zener voltage is shorted directly to the potentiometer. The output voltage is still very high and varying the potentiometer has little effect. The output voltage measures within 2 V of the input voltage.

Troubleshooting Switching Regulators   First of all, very high currents and voltages may be involved with switching regulators, so use extreme caution! The pass transistor acts as a switch; therefore, you should see a series of pulses at its base. The collector–emitter region is also pulsing on and off. These measurements can be made with an oscilloscope. Disconnect the *LC* circuit shown in Fig. 7-7, if possible, to prevent interference from a shorted component towards the output.

If the voltage at V_{IN} measures at its rated value, then the transformer and bridge rectifier are working properly. A discrepancy in the voltage level or excessive ripple is your signal to check this section. This includes the filter capacitors.

If there are no pulses at the base of the pass transistor, then the problem lies somewhere between the oscillator and the base. Each of these subcircuits should be checked for an appropriate waveform or voltage level.

In the case where V_{IN} and the base of the pass transistor check out, the process of elimination should lead you to the *LC* circuit and output connectors. The problem may be anything from an open in the line or coil to a shorted capacitor or diode. A resistance check should be made on the appropriate component.

Troubleshooting IC Regulators   IC regulators conserve space and simplify the design of a power supply. These benefits come at a price. The previous regulators, which use **discrete components,** allow you to replace just the pass transistor if necessary. The IC regulator has all the components **fabricated** within the casing. The entire regulator must be replaced even if only the internal pass transistor is bad. This increases the cost of repairs.

Discrete Components: Individual components, such as resistors, capacitors, inductors and transistors, that have not been fabricated in IC form.
Fabricate: To manufacture.

To troubleshoot an IC regulator, simply measure the input and output voltages. Disconnect the output of the regulator to prevent interference from other components, but use a characteristic load resistance. If the input voltage is correct and the output voltage is not, replace the IC regulator.

REVIEW QUESTIONS

53. Where should you normally begin troubleshooting?
54. What effect does an open zener diode have on the output voltage of the circuit of Fig. 7-15?

55. What effect does a shorted zener diode have on the output voltage of the circuit of Fig. 7-15?

56. What is the resulting output voltage of the circuit of Fig. 7-15 if the base–emitter junction of Q_1 were to open?

57. What is the resulting output voltage of the circuit of Fig. 7-15 if the base–emitter junction of Q_1 were to short?

58. What is the resulting output voltage of the circuit of Fig. 7-15 if the base–emitter junction of Q_3 were to open?

59. What is the resulting output voltage of the circuit of Fig. 7-15 if the base–emitter junction of Q_3 were to short?

60. What is the resulting output voltage of a switching regulator if the pass transistor is not pulsed on?

SUMMARY

1. A regulator provides a constant or regulated dc voltage.

2. A zener diode provides a regulated voltage to a load in parallel to it.

3. Smaller load resistances draw more current.

4. The value of the current-limiting resistor in a zener regulator circuit can be designed by

$$R_1 = \frac{V_{IN} - V_{ZD}}{V_{ZD}/R_{L(MIN)} + I_{DZ(MIN)}}$$

5. The output voltage of the emitter-follower regulator follows the voltage at the base.

6. The series pass transistor passes all of the load current and drops the difference between input and output voltages across its collector–emitter region.

7. Important formulas for the emitter-follower regulator are

$$V_{OUT} = V_{D1} - V_{BE}$$
$$V_{CE} = V_{IN} - V_{OUT}$$
$$I_E = I_{R_L} = \frac{V_{OUT}}{R_L}$$
$$P_{Q1} = I_E \times V_{CE}$$

8. The feedback transistor helps to provide and sample output voltage. It aids in offsetting changes in output voltage.

9. Important formulas for the variable-feedback regulator are

$$V_{OUT} = V_{R3(wiper)} + V_{BE2}$$
$$I_{R_L} = \frac{V_{OUT}}{R_L}$$
$$V_{CE1} = V_{IN} - V_{OUT}$$
$$P_{Q1} = I_{E1} \times V_{CE1}$$

10. The Darlington feedback regulator uses a Darlington-pair arrangement of transistors in order to achieve a high current gain.

11. Important formulas for the Darlington feedback regulator are

$$V_{OUT} = \frac{R_{VD}}{R_W} V_{B3}$$

$$I_{R_L} = \frac{V_{OUT}}{R_L}$$

$$I_{VD} = \frac{V_{OUT}}{R_{VD}}$$

$$I_{E1} = I_L + I_{VD}$$

$$V_{R3} = I_{E1} \times R_3$$

$$V_{CE1} = V_{IN} - (V_{OUT} + V_{R3})$$

$$P_{Q1} = I_{E1} \times V_{CE1}$$

$$V_{E2} = V_{E1} + V_{BE1}$$

$$I_{E2} = \frac{I_{E1}}{\beta_1}$$

$$V_{CE2} = V_{IN} - V_{E2}$$

$$P_{Q2} = I_{E2} \times V_{CE2}$$

12. The current-sensing resistor monitors the current that flows to the pass transistor.

13. Current limiting is used to protect the pass transistor.

14. Linear regulators operate in the active region (between cutoff and saturation).

15. Efficiency for linear regulators is usually less than 50%.

16. Switching regulators operate at cutoff and saturation.

17. Efficiency for switching regulators is upwards of 90%.

18. The pass transistor of a switching regulator operates like a switch that closes every time it receives a forward pulse at its base.

19. The *LC* filter of the switching regulator maintains the output voltage during the time that the pass transistor is cut off.

20. The duty cycle is the percentage of time that a pulse exists within a cycle.

21. Important formulas for the switching regulator are

$$D = \frac{W}{P}$$

$$V_{OUT} = V_{IN} \times D$$

22. IC regulators contain a regulator circuit fabricated within a single package.

23. Fixed regulator ICs provide a specific value of output voltage.

24. Variable regulator ICs allow for an adjustable output voltage.

25. The unregulated voltage entering an IC regulator must be greater than the rated output voltage.

26. The zener diode and pass transistor are generally the first components to be destroyed within a regulator circuit.

27. Components that are shorted measure 0 V and open components measure higher than normal voltages.

28. A short measures zero ohms and an open measures infinite resistance.

GLOSSARY

Darlington Feedback Regulator: A regulator that uses the Darlington pair as the pass transistor in order to enhance the current capabilities.

Discrete Components: Individual components, such as resistors, capacitors, inductors and transistors, that have not been fabricated in IC form.

Emitter Follower: A regulator in which the output from the emitter follows the value of the zener diode at the base.

Fabricate: To manufacture.

Feedback: A condition where output information is returned back to the input.

Feedback Transistor: Also called the error amplifier, it monitors the output voltage, using feedback to make any necessary adjustments.

Linear Regulator: Regulators that operate between cutoff and saturation.

Noise: An unwanted signal or signals.

Oscillator: A circuit that produces ac from dc.

Pass Transistor: A transistor through which total load current passes and is controlled in a regulator circuit.

Regulator: A circuit that maintains a constant output voltage.

Transient: A sudden surge or spike of high voltage.

Unregulated Voltage: A fluctuating dc voltage as it pertains to regulator circuits. The voltage from a bridge rectifier is unregulated.

PROBLEMS

SECTION 7-1

The following problems refer to the circuit of Fig. 7-17.

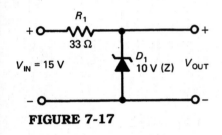

FIGURE 7-17

7-1. What is the value of V_{OUT}?

7-2. Calculate the total current.

7-3. Determine the zener current when a 1-kΩ load is connected to the output.

7-4. How much power is dissipated by the zener diode when a 1-kΩ load is connected to the output?

7-5. What is the minimum load resistance in order to maintain a zener current of at least 25 mA?

7-6. What would be a safe wattage rating for R_1?

7-7. Determine the size of the current-limiting resistor necessary to service an 8-Ω load while providing 25 mA of zener current.

SECTION 7-2

The following problems refer to the circuit of Fig. 7-18.

7-8. What is the value of the output voltage?

7-9. Calculate the current of the pass transistor for a 500-Ω load.

7-10. Determine the collector–emitter voltage of the pass transistor.

7-11. How much power will the pass transistor dissipate when a 500-Ω load is connected?

7-12. What value of voltage is dropped across R_1?

7-13. Is the zener diode forward or reverse biased?

7-14. Does the power of the pass transistor increase or decrease as the resistance of the load decreases?

7-15. How does an increase in V_{IN} affect the output voltage?

7-16. What is the load current for a 100-Ω load resistance?

7-17. Calculate the power of the pass transistor when a 100-Ω load resistance is connected across the output.

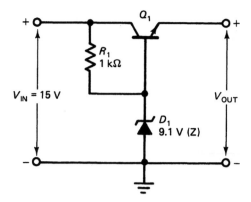

FIGURE 7-18

SECTION 7-3

The following problems refer to the circuit of Fig. 7-19.

7-18. What is the minimum output voltage?

7-19. What is the maximum output voltage?

7-20. Determine the current through a 1-kΩ load at the maximum output voltage.

7-21. Calculate the power dissipated by Q_1 for a 1-kΩ load at the maximum output voltage.

7-22. Describe the function of Q_2.

7-23. Calculate the power dissipated by Q_1 for a 1-kΩ load at the minimum output voltage.

7-24. What is the voltage across R_2?

7-25. What is the voltage at the collector of Q_2?

7-26. Determine the output voltage when the wiper of the potentiometer is adjusted half way.

7-27. Calculate the power dissipated by Q_1 when the wiper of the potentiometer is adjusted half way and a 100-Ω load connected to the output.

7-28. What is the purpose of a heat sink?

7-29. What is the difference between regulated and unregulated voltage?

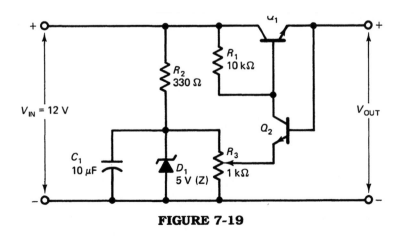

FIGURE 7-19

The following problems refer to the circuit of Fig. 7-20.

7-30. What is the minimum V_{OUT}?

7-31. What is the maximum V_{OUT}?

7-32. Find the current through a 1-kΩ load resistance at the maximum output voltage.

7-33. Determine the voltage across R_3 when a 1-kΩ load resistance is connected across the output at the maximum V_{OUT}.

7-34. Does Q_4 conduct as a result of the previous problem?

7-35. Calculate the power dissipated by Q_1 when a 1-kΩ load is connected across the maximum output voltage.

7-36. Calculate the power dissipated by Q_2 when a 1-kΩ load is connected across the maximum output voltage.

7-37. What is the name given to the configuration created by Q_1 and Q_2?

7-38. What is the purpose of each transistor?

7-39. Determine the current limit created by Q_4 and R_3.

7-40. What value of load resistance at the maximum output voltage causes Q_4 to begin conducting?

7-41. What value of R_3 provides a current limit of 120 mA for the pass transistor?

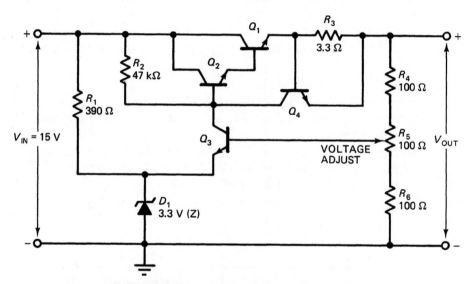

FIGURE 7-20 (Note: Beta of Qs = 125.)

SECTION 7-5

The following problems refer to the circuit of Fig. 7-21.

7-42. What is the period of the waveform?

7-43. What is the time of the pulse width?

7-44. Determine the duty cycle.

7-45. Find the value of the output voltage.

7-46. Calculate the output voltage for a pulse width of 20 μs.

184

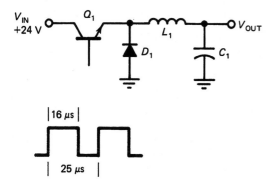

FIGURE 7-21

7-47. Explain the operation of the pass transistor along with the required polarity on its base.

7-48. Explain the action of the LC filter, including the magnetic field.

SECTION 7-6

The following problems refer to the circuit of Fig. 7-22.

7-49. What type of voltage regulator is used?

7-50. Determine the output voltage when the rheostat is adjusted to 0 Ω.

7-51. Determine the output voltage when the rheostat is adjusted to 100 Ω.

7-52. Does decreasing R_2 increase or decrease the maximum output voltage?

7-53. List the various regulator ICs along with a brief description of each. Refer to a parts catalog, if available.

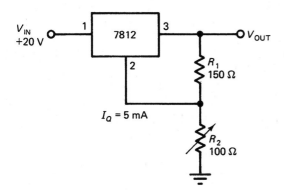

FIGURE 7-22

SECTION 7-7

7-54. What would be the output voltage of the circuit of Fig. 7-17 if the zener diode were to open?

7-55. The output voltage of the circuit of Fig. 7-20 is 0 V and adjusting the potentiometer has no effect. What is the probable cause?

7-56. The output voltage of the circuit of Fig. 7-20 is nearly equal to V_{IN} and adjusting the potentiometer has no effect. What is the probable cause?

7-57. The output voltage of the circuit of Fig. 7-20 is less than 3 V and adjusting the potentiometer varies the voltage down to about 1 V. What is the probable cause?

PROBLEMS

7-58. The output voltage of the circuit of Fig. 7-20 is about half the value of V_{IN} and adjusting the potentiometer has no effect. What is the probable cause?

7-59. Describe some important points in troubleshooting switching regulators.

7-60. How would you troubleshoot a regulator IC?

ANSWERS TO REVIEW QUESTIONS

1.	regulation	**2.**	dc	**3.**	reverse
4.	unregulated	**5.**	current-limiting	**6.**	40 mA
7.	It remains constant	**8.**	more	**9.**	minimum
10.	maximum	**11.**	3 mA	**12.**	Common collector
13.	False	**14.**	Pass transistor	**15.**	load
16.	5.3 V	**17.**	16.1 mA	**18.**	8.4 V
19.	Heat sink	**20.**	445 mW	**21.**	Q_1
22.	0.7 V	**23.**	700 μA	**24.**	24.3 V
25.	17 mW	**26.**	increases	**27.**	decreases
28.	remains the same	**29.**	increases	**30.**	unregulated
31.	It can drive smaller loads.	**32.**	down	**33.**	9.77 V
34.	No	**35.**	428 mW	**36.**	R_3
37.	0.7 V	**38.**	increasing	**39.**	4.67 Ω
40.	Linear	**41.**	switching	**42.**	closed
43.	saturate	**44.**	0.429	**45.**	108 V
46.	increases	**47.**	decrease	**48.**	5-V regulator IC
49.	fixed	**50.**	12 V	**51.**	potentiometer, rheostat
52.	−12-V	**53.**	At the power supply	**54.**	Increases to V_{IN}
55.	Decreases to <2.2 V	**56.**	0 V	**57.**	Between 0 V and V_{IN}
58.	Increases almost to V_{IN}	**59.**	Increases almost to V_{IN}	**60.**	0 V

Transistor Logic

8

When you have completed this chapter, you should be able to:

- Apply the switching capability of the transistor.
- Analyze the application and operation of TTL and RTL logic.
- Recognize the different logic functions of the various gates.
- Investigate the use of digital logic with ordinary applications.
- Troubleshoot digital circuits.

INTRODUCTION

Bipolar junction transistors have been studied from a variety of approaches. The construction and biasing of the BJTs were examined first. Next, their use in small-signal amplifier applications was examined. The ac applications continued with power amplifiers and the last chapter concluded by using the transistor in regulated dc power supplies.

Up to this point, the applications of the transistor have been **analog** in nature. The BJT has been operated in a continuous fashion throughout its active region. Now we use the transistor in **digital** applications. It will be operated only at its two extremes: cutoff and saturation. It is this switchlike capability of the transistor that occupies our attention throughout this chapter.

Analog: Provides a continuous range of values.

Digital: Provides a discrete range of values.

Digital IC.

Transistor logic comes in a variety of forms of which two are discussed in the following sections. Logic is composed of two basic states, the logic-1 state and the logic-0 state. The logic-1 state is also referred to as high, set, on, or true; and the logic-0 state as low, clear, off, or false. In terms of voltage, a logic-1 state falls between 2 and 5 V for bipolar circuits. A logic-0 state falls between 0 and 0.8 V. These voltage levels vary slightly from input to output.

Logic Levels: A logic 1 is indicated by a voltage greater than 2 V and a logic 0 by a voltage less than 0.8 V.

8-1   TRANSISTOR–TRANSISTOR LOGIC

Transistor–transistor logic (TTL) is one of the most popular forms of logic today. It offers medium- to high-speed capability in switching states. High noise immunity along with a high output-current capability are also advantages. The disadvantages are the larger space requirement and power consumption as compared with the more efficient **metal-oxide semiconductor** (MOS) technology. This technology is discussed in the next chapter.

Transistor–Transistor Logic: TTL is a widely used form of creating digital devices. It provides speed and good current capability.

Metal-Oxide Semiconductors: These semiconductors use less power and space than TTLs, but are not as fast at changing logic states.

There are a variety of logic gates. A logic gate is a device that produces a certain output based on a certain input or combination of inputs. The **NAND gate** is used in our example. Each gate can be recognized by its **truth table.** A truth table lists all the possible input conditions along with the resulting output conditions.

TTL NAND Gate with a Low Output   Figure 8-1 is a representation of the TTL logic NAND gate. The inputs are labeled A and B. The output is labeled X. A V_{CC} of +5 V is used to power the circuit. The entire circuit will be discussed component by component, but, first, the truth table for a NAND gate:

NAND Gate Truth Table

A	B	X
L	L	H
L	H	H
H	L	H
H	H	L

Notice that all input conditions, with the exception of two highs, produce a logic 1, or high, output. A truth table is easier to remember if you concentrate on the unique output condition. Apply this truth table to the circuit of Fig. 8-1, which shows that both the A and B inputs are receiving a logic 1, or high.

Diodes D_1 and D_2 are negative clamping diodes. They protect the circuit from an accidental negative voltage at the inputs. If the input voltage were to reach -0.7 V, the associated diode would conduct.

The input transistor Q_1 has multiple emitters. A transistor of this type can have as many as eight emitters. Forward biasing any of the emitters causes the transistor to conduct. Since this is an NPN transistor, the emitter would need a logic 0, or low, in order to be forward biased. The inputs are both high; therefore,

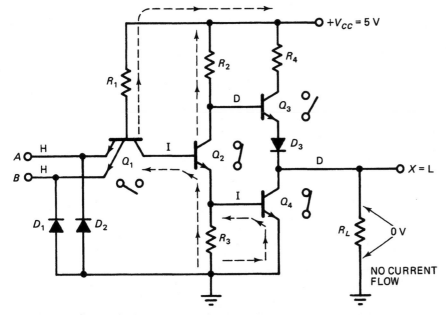

FIGURE 8-1   TTL NAND gate with a low output.

the transistor is not conducting and acts as an open switch, as illustrated. The open collector–emitter region of Q_1 causes an increase (I) in voltage at the collector. The base–collector junction of this transistor is forward biased and acts like an ordinary silicon diode. Figure 8-1 shows current flowing through this junction.

Q_2 acts as a **phase splitter.** An increase in voltage at its base causes it to conduct. The collector and emitter of Q_2 are 180° out of phase with each other. An increase on the base results in an increase at the emitter and a decrease at the collector. The emitter resistor R_3 is in parallel with the base–emitter junction of Q_4 and can only drop 0.7 V. Q_2 is saturated, as indicated by the closed switch, and drops about 0.1 V across its collector–emitter region.

The **totem-pole** output consists of R_4, Q_3, D_3, and Q_4. Either Q_3 or Q_4 conducts. The increase at Q_4's base causes it to conduct. The combination of Q_{BE4} and Q_{CE2} places a 0.8-V potential at the base of Q_3. This is not enough to turn on both D_3 and Q_{BE3}, so Q_3 remains cut off, as indicated by the open switch. Because current cannot flow through the load resistance and up through D_3 and Q_3, the output logic state is a 0, or low.

The purpose of D_3 is, as explained earlier, to prevent Q_3 from being turned on by the 0.8 V at its base. The cut-off Q_3 prevents current flow through the load when the output is low, which, in turn, reduces the power consumption of the circuit.

Try following the switching action through the circuit once more. Use the increase (I) and decrease (D) symbols as your guide. An open collector–emitter switch usually increases the voltage at the collector and a closed collector–emitter switch decreases the voltage at the collector.

TTL NAND Gate with a High Output
Figure 8-2 shows a NAND gate with a high output. This is the result of a low logic level on one of the emitter inputs. The truth table for a NAND gate indicates that a low on any or all inputs produces a high output.

The switching action begins with Q_1, which is forward biased due to the low on one of its emitter inputs. The collector–emitter switch closes, causing a de-

Phase Splitter: A transistor that is tapped on both its collector and emitter to provide signals of opposite phase.

Totem Pole: A method of stacking transistors and diodes to conserve power.

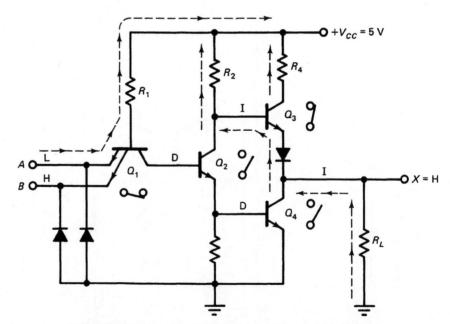

FIGURE 8-2 TTL NAND gate with a high output.

crease at the collector of Q_1. Current flows through the base–emitter junction of Q_1 and saturates the transistor.

The decrease in voltage experienced at the base of Q_2 opens its collector–emitter switch. The phase-splitting action produces an increase at the collector of Q_2 and a decrease at its emitter. The open collector–emitter switch of Q_2 prevents it from conducting.

The decrease at the base of Q_4 opens its collector–emitter switch. Also, the path for its base current through Q_2 has been removed. Therefore, Q_4 is cut off.

The base voltage of the emitter follower, Q_3, has been increased. This closes the collector–emitter switch and provides a path for the load current. As current flows through the load resistance, a high logic level is developed at the output.

EXAMPLE 8-1

What logic level would appear at the output of the circuit of Fig. 8-2 if 0 V were applied to both the A and B inputs?

Solution: The logic-0 voltage levels on inputs A and B cause Q_1 to conduct. A decrease in voltage at the collector of Q_1 is experienced and, in turn, cuts off Q_2. If Q_2 does not conduct, then Q_4 has no path for the base current and also cuts off. Cutting off Q_2 increases its collector voltage, which turns on Q_3. Current flows through the load resistor and up through Q_3. This produces a voltage drop across R_L equivalent to a logic-1 level at the output. ∎

REVIEW QUESTIONS

1. What term describes a continuous range of values?
2. How many defined logic states are there in digital applications?
3. What are the two extremes of transistor operation?
4. The term "high" is equivalent to a logic _____.
5. The term "set" is equivalent to a logic _____.
6. What voltage range is defined by a logic 0?
7. What does TTL represent?
8. What are the advantages of MOS technology over TTL?
9. What is the name given to a list of all possible input and output conditions for a given gate?
10. A condition of high and low on the inputs of a NAND gate produces a _____ output.
11. Which diode(s) provide negative clamping in the circuit of Fig. 8-1?
12. Name an advantage of the totem-pole output of the circuit of Fig. 8-1.
13. Forward biasing the base–emitter junction of a transistor (opens or closes) its collector–emitter switch.
14. A closed collector–emitter switch (increases or decreases) the voltage at the collector.
15. An NPN transistor requires a logic (high or low) at its emitter in order to forward bias it.
16. Which transistor in the totem pole must conduct for load current to flow in the circuit of Fig. 8-2?
17. What is the name given to transistor Q_2 in the circuit of Fig. 8-2?

Resistor-Transistor Logic: RTL is an earlier form of designing digital gates that provides simplicity at the expense of power.

Resistor-transistor logic (RTL) is an older method of creating gates. It offers moderate switching speed and a simpler design. The disadvantages are the space requirement, power consumption, and lower noise immunity.

RTL NAND Gate with a Low Output Figure 8-3 shows the RTL configuration of a NAND gate. The logical operation of this gate is the same as that of the TTL version. All NAND gates have the same truth table.

NAND Gate Truth Table

A	B	X
0	0	1
0	1	1
1	0	1
1	1	0

The operation of the circuit of Fig. 8-3 is straightforward. A high on the base of an NPN transistor closes the collector–emitter switch. A low on the base of an NPN transistor opens the collector–emitter switch.

Both inputs of the circuit of Fig. 8-3 are high, so both switches are closed. Voltage across a closed switch is 0 V. The output is taken across the two closed switches, which results in a low output.

Both transistors are conducting and provide a low resistance path for current. Current is consequently redirected away from the load. The current through R_3 produces a voltage drop that is just about equal to V_{CC}. The base resistor of each transistor helps to limit base current flow.

RTL NAND Gate with a High Output There are a number of input conditions that produce a high output. Figure 8-4 illustrates one such possibility. Notice that the low on Q_1 opens its collector–emitter switch. This removes the path of current through the two transistors. Current is now forced to flow through the load. The current through the load produces a voltage drop that is equal to a logic 1, or high.

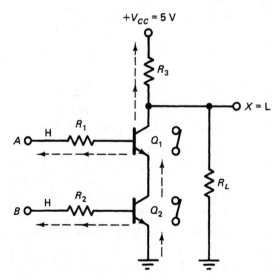

FIGURE 8-3 RTL NAND gate with a low output.

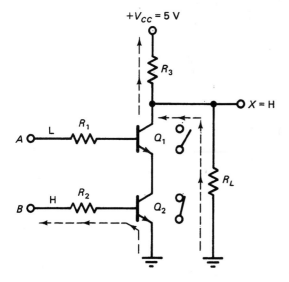

$+V_{CC} = 5$ V

R_3

R_1

A

L

Q_1

$X = $ H

R_L

B

H

R_2

Q_2

FIGURE 8-4 RTL NAND gate with a high output.

EXAMPLE 8-2

What logic level would appear at the output of the circuit of Fig. 8-4 if 5 V were applied to input A and 0 V to input B?

Solution: Since Q_1 and Q_2 are in series with each other, a low on the base of either prevents both transistors from conducting. The 0 V on input B results in current being forced to flow through the load resistance, thus causing a logic-1 level to appear at the output. ■

REVIEW QUESTIONS

18. What does RTL represent?
19. Two lows on the inputs of a NAND gate causes a _____ output.
20. A high on the base of an NPN transistor (opens or closes) its collector–emitter switch.
21. The voltage across a closed switch is equal to _____.
22. What is the voltage level at the collector of Q_2 in the circuit of Fig. 8-4?
23. What is the purpose of the base resistors in the circuit of Fig. 8-4?

8-3 LOGIC GATES

Digital logic includes a variety of gates. The NAND gate has been used in all of the previous examples. Each example examined the NAND gate in terms of its discrete components. Actually, these circuits are packaged in IC form. The overall circuit is then represented by the schematic symbols of Fig. 8-5.

The purpose of this chapter is to understand the concept of using a transistor as a switch. Digital logic gates are one application. We use the simpler design of RTL in reviewing these circuits.

NOT Gate   The **NOT gate** is also called an inverter because the logic level at its input is inverted (changed to its opposite state). Figure 8-6 shows a transistor connected as a NOT gate. It has a single input and output. The truth table for a NOT gate has only two conditions.

NOT Gate: Also called an inverter, it produces an output logic level that is the opposite state of the input.

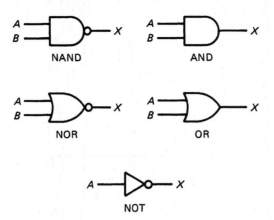

NOT

FIGURE 8-5 Digital gates.

NOT Gate Truth Table

A	X
L	H
H	L

The output is "not" the input. In other words, the output is the opposite logic level of the input. A low at the input of the circuit of Fig. 8-6 opens the collector–emitter switch and current is forced to flow through the load. A high logic level develops across the load at the output.

If a high is applied to the input of the inverter, then the collector–emitter switch closes. This shorts the load and maximum current flows through the transistor. The voltage at the output drops to the value of the closed collector–emitter switch, which is nearly 0 V. This is a low logic output.

AND Gate The NAND gate discussed earlier in this chapter was actually a combination of two logic functions: the NOT and AND functions. The NOT gate inverts a logic level and the **AND gate** performs logical multiplication. A better understanding of the AND gate can be gained by considering its truth table.

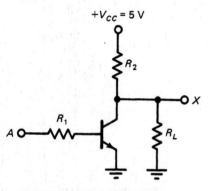

FIGURE 8-6 NOT gate.

AND Gate: A digital device in which any low input causes a low output.

AND Gate Truth Table

A	B	X
L	L	L
L	H	L
H	L	L
H	H	H

If this truth table were compared to the NAND gate truth table, it would be found to be the exact opposite. This occurs because the NAND gate is an inverted AND gate. The truth table shows us that a low ANDed with anything produces a low output. Thinking along the lines of multiplication, we know that 0 times anything is 0. This concept can be applied to logical multiplication.

Figure 8-7 shows a RTL AND gate used in a computer printer application. The printer must have paper and be selected in order to be ready for use. Switches are employed to demonstrate these conditions. A small light is used to indicate that the printer is ready.

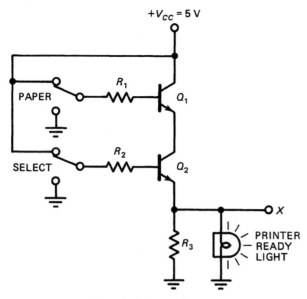

FIGURE 8-7 AND gate.

The switches have the capability of being thrown into one of two positions. They are currently positioned to provide $V_{CC} = +5$ V to the inputs of both transistors. Each transistor closes its collector–emitter switch and a complete path is provided for current to flow. It flows from ground through the bulb, R_3, and the transistors up to V_{CC}. The printer ready light is now lit.

If the printer were to run out of paper or be deselected, one or both of the switches would change its position to ground. This applies a logic low to the base of the transistor and causes the collector–emitter switch to open. When either transistor stops conducting, the light goes out, indicating that the printer is not ready.

OR Gate   The OR gate provides the function of logical addition. A logic 1 on any input results in a logic-1 output. You must be careful in drawing a parallel with everyday addition. In digital logic, there are only two states, or conditions. It is **binary** in nature. The two possible states are a logic 1 and a logic 0. This is why a logic 1 ORed with anything results in a logic 1.

Binary: A number system consisting of two symbols: 0 and 1.

OR Gate Truth Table

A	B	X
L	L	L
L	H	H
H	L	H
H	H	H

Figure 8-8 illustrates an OR gate used in the context of an automotive application. The courtesy light in a car will light if a door is opened. The doors are represented by switches and the courtesy light by an LED. The LED lights when forward biased. Therefore, a logic high is required at the output of the circuit to light the LED.

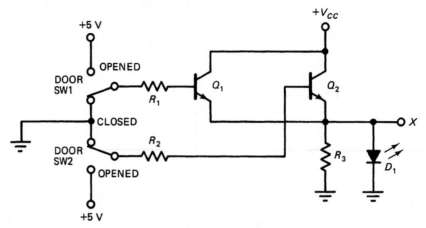

FIGURE 8-8 OR gate.

The current condition of the circuit of Fig. 8-8 shows the doors closed. A low is applied to the base of each transistor. Both transistors act as open switches and do not conduct. Current cannot flow through R_3 or D_1 and so the LED remains unlit.

If either of the doors were opened, a logic 1 would be applied to the base of the associated transistor. This transistor conducts and thereby provides a path for current through the LED. The LED lights and remains lit for as long as either or both doors are open.

NOR Gate   The NOR gate provides a combination of NOT and OR functions. The truth table is the exact opposite of that of an OR gate. A logic high on any input causes a low output.

NOR Gate Truth Table

A	B	X
L	L	H
L	H	L
H	L	L
H	H	L

The circuit of Fig. 8-9 causes a buzzer to sound if the key is in the ignition and the door is open. The current condition of the circuit shows that the key is in the ignition and the car door is open. A logic 0 is applied to the bases of both transistors. Neither transistor conducts, so the current has no alternative but to flow through the buzzer. As current flows through the buzzer, a sound is emitted.

Closing the car door or removing the key from the ignition turns on one of the transistors. When the transistor conducts, it provides a low-resistance path for current. Current is then redirected away from the buzzer and through the transistor. Withdrawing current from the buzzer silences it.

EXAMPLE 8-3 _____

(a) Which logic gate produces a high output if any input is high and (b) which logic gate produces a low output if any input is low?

Solution:

(a) The OR logic gate produces a high output if any of its inputs are high. This can be proven by its truth table or by considering the circuit of Fig. 8-8. The transistor switches in this circuit are essentially in parallel with each other.

(b) The AND logic gate produces a low output if any of its inputs are low. This can be proven by its truth table or by considering the circuit of Fig. 8-7. The transistor switches in this circuit are essentially in series with each other. ■

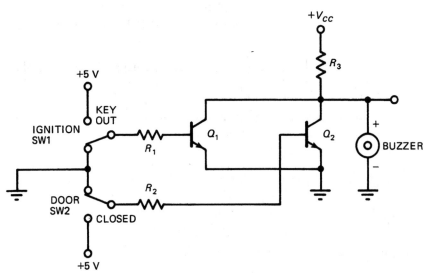

FIGURE 8-9 NOR gate.

REVIEW QUESTIONS

24. What is another name for a NOT gate?
25. A logic 1 at the input of a NOT gate causes a _____ output.
26. Does a low at the input of the circuit of Fig. 8-6 cause the transistor to conduct?
27. Which gate performs logical multiplication?
28. If the paper switch of the circuit of Fig. 8-7 were positioned at ground, would the light be lit or unlit?
29. When the printer is deselected in the circuit of Fig. 8-7, is the collector–emitter region of Q_2 saturated or cut off?
30. What is the logic level of the output when both inputs of an AND gate are low?
31. What is the logic level of the output of an OR gate that has a logic 1 on both inputs?
32. The OR gate performs the function of logical _____ .
33. The LED of Fig. 8-8 must be _____ biased to conduct.
34. Which transistor conducts if SW1 is opened in the circuit of Fig. 8-8?
35. Will the LED light if SW2 is opened in the circuit of Fig. 8-8?

36. What is the output level of a NOR gate with a low on one input and a high on the other?

37. What effect does closing SW2 have on the buzzer in the circuit of Fig. 8-9?

38. Which transistor(s) would conduct in the circuit of Fig. 8-9 if the key were removed from the ignition and the car door opened?

8-4   COMBINATIONAL LOGIC

Prime: The prime mark (') is sometimes used in place of the overbar to indicate a NOT function.

Flip-Flop: A memory device that can maintain the logic level of its two outputs.

Logic gates can be combined to create other digital devices. The $S'C'$ flip-flop is an example of a very useful device that is composed of two NAND gates. It is pronounced as either "set prime clear prime flip-flop" or "set not clear not flip-flop." The **prime** mark (') denotes the NOT function. An overbar is also used to signify the NOT function as in $\overline{S}\,\overline{C}$ flip-flop. The name **flip-flop** is descriptive of the way the outputs flip to a high logic state and then flop to a low.

Earlier, different designations for the same logic state were discussed. Set indicates a logic 1, or high, and clear indicates a logic 0, or low. The inputs are named according to their effect on the Q output. From Fig. 8-10, you notice that there are two inputs and two outputs. The inputs are S' and C'. The outputs are Q and Q'; however, Q is the primary output. Whatever logic state Q is in, Q' will be the opposite.

When an input is labeled with a prime mark or overbar, it means that the input is activated by a low. The absence of a prime mark or overbar indicates it is activated by a high. The flip-flop of Fig. 8-10 has active low inputs. If a low is applied to the S' input, it will set the Q output. On the other hand, if a low is applied to the C' input, Q will be cleared. Notice that the focus is upon the Q output. Q' is always the opposite state of Q. Consider the following truth table.

S'C' Flip-Flop Truth Table

S'	C'	Q	Q'
L	L	Illegal	
L	H	H	L
H	L	L	H
H	H	No change	

The first set of input conditions results in an illegal output. This is because an attempt is made to activate both the set and clear inputs. The Q output can be either set or cleared. It cannot be set and cleared at the same time.

The next two sets of input conditions are straightforward. A low on S' sets Q and a low on C' clears Q. The emphasis is on the input with the active logic level, which, in this case, is low.

The final set of input conditions shows a high on both inputs. Since neither input is activated, the output remains unchanged. In other words, it maintains the same logic level that existed before the two highs were applied. It is this set of input conditions that allows the flip-flop to act as a memory device. It can remember or store one bit of information at its output after the input conditions that caused it have been removed. Bit is a conjunction of *bi*nary digi*t*, which simply means a 1 or 0.

Begin by visually isolating the two NAND gates in the circuit of Fig. 8-10. Look for an active logic level at the inputs. For example, S' receives the low it needs to set the Q output. The low at the base of Q_1 opens its collector–emitter

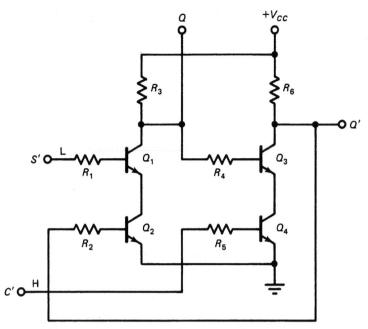

FIGURE 8-10 S'C' flip-flop from NAND gates.

switch and a high is measured at the collector with reference to ground. The high output of Q_1 is tied to the base of Q_3. This high allows Q_3 to conduct.

The C' input provides a high to the base of Q_4, which closes its collector–emitter switch. The combination of Q_3 and Q_4 conducting reduces the Q' output to a low logic level with reference to ground. This low is tied back to the base of Q_2 and opens the collector–emitter switch of this transistor. A comparison of logic states at this point shows that Q and Q' are the opposite of each other.

How No Change Occurs   At this point, it should be understood that Q is set and Q' is clear in the circuit of Fig. 8-10. Now, what would happen if the S' input were changed to a high logic level? A no-change output condition occurs according to the truth table. Examine why the outputs maintain their current logic levels.

Changing S' to a high closes the collector–emitter switch of Q_1; however, the collector–emitter switch of Q_2 is still open due to the low it has been receiving from the Q' output. Current cannot flow through Q_2 and Q_1 due to this open. The logic level measured across these two transistors results in a high at Q.

If the Q output remains the same, then the Q' output must be the same because no changes have occurred at the bases of Q_3 and Q_4. Although the input conditions were changed, the logic 1 is remembered at the Q output.

EXAMPLE 8-4 _____

Determine the output for the circuit of Fig. 8-10 when a low is applied to C' and a high is applied to S'.

Solution: The low on C' causes Q_4 to cut off. Opening the Q_4 transistor switch forces the Q' output to a high logic level. This high is applied to the base of Q_2 and causes it to conduct. Meanwhile, the high on S' causes Q_1 to conduct. Since both Q_1 and Q_2 are saturated, the Q output is low. This low is also applied to the base of Q_3 and causes it to cut off, thus reinforcing the high output of Q'. ■

REVIEW QUESTIONS

39. What does $S-C$ represent in the name of a flip-flop?
40. How are the outputs of a flip-flop labeled?
41. What logic level does set represent?
42. Which set of input conditions is considered illegal for a $S'C'$ flip-flop?
43. What does the term "bit" represent?
44. Which transistors conduct in the circuit of Fig. 8-10 when S' is high and C' is low?
45. What type of gate is used to make the flip-flop of Fig. 8-10?

Table 8-1 compares the output conditions of the various gates under the same input conditions.

TABLE 8-1 Logic Gates

INPUTS		OUTPUTS			
A	B	OR	NOR	AND	NAND
L	L	L	H	L	H
L	H	H	L	L	H
H	L	H	L	L	H
H	H	H	L	H	L

8-5 TROUBLESHOOTING

Troubleshooting digital circuits requires a working knowledge of the various gates and other logic devices. The focus has been on the use of transistors as switches. A transistor applied as a switch offers two basic conditions: a forward-bias voltage applied to the base closes the collector–emitter switch and zero volts applied to the base opens the collector–emitter switch. Opening and closing the transistor switches redirect current toward or away from the load at the output. The result is either a high or low output.

There are different ways to observe the logic levels throughout a circuit. A digital multimeter or oscilloscope can be used to measure the dc voltage levels. We mentioned earlier that any voltage below 0.8 V is a low and a voltage above 2 V is considered to be a high for bipolar circuits. Any voltage that falls between these values is considered to be **floating**. It is an indeterminate state and indicates a problem.

Floating: The indeterminate condition that exists between a logic 1 and 0.

Logic Probe: A digital measuring tool that indicates the logic level through a set of LEDs.

The **logic probe** is another measuring tool used to measure and troubleshoot digital circuits. It consists of a probe tip for taking a measurement and three LEDs that indicate high, low, and pulse. It is powered by the circuit under test by means of a pair of clip leads. Some logic probes also emit different sounds to indicate different logic conditions. This enables you to concentrate on the circuit without having to glance at a meter for a reading. If none of the LEDs light and no sound is heard, then a floating condition exists.

Digital logic probe indicates logic levels.

Locating Bad Transistor Gates   A bad transistor certainly affects the output of the circuit in which it is used. In gate applications, it may be necessary to force a particular output condition to occur in order to detect the bad transistor. This is accomplished by connecting ground or V_{CC} directly to the input while monitoring the output.

Figure 8-11 shows an AND gate connected to a NOT gate. The load is an LED that remains lit as long as Q_3 does not conduct. Only a high on inputs A and B of the AND gate can provide the necessary voltage at the base of Q_3 to make it conduct. This means that three out of the four sets of input conditions on the AND result in a lit LED.

If an open exists at the collector–emitter switch of Q_3, the LED would be lit and the circuit would appear in proper working order. Only when highs are applied to the inputs of the AND gate would the problem be visible: a lit LED when it should be off. This point is stressed to emphasize that although a piece of equipment may appear to work correctly, all possible conditions must be tested to verify it. This example takes advantage of a visual indicator. When two highs are applied to the inputs of the AND gate and the light remains lit, the problem is apparent. It is not in the power supply since the light is drawing power.

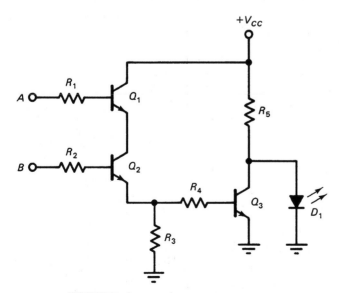

FIGURE 8-11   Troubleshooting.

The first step in troubleshooting is to measure the output of the AND gate while flipping the input switches to highs. A measurement at this point cuts the circuit in half. If the output state of the AND gate changes to a high, then the AND gate is good. A high on the base of Q_3 should cause it to conduct. A lit LED indicates that it is not conducting, so disconnect the load and measure the voltage at the collector. If it equals V_{CC}, then we know the transistor is bad.

The opposite extreme is an LED that is always unlit. In this case, begin by ensuring that there is V_{CC}. If not, check the power supply. Assuming V_{CC} is present, a quick check of the LED can be made by **piggybacking** a good LED across it. Piggybacking is a troubleshooting method in which good components are seated on the backs of suspected bad components. If this causes the system to operate again, then the problem lies in the component that was piggybacked. Piggybacking does not work in the case of shorted components. An ohmmeter check should be made to confirm a suspected fault.

If the power supply and LED provide correct voltage readings, the circuit can be split in half again with a measurement at the output of the AND gate in Fig. 8-11. Changing the input conditions from two highs to two lows should cause a change at the output. Disconnect the NOT gate to avoid any interference. If the output does not change, the problem lies in one of the AND gate transistors. Use an ohmmeter to check them individually.

A final cause for an unlit LED in the circuit of Fig. 8-11 is a shorted NOT gate. This can be tested by using a clip lead to short its base to ground. If the LED still does not light, then Q_3 is bad.

The resistors have been purposely left out of this discussion because they are easily tested with an ohmmeter. If a resistor is open, it may show a crack or burn mark. A shorted resistor may be due to a solder bridge on the printed-circuit board. A close visual inspection should reveal most of these faults.

Piggybacking: The troubleshooting method of placing a good component on top of a bad component to check for faults.

REVIEW QUESTIONS

46. A forward-bias voltage on the base of a transistor (opens or closes) its collector–emitter switch.

47. A(n) (open or closed) collector–emitter switch provides a low-resistance path for current.

48. What is the upper voltage limit for a logic low?

49. What is the lower voltage limit for a logic high?

50. What is the name given to the voltage level between a high and a low?

51. What is the name of a digital measuring tool that indicates logic levels on a set of LEDs?

52. What term describes the procedure of placing a good component on top of a bad component when troubleshooting?

53. What is the logic level at the input of Q_3 in the circuit of Fig. 8-11 if the base–emitter junction of Q_2 is open?

54. Should the LED light in the circuit of Fig. 8-11 if the base of Q_3 is purposely shorted to ground?

55. What should the forward resistance of a PN junction measure on an ohmmeter?

SUMMARY

1. Digital applications center around two logic or voltage levels.
2. Logic 1 (high, set, on, or true) represents all voltages between 2 and 5 V for bipolar circuits.
3. Logic 0 (low, clear, off, or false) represents all voltages between 0 and 0.8 V for bipolar circuits.
4. TTL stands for transistor–transistor logic.
5. A positive voltage applied to the base of an NPN BJT closes its collector–emitter switch, causing conduction.
6. Zero volts applied to the base of an NPN BJT opens its collector–emitter switch, preventing conduction.
7. Clamping diodes protect a circuit from accidental negative voltages at the inputs.
8. A phase splitter provides two output signals of opposite phase from each other.
9. RTL stands for resistor–transistor logic.
10. The output of a NAND gate is a logic 1 only when all of its inputs are logic 0.
11. The output of an AND gate is a logic 1 only when all of its inputs are a logic 1.
12. The output of a NOR gate is a logic 0 only when all of its inputs are logic 1.
13. The output of an OR gate is a logic 0 only when all of its inputs are logic 0.
14. The output logic level of an inverter gate is opposite that of its input.
15. The Q output of an $S'C'$ flip-flop is set or cleared by applying a low to either S' or C'. A high on both inputs results in no change at the Q output.
16. An NPN transistor used as a switch can be forced on or off by applying +5 V or ground to its base.

GLOSSARY

Analog: Provides a continuous range of values.

AND Gate: A digital device in which any low input causes a low output.

Binary: A number system consisting of two symbols: 0 and 1.

Digital: Provides a discrete range of values.

Flip-Flop: A memory device that can maintain the logic level of its two outputs.

Float: The indeterminate condition that exists between a logic 1 and 0.

Logic Levels: A logic 1 is indicated by a voltage greater than 2 V and a logic 0 by a voltage less than 0.8 V.

Logic Probe: A digital measuring tool that indicates the logic level through a set of LEDs.

Metal-Oxide Semiconductors: These semiconductors use less power and space than TTLs, but are not as fast at changing logic states.

NAND Gate: A digital device in which any low input causes a high output.

NOT Gate: Also called an inverter, it produces an output logic level that is the opposite state of the input.

Phase Splitter: A transistor that is tapped on both its collector and emitter to provide signals of opposite phase.

Piggybacking: The troubleshooting method of placing a good component on top of a bad component to check for faults.

Prime: The prime mark (') is sometimes used in place of the overbar to indicate a NOT function.

Resistor–Transistor Logic: RTL is an earlier form of designing digital gates that provides simplicity at the expense of power.

Totem Pole: A method of stacking transistors and diodes to conserve power.

Truth Table: A map of how a device works.

Transistor–Transistor Logic: TTL is a widely used form of creating digital devices. It provides speed and good current capability.

PROBLEMS

SECTION 8-1

8-1. List all terms that describe a logic 0 or 1.

8-2. List the advantages and disadvantages of TTL.

8-3. Draw the truth table for a NAND gate.

The following problems refer to the circuit of Fig. 8-12.

8-4. Give the name and function of each component in the TTL NAND gate.

8-5. What value of voltage is found at the base of Q_2 when both inputs are high?

8-6. What value of voltage is found at the output when both inputs are high?

8-7. Which transistors conduct when both inputs are high?

8-8. Determine the approximate voltage at the collector of Q_2 when both inputs are high.

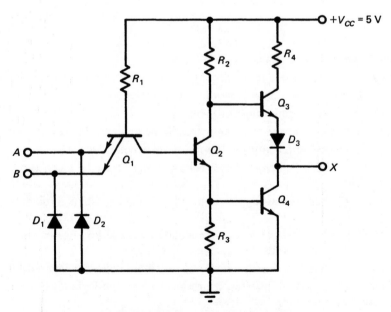

FIGURE 8-12 TTL NAND gate.

8-9. Which transistors conduct when one input is low?

8-10. What value of voltage is found at the base of Q_2 when one input is low?

8-11. Determine the voltage at the output when one input is low. Hint: $V_{R2} = 0.2$ V.

8-12. What are the voltage ranges for each logic level?

8-13. Describe the operation of a transistor in both saturation and cutoff.

SECTION 8-2

8-14. List the advantages and disadvantages of RTL.

8-15. Draw the RTL transistor configuration for each input condition of the truth table along with the logic levels and collector–emitter switch conditions.

The following problems refer to the circuit of Fig. 8-13.

8-16. Which transistor acts like an open switch as a result of the logic level at its base?

8-17. What is the voltage level at the collector of Q_2?

8-18. Outline the path of current.

8-19. What is the output logic level?

8-20. Explain the reason for the conduction or nonconduction of each transistor?

SECTION 8-3

The home security system of Fig. 8-14 is made up of three different logic gates. An alarm is represented by an LED. The LED is unlit because the door is shown closed. If the door is opened before the correct code is entered, the alarm will go on (the LED will light). The switches are currently set to an incorrect code. One of the goals is to determine the correct code before opening the door so that the LED does not light. The following problems refer to the circuit of Fig. 8-14.

8-21. Redraw each transistor gate with the appropriate truth table and digital schematic symbol next to it.

8-22. What type of gate is depicted by Gate A?

8-23. Is Q_1 conducting?

8-24. Is Q_2 conducting?

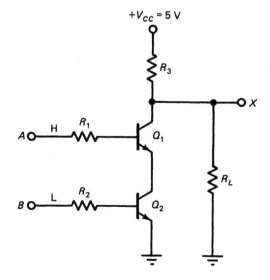

FIGURE 8-13

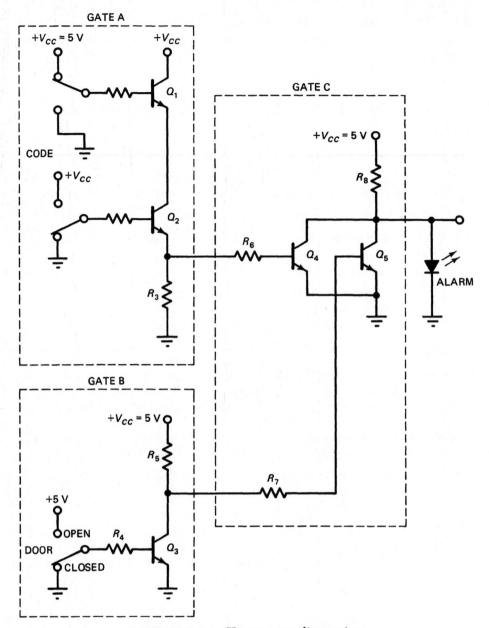

FIGURE 8-14 Home security system.

8-25. What is the output of Gate *A*?

8-26. What type of gate is depicted by Gate *B*?

8-27. Is Q_3 conducting?

8-28. What is the output of Gate *B*?

8-29. What type of gate is depicted by Gate *C*?

8-30. Is Q_4 conducting?

8-31. Is Q_5 conducting?

8-32. What is the output of Gate *C*?

8-33. Is the LED lit?

8-34. If the door is opened, what will be the logic level of the output?

8-35. Describe the code necessary to open the door without the LED going on.

SECTION 8-4

Figure 8-15 shows an *S–C* flip-flop constructed from NOR gates. The *S–C* flip-flop works in the same fashion as the *S'C'* flip-flop with the exception that its inputs are activated by a high. This means that the truth table is the exact opposite of the one studied earlier. Hint: always trace through the circuit starting at the activated input. The following problems refer to the circuit of Fig. 8-15.

8-36. Draw the truth table for the *S–C* flip-flop shown in this figure.

8-37. What is the logic level of the Q output if S is high and C is low?

8-38. Describe, step by step, how no change occurs at the output for a certain pair of input conditions.

8-39. What input conditions are necessary to cause a logic 1 to appear at the Q' output?

8-40. Which transistors conduct when S is low and C is high?

SECTION 8-5

8-41. What are three types of meters that can be used to troubleshoot transistor gate circuits?

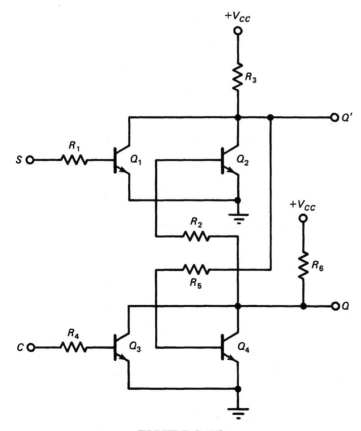

FIGURE 8-15

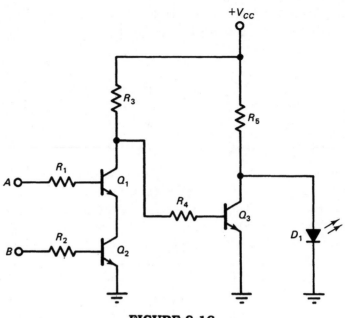

FIGURE 8-16

The following problems refer to the circuit of Fig. 8-16.

8-42. What gates are represented in this circuit?

8-43. List all input conditions that turn on the LED.

8-44. Would a logic 1 at the collector of Q_1 indicate a problem if both inputs were low?

8-45. Would a logic 1 at the collector of Q_3 indicate a problem if inputs A and B were low?

8-46. What logic level should be measured at the collector of Q_2 if input A were high and input B were low?

8-47. Should the LED light if the base of Q_3 were shorted to ground with a clip lead?

8-48. Will the LED light if the base–emitter junction of Q_2 opens?

8-49. What will be the logic level at the collector of Q_2 if its base–emitter junction opens?

8-50. Will the LED light if the collector–emitter region of Q_3 shorts?

ANSWERS TO REVIEW QUESTIONS

1. Analog	**2.** Two	**3.** Cutoff and saturation
4. 1	**5.** 1	**6.** Less than 0.8 V
7. Transistor–transistor logic	**8.** Less space and power	**9.** Truth table
10. High	**11.** D_1 and D_2	**12.** Reduces power
13. closes	**14.** decreases	**15.** low
16. Q_3	**17.** Phase splitter	**18.** Resistor–transistor logic
19. high	**20.** closes	**21.** 0 V
22. 0 V	**23.** To limit current	**24.** An inverter
25. logic-0	**26.** No	**27.** AND

28.	Unlit	29.	Cut off	30.	Low
31.	High	32.	addition	33.	forward
34.	Q_1	35.	Yes	36.	Low
37.	Silences it	38.	Q_1	39.	Set-clear
40.	Q and Q'	41.	Logic 1, or high	42.	Low, low
43.	Binary digit	44.	Q_1 and Q_2	45.	NAND
46.	closes	47.	closed	48.	0.8 V
49.	2 V	50.	A float	51.	A logic probe
52.	Piggybacking	53.	Low	54.	Yes
55.	Low resistance				

9

Field-Effect Transistors

When you have completed this chapter, you should be able to:

- Assess the differences between FETs and BJTs.
- Study the construction and operation of JFETs and MOSFETs.
- Investigate the similarities and differences of biasing methods as compared to those of BJTs.
- Analyze and design FET amplifiers.
- Apply FETs as switches.
- Develop a sound procedure for troubleshooting FET circuits.

INTRODUCTION

Field-effect transistors are often referred to by the abbreviation FET. They differ from bipolar junction transistors (BJTs) in that FETs are **unipolar** devices. FETs rely on majority current carriers only. It is this characteristic that defines FETs as unipolar devices. BJTs, on the other hand, are bipolar devices, relying on two types of current carriers to enable conduction, majority and minority current carriers.

There are two main types of FETs: JFETs and MOSFETs. Each is discussed in the following sections. Whereas BJTs rely on base current for operation, FETs are voltage-controlled devices.

Field-Effect Transistor (FET): A transistor whose field around the gate is affected by the gate voltage.

Unipolar: A device that relies only on majority carriers for current flow.

9-1   JFET CHARACTERISTICS

Construction   The construction of a **junction field-effect transistor** (JFET) is different from that of a BJT. Figure 9-1 shows a block diagram of a JFET. There are three leads: **source, gate,** and **drain.** The N-type source and drain are connected by a channel between the gate regions. The **channel** provides the path for current flow from source to drain.

The two P-type regions form the gate and are connected internally. A PN junction is formed by the gate and channel. The importance of this junction is brought out when biasing is discussed.

The source and drain regions are of equal size and shape. For this reason, they are called **symmetrical.** Being symmetrical, their leads can be interchanged with little or no difference in operation. However, treat each lead as designated by the manufacturer's spec sheet.

A schematic symbol for the N-channel JFET follows the block diagram in Fig. 9-1. The gate lead contains the arrow that points inward. This is what differentiates an N-channel from a P-channel JFET schematic symbol. Imagine the

Junction Field-Effect Transistor (JFET): A unipolar device consisting of a source, drain, and gate. Its drain current (I_D) is controlled by the gate-to-source voltage (V_{GS}).

Source: An area of a FET through which electron current normally enters the device. It is symmetrical with the drain.

Gate: An area of a FET that is insulated from the channel in MOSFETs and made of the opposite type of material from the channel for JFETs.

Drain: An area of the FET through which electron current normally exits. It is symmetrical with the source.

Channel: The area joining the source to the drain.

Symmetrical: Same in size and shape.

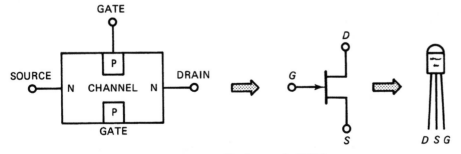

FIGURE 9-1   N-channel JFET.

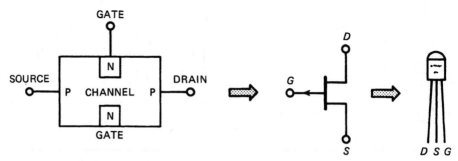

FIGURE 9-2 P-channel JFET.

center bar as being the channel and recall that the arrow points toward the N-type material; this schematic symbol can then be recognized as an N-channel device.

Figure 9-1 also shows the physical packaging of a JFET. It is similar to the packaging of a BJT. Only the part number allows you to distinguish one from another. Once again, it is important to follow the information provided with the particular FET's spec sheet.

The P-channel JFET is depicted in Fig. 9-2. It shows some subtle differences from the N-channel device. Polarities within the various regions of the block diagram have changed. The arrow on the gate of the schematic symbol is pointing outward, but the physical package remains the same except for the part number.

Operation   Setting up the JFET for proper operation requires that it be biased correctly. As was mentioned earlier, the JFET is a voltage-controlled device. The gate–source voltage, V_{GS}, determines the amount of current flow through the device.

Figure 9-3 shows the proper biasing voltages being applied to the appropriate leads. The positive supply, $+V_{DD}$, provides the current that enters the source and exits the drain. This current is referred to as drain current (I_D) and is equal to the source current.

Notice that no current is flowing into or out of the gate. The negative gate voltage, $-V_{GG}$, reverse biases the PN junction formed by the P-type gate and N-type channel. A discussion earlier in the book described the effect of reverse biasing a PN junction. An area called a depletion region forms. This area is depleted of current carriers. The extent of this region is determined by the amount of $-V_{GS}$. As the negative gate–source voltage increases, so does the depletion region. Eventually, the two regions meet and cut off the current flow through the channel. This value of voltage is called $-V_{GS(OFF)}$.

The opposite extreme is maximum current flow. Drain current reaches its maximum value when $V_{GS} = 0$ V. A V_{GS} of zero volts is the equivalent of shorting the gate and source leads together. For this reason, maximum drain current is labeled as I_{DSS} (drain current when the source is shorted to the gate).

The thought of placing a positive potential on the gate to increase drain current may have crossed your mind, but doing so creates a large gate current and damages the JFET. Caution should be used when connecting this device in a circuit. Remember that the polarity of the voltage at the gate should be negative for N-channel JFETs and positive for P-channel JFETs relative to the source.

Drain Curves   The **drain curves** in Fig. 9-4 graphically show the operation of an N-channel JFET. Observing the vertical axis, we note that drain current decreases as V_{GS} becomes more negative and increases as a V_{GS} of 0 V is

V_{GS}: Gate–source voltage is used to control I_D in FETs.

$V_{GS(OFF)}$: The value of V_{GS} that cuts off JFETs and D-type MOSFETs.

I_{DSS}: The drain current that flows when the source is shorted to the gate.

Drain Curves: A pictorial representation of the relationship between I_D, V_{GS}, and V_{DS}.

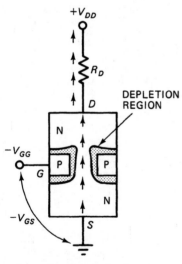

FIGURE 9-3 Biasing an N-channel JFET.

CHAP. 9 / FIELD-EFFECT TRANSISTORS

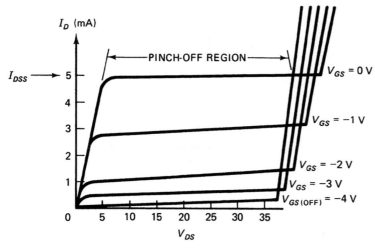

FIGURE 9-4 Drain curves for an N-channel JFET.

approached. At the point where $V_{GS} = 0$ V, maximum drain current flows; it is labeled I_{DSS}. As V_{GS} is made more negative, the value of $V_{GS(OFF)}$ is reached. For all practical purposes, no drain current flows when $V_{GS(OFF)}$ is applied from gate–source.

The drain current for each value of V_{GS} increases in a linear fashion and then levels off. The point at which the drain current levels off is called the pinch-off region. This is because the drain current is pinched off or prevented from further increase. The value of drain–source voltage at this point is called the **pinch-off voltage.** Along the horizontal axis are a series of values for the drain–source voltage (V_{DS}). The drain–source voltage must be maintained within the pinch-off region in order to maintain a constant or stable drain current. A curious fact is that the pinch-off voltage always equals the value of $V_{GS(OFF)}$ (but with opposite polarity) when $V_{GS} = 0$ V. In Fig. 9-4, the value of the pinch-off voltage is 4 V.

At the right side of the drain curves, the drain current increases dramatically. This occurs when V_{DS} is increased to the point of actually causing a breakdown within the JFET. Permanent damage results from too high a value of V_{DS}.

Gate current is purposefully absent from the drain-curve chart because the gate region is reverse biased and no gate current flows. This is similar to a reverse-biased diode. Also, no positive value of V_{GS} is shown since this would damage the device.

Pinch-Off Voltage: The value of V_{DS} right at the knee of the drain curves. A V_{DS} above the pinch-off voltage maintains a steady I_D.

Input Impedance The fact that there is no gate current brings us to the topic of input impedance. A small value of current naturally leads one to the conclusion that there is a large value of resistance or impedance. No current would indicate an infinite resistance. Although we speak of there being no gate current, there is a small reverse leakage current. It is so small, however, that it is not taken into consideration. This extremely small current indicates an extremely high input resistance (hundreds of megohms). The extremely large input impedance of JFETs is one of their chief advantages.

Transconductance **Transconductance** is defined as the ratio of the change in drain current to the corresponding change in gate–source voltage. Thus, it shows the effect that a change in V_{GS} has on I_D. It is a parameter that can also show the control that an ac input signal has on the output signal.

Transconductance: The control that V_{GS} exercises over the drain current. A g_m of 2000 μS predicts an average change of 2000 μA in drain current for every volt of gate–source voltage.

TABLE 9-1

Type No.	BV$_{gss}$ (V) Min	V$_p$ (V) @ Min	V$_p$ (V) @ Max	V$_{ds}$ (V)	I$_d$ (nA)	G$_{fs}$ @ (mS) Min	G$_{fs}$ @ (mS) Max	C$_{iss}$ (pF) Max	C$_{rss}$ (pF) Max	e$_n$ @ F (Hz)		Package
General Purpose N-Channel JFETs by BV$_{gss}$												
MPF110	20	0.5	10	10	1	0.5						TO-92
MPF111	20	0.5	10	10	1000	0.5						TO-92
2N5103	25	0.5	4	15	1	2	8	5	1	100	10	TO-72
2N5104	25	0.5	4	15	1	3.5	7.5	5	1	50	10	TO-72
2N5105	25	0.5	4	15	1	5	10	5	1			TO-72
2N5457	25	0.5	6	15	10	2	5	7	3			TO-92
2N5458	25	1	7	15	10	1.5	5.5	7	3			TO-92
2N5459	25	2	8	15	10	2	6	7	3			TO-92
J210	25	1	3	15	1	4	12	15	11.5	110	999	TO-92
J211	25	2.5	4.5	15	1	7	12	15	11.5	110	999	TO-92
J212	25	4	6	15	1	7	12	15	11.5	110	999	TO-92
MPF103	25		6	15	1	1	5	7	3			TO-92
MPF104	25		7	15	1	1.5	5.5	7	3			TO-92
MPF105	25		8	15	1	2	6	7	3			TO-92
MPF109	25	0.2	8	15	10	0.8	6	7	3	115	999	TO-92
MPF112	25	0.5	10	10	1000	1	7.5					TO-92
PN5163	25	0.4	8	15	1000	2	9	12	3	50	999	TO-92
TIS58	25	0.5	5	15	20	1.3	4	6	3			TO-92
TIS59	25	1	9	15	20	1.3		6	3			TO-92
2N3967	30	2	5	20	1	2.5		5	1.3	84	100	TO-72
2N3967A	30	2	5	20	1	2.5		5	1.3	160	10	TO-72
2N3968	30		3	20	1	2		5	1.3	84	100	TO-72
2N3968A	30		3	20	1	2		5	1.3	160	10	TO-72
2N3969	30		1.7	20	1	1.3		5	1.3	84	100	TO-72
2N3969A	30		1.7	20	1	0.3		5	1.3	160	10	TO-72
2N4220	30		4	15	0.1	1	4	6	2			TO-72
2N4220A	30		4	15	0.1	1	4	6	2	115	100	TO-72
2N4221	30		6	15	0.1	2	5	6	2			TO-72
2N4221A	30		6	15	0.1	2	5	6	2	115	100	TO-72
2N4222	30		8	15	0.1	2.5	6	6	2			TO-72
2N4222A	30		8	15	0.1	2.5	6	6	2	115	100	TO-72
2N5556	30	0.2	4	15	1	1.5	6.5	6	3	35	10	TO-72
2N5557	30	0.8	5	15	1	1.5	6.5	6	3	35	10	TO-72
2N5558	30	1.5	6	15	1	1.5	6.5	6	3	35	10	TO-72
PN4220	30		4	15	1	1	4	6	2			TO-92
PN4221	30		6	15	1	2	5	6	2			TO-92
PN4222	30		8	15	1	2.5	6	6	2			TO-92
PN4302	30		4	20	10	1		6	3	100	999	TO-92
PN4303	30		6	20	10	2		6	3	100	999	TO-92
PN4304	30		10	20	10	1		6	3	125	999	TO-92
2N3369	40		6.5	20	1000	0.6	2.5	20	3			TO-18
2N3370	40		3.2	20	1000	0.3	2.5	20	3			TO-18
2N5358	40	0.5	3	15	100	1	3	6	2	115	100	TO-72
2N5359	40	0.8	4	15	100	1.2	3.6	6	2	115	100	TO-72
2N5360	40	0.8	4	15	100	1.4	4.2	6	2	115	100	TO-72
2N5361	40	1	6	15	100	1.5	4.5	6	2	115	100	TO-72
2N5362	40	2	7	15	100	2	5.5	6	2	115	100	TO-72
2N5363	40	2.5	8	15	100	2.5	6	6	2	115	100	TO-72
2N5364	40	2.5	8	15	100	2.7	6.5	6	2	115	100	TO-72
J201	40	0.3	1.5	20	10	0.5		15	12	110	999	TO-92
J202	40	0.8	4	20	10	1		15	12	110	999	TO-92
J203	40	2	10	20	10	1.5		15	12	110	999	TO-92
2N3458	50		7.8	20	1000	2.5	10	18	5	225	20	TO-18
2N3459	50		3.4	20	1000	1.5	6	18	5	155	20	TO-18
2N3460	50		1.8	20	1000	0.8	4.5	18	5	155	20	TO-18

Transconductance is mentioned at this point so that we can create a transconductance curve. This curve offers another view of the relationship between V_{GS} and I_D. It also aids in designing the self-bias circuit that will soon be discussed. (See Table 9-1.)

Let us begin by assuming that the measured values of a JFET are $I_{DSS} = 12.8$ mA and $V_{GS(OFF)} = -6.6$ V. These values are marked as the vertical and horizontal axes, respectively, of a graph, as shown in Fig. 9-5(A). Next, the range of V_{GS} and I_D values are filled in, as shown in Fig. 9-5(B).

Now we can begin plotting the points of drain current for the different values of V_{GS}. This requires a special formula. It is called the square-law formula because a portion of the formula is squared.

$$I_D = \left(1 - \frac{V_{GS}}{V_{GS(OFF)}}\right)^2 \times I_{DSS}$$

where I_D = drain current

V_{GS} = any value of gate–source voltage

$V_{GS(OFF)}$ = cutoff voltage of the JFET

I_{DSS} = maximum value of drain current

Apply the square law to each of the values of V_{GS} in order to plot the points shown in Fig. 9-5(C). Beginning with a V_{GS} of −6 V,

$$I_D = \left(1 - \frac{-6\ \text{V}}{-6.6\ \text{V}}\right)^2 \times 12.8\ \text{mA}$$

$$= 106\ \mu\text{A}$$

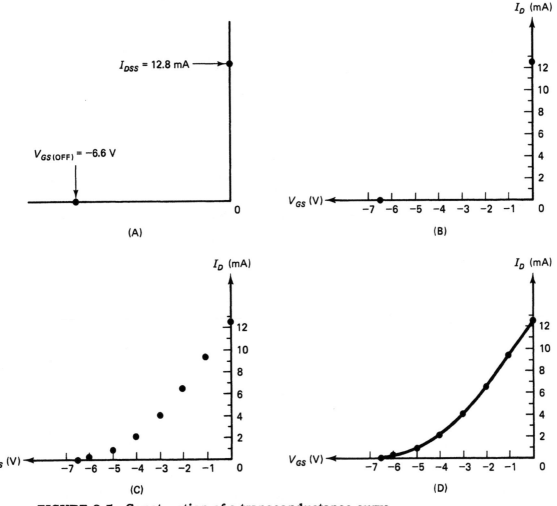

FIGURE 9-5 Construction of a transconductance curve.

The drain current in this example is about 0.10 mA, therefore, the first point is plotted directly above the −6-V mark at about 0.1 mA according to the vertical axis. Continuing with a value of −5 V for V_{GS},

$$I_D = \left(1 - \frac{-5 \text{ V}}{-6.6 \text{ V}}\right)^2 \times 12.8 \text{ mA}$$

$$= 752 \ \mu\text{A}$$

This value of current is plotted above the −5-V mark and at about the 0.75-mA level. If this process is repeated with the remaining values of V_{GS}, the drain currents should be

for $V_{GS} = -4$ V, $I_D = 1.99$ mA
for $V_{GS} = -3$ V, $I_D = 3.81$ mA
for $V_{GS} = -2$ V, $I_D = 6.22$ mA
for $V_{GS} = -1$ V, $I_D = 9.22$ mA
for $V_{GS} = 0$ V, $I_D = 12.8$ mA

The remaining values of drain current are plotted above the associated V_{GS}, as shown in Fig. 9-5(C). Once this is completed, a parabolic curve is created by connecting the dots. The result is the transconductance curve of Fig. 9-5(D).

Examine the curve just created and observe how a change in V_{GS} from −5 to −6 V causes less of a change in drain current than a change in V_{GS} from −1 to −2 V. In other words, certain portions of the curve allow for a greater influence of I_D by V_{GS} than other portions. This relationship, or ratio, between these two sets of values is expressed in terms of the device's transconductance.

The symbol for transconductance is either g_m or y_{fs} (or G_{fs} as in the spec sheet shown in Table 9-1). It is expressed in units of measurement called the siemen, which is symbolized by the letter S. The formula for transconductance is expressed as

$$g_m = \frac{\Delta I_D}{\Delta V_{GS}}$$

The triangle in the formula is the Greek symbol denoting change. By incorporating some values from Fig. 9-5(D), we can examine the transconductance of this particular JFET. Begin by assuming an operating range that includes the entire curve. The change in I_D is the difference between the maximum current and zero.

$$\Delta I_D = 12.8 \text{ mA} - 0 \text{ mA}$$

$$= 12.8 \text{ mA}$$

Likewise, the change in V_{GS} is the difference between the cutoff voltage and zero.

$$\Delta V_{GS} = -6.6 \text{ V} - 0 \text{ V}$$

$$= -6.6 \text{ V}$$

Inserting these values into the formula reveals the transconductance of the device. Polarities are irrelevant and so they are omitted.

$$g_m = \frac{12.8 \text{ mA}}{6.6 \text{ V}}$$

$$= 1939 \ \mu\text{S}$$

The transconductance of this device using the entire operating range available is 1939 microsiemens. The larger the number, the greater the ability of V_{GS} to change I_D. For instance, if the operating range is from a V_{GS} of -2 to -3 V, the value of transconductance changes. Prove this by determining the corresponding change in drain current.

$$\Delta V_{GS} = -3 \text{ V} - (-2 \text{ V})$$

$$= -1 \text{ V}$$

and

$$\Delta I_D = 6.22 \text{ mA} - 3.81 \text{ mA}$$

$$= 2.41 \text{ mA}$$

Finally,

$$g_m = \frac{2.41 \text{ mA}}{1 \text{ V}}$$

$$= 2410 \ \mu\text{S}$$

A circuit that is biased to operate in this range has a greater influence or control over I_D than the example with a transconductance of 1939 μS. The ideal is a maximum change of the drain current for a minimum change in V_{GS}.

EXAMPLE 9-1

Using Fig. 9-5(d), determine the transconductance when the operating range extends from a V_{GS} of -2 to -1 V.

Solution: The change in V_{GS} for this example is

$$\Delta V_{GS} = -2 \text{ V} - (-1 \text{ V})$$

$$= -1 \text{ V}$$

and the corresponding change in drain current is

$$\Delta I_D = 9.22 \text{ mA} - 6.22 \text{ mA}$$

$$= 3 \text{ mA}$$

The square law was used to determine the specific current values. $V_{GS(\text{OFF})} = -6.6$ V and $I_{DSS} = 12.8$ mA.

Therefore, the transconductance in this case is

$$g_m = \frac{3 \text{ mA}}{1 \text{ V}}$$

$$= 3000 \ \mu\text{S} \qquad \blacksquare$$

Practice Problem: What is the transconductance for an operating range of -4 to -2 V using Fig. 9-5(d)? *Answer:* $g_m = 2115 \ \mu\text{S}$.

1. BJTs are bipolar devices and FETs are _____ devices.
2. FETs rely on _____ carriers for conduction.
3. Unipolar means "one _____."
4. FETs are (voltage- or current-) controlled devices.
5. What are the names of the three leads?
6. If the gate is made of P-type material, then the JFET is a(n) _____-channel device.
7. An outward pointing arrow in a JFET schematic symbol indicates that it is a(n) _____-channel JFET.
8. What is the term used to describe the fact that the source and drain are the same size and shape?
9. What polarity of voltage should be applied to the gate of an N-channel JFET?
10. What label is used to denote maximum drain current?
11. What is the name of the region formed around the gate when a reverse-bias voltage is applied?
12. What label is used to denote the voltage at which a JFET ceases to conduct?
13. Drain current is equal to (gate or source) current.
14. As gate voltage is increased, drain current _____ .
15. Maximum current flow occurs when $V_{GS} =$ _____ .
16. JFETs have _____ input impedances.
17. The V_{DS} at which the drain current levels off is referred to as the _____ _____ .
18. What term is used to describe the control that a change in V_{GS} has over a corresponding change in I_D?
19. The value of g_m is expressed in _____ .
20. The smaller the value of g_m, the more control V_{GS} has over I_D. True or false?

9-2 JFET BIASING

Like the BJT, the JFET must be biased in order to achieve a particular mode of performance. The goal of biasing a device is to achieve stable operation. Unfortunately, JFETs do not provide the consistent values that BJTs offer. The base–emitter voltage of any silicon BJT is 0.7 V. On the other hand, the V_{GS} of JFETs can vary between −2 and −6 V and the I_{DSS} can range from 5 to 15 mA. This wide range of values makes JFETs difficult to bias.

Self-Bias   Before describing the action of the self-bias circuit, it is necessary to realize that V_{GS} controls drain current. Reexamine the drain curves of Fig. 9-4 to confirm this observation. As V_{GS} is increased, drain current decreases.

Self-biasing is illustrated in Fig. 9-6. The output is taken off the drain; therefore, the dc operating voltage at the drain is the focus of our attention.

If drain current attempts to increase, the voltage across R_S will also increase. The polarities of this voltage drop are indicated. An increase in drain current also increases the voltage across R_D. Yet, before a rise in drain current has a chance to

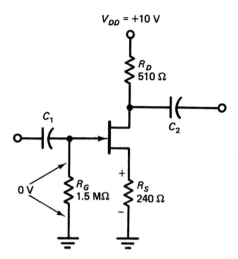

FIGURE 9-6 Self-bias.

occur, the increase in voltage across R_S expands the depletion regions within the JFET, causing the drain current to decrease. To see how this happens, first observe that R_G has 0 V dropped across it because no gate current exists. Next, follow the path from the gate through ground and back up to the source. We pass through both R_G and R_S. In other words, the voltage drops between the gate and source are

$$V_{GS} = V_{R_G} + (-V_{R_S})$$

A negative value is used for V_{R_S} because the negative polarity of R_S is tied to the gate through ground and R_G. The overall effect is that as the drain current attempts to increase, V_{R_S} increases, which consequently increases $-V_{GS}$. According to the drain-curve chart, an increase in $-V_{GS}$ causes a decrease in the drain. The original attempted increase in drain current is offset by the self-bias circuit.

Assume a drain current of 7 mA for the sake of working through an example of circuit operation. The drain current flows through both the source and drain resistors. A voltage is developed across R_S and can be found by

$$V_{R_S} = I_D \times R_S$$
$$= 7 \text{ mA} \times 240 \ \Omega$$
$$= 1.68 \text{ V}$$

According to the explanation earlier, the value of V_{R_S} is equal in value but opposite in polarity to V_{GS}.

$$V_{GS} = V_{R_G} + (-V_{R_S})$$
$$= 0 \text{ V} + (-1.68 \text{ V})$$
$$= -1.68 \text{ V}$$

The operating voltage at the output is labeled V_D. The voltage at the drain is found in the same way as the voltage at the collector for BJTs. Let us begin by finding the voltage across R_D. Subtract this value from the supply voltage and the remainder is found from drain to ground. Notice that the assumed source-current value of 7 mA is used for the drain current.

219

$$V_D = V_{DD} - V_{R_D}$$
$$= 10 \text{ V} - (7 \text{ mA} \times 510 \text{ }\Omega)$$
$$= 6.43 \text{ V}$$

EXAMPLE 9-2 _____

Determine the V_D in Fig. 9-6 if $V_{GS} = -2$ V.

Solution: We start by realizing that $V_{R_s} = V_{GS}$, but with the opposite polarity. Knowing the voltage across the source resistor allows us to find the source current and, therefore, the drain current.

$$I_D = \frac{2 \text{ V}}{240 \text{ }\Omega}$$
$$= 8.33 \text{ mA}$$

Drain current flows through the drain resistor, developing a voltage drop. V_D is the difference between V_{R_D} and V_{DD}.

$$V_D = 10 \text{ V} - (8.33 \text{ mA} \times 510 \text{ }\Omega)$$
$$= 5.75 \qquad \blacksquare$$

Designing a Self-Bias Circuit All of the values in the previous discussion were provided without explanation. This was done to focus attention on the operation of the circuit rather than on the design. To design a self-bias circuit, a transconductance curve is helpful. Refer to the curve of Fig. 9-5(d).

Start by choosing a desired drain current. A value of 4 mA is used since it falls at about the center of the curve in Fig. 9-5(d). Trace a horizontal line from the 4-mA mark across so that it meets the curve. Look down and read the V_{GS} from the point at which it meets the curve. This value is slightly less than -3 V. We estimate a V_{GS} of -2.9 V. The value does not have to be exact since JFETs vary quite a bit from one to another. The value of V_{GS} can be checked to determine whether it produces the desired value of drain current by using the square-law formula.

$$I_D = \left(1 - \frac{-2.9 \text{ V}}{-6.6 \text{ V}}\right)^2 \times 12.8 \text{ mA}$$

$$= 4.02 \text{ mA}$$

This is very close to the desired value of drain current. All that remains to do is to determine a value for the new R_S in Fig. 9-6. The value of R_S is determined by applying Ohm's law to the values that were just found. In a self-bias circuit, V_{GS} is equal to the voltage across the source resistor and drain current is equal to source current. Therefore,

$$R_S = \frac{V_{GS}}{I_D}$$

$$= \frac{-2.9 \text{ V}}{4.02 \text{ mA}}$$

$$= -721$$

The polarity of the resistor is irrelevant. The nearest value for a single resistor is either 680 or 750 Ω. Two resistors can be combined in series or in parallel to create this value, or an adjustable resistor can be used.

The voltage at the drain can now be determined by providing the proper value of R_D. If we desire a V_D of 7 V in the newly designed version of Fig. 9-6, then 3 V must be dropped across R_D. The drain current has already been established at 4.02 mA; therefore,

$$R_D = \frac{V_{R_D}}{I_D}$$

$$= \frac{3 \text{ V}}{4.02 \text{ mA}}$$

$$= 746 \ \Omega$$

Determining the value of transconductance in this circuit requires knowledge of the value of the input signal. This topic is discussed in Section 9-3, "JFET Amplifiers."

Practice Problem: What value of R_S is necessary to produce a drain current of 6 mA? Use the transconductance curve of Fig. 9-5(d) for reference. *Answer:* If V_{GS} is estimated at 2.1 V, then $I_D = 5.95$ mA and an R_S valued at 353 Ω would be needed. Your answer may vary slightly.

Voltage-Divider Bias The voltage-divider bias of Fig. 9-7 should look familiar; R_1 and R_2 make up the voltage divider. The voltage across R_2 is the same as V_G, the gate voltage. It is found by using the voltage-divider formula.

$$V_G = \frac{R_2}{R_1 + R_2} V_{DD}$$

$$= 3.72 \text{ V}$$

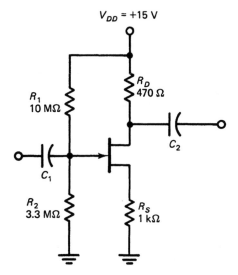

FIGURE 9-7 Voltage-divider bias.

If the measured value of V_{GS} in Fig. 9-7 were -2 V, then the voltage at the source can be calculated as

$$V_S = V_G - (-V_{GS})$$
$$= 3.72 \text{ V} - (-2 \text{ V})$$
$$= 5.72 \text{ V}$$

Notice that subtracting a negative voltage results in the minus signs canceling, producing simple addition. The value of V_{GS} can vary from JFET to JFET and this makes designing a consistent circuit difficult. A very large gate voltage is needed to swamp out any changes in the value of V_{GS}. Swamping means to overwhelm the changes in V_{GS} with such a large value of V_G that it is hardly noticeable.

The voltage at the source allows the current through the source resistor to be determined. This current is the same as the drain current.

$$I_D = \frac{V_S}{R_S}$$
$$= \frac{5.72 \text{ V}}{1 \text{ k}\Omega}$$
$$= 5.72 \text{ mA}$$

As this current exits the drain, it flows through R_D, which results in a voltage drop across R_D.

$$V_{RD} = I_D \times R_D$$
$$= 5.72 \text{ mA} \times 470 \text{ }\Omega$$
$$= 2.69 \text{ V}$$

The operating voltage at the drain is the difference between the voltage across R_D and the supply voltage.

$$V_D = V_{DD} - V_{R_D}$$
$$= 15 \text{ V} - 2.69 \text{ V}$$
$$= 12.3 \text{ V}$$

The only remaining value of any importance is the voltage from drain to source. This value is the difference between V_D and V_S.

$$V_{DS} = V_D - V_S$$
$$= 12.3 \text{ V} - 5.72 \text{ V}$$
$$= 6.58 \text{ V}$$

The procedures are almost identical to those used with the BJTs earlier. Although the voltage-divider bias used with BJTs eliminated the fluctuating value of beta, the uncertainty of V_{GS} still remains with JFETs.

EXAMPLE 9-3

Determine V_D in the circuit of Fig. 9-7 if $V_{GS} = -3$ V.

Solution: The voltage at the gate in this circuit is equal to the voltage across R_2.

$$V_G = \frac{3.3 \text{ M}\Omega}{(10 \text{ M}\Omega + 3.3 \text{ M}\Omega)} 15 \text{ V}$$

$$= 3.72 \text{ V}$$

The voltage at the source is 3 V more positive than the gate voltage due to $-V_{GS}$.

$$V_S = 3.72 \text{ V} - (-3 \text{ V})$$

$$= 6.72 \text{ V}$$

Source current, which is equal to drain current, can now be determined.

$$I_D = \frac{6.72 \text{ V}}{1 \text{ k}\Omega}$$

$$= 6.72 \text{ mA}$$

Finally, the voltage at the drain is the difference between V_{DD} and V_{R_D}.

$$V_D = 15 \text{ V} - (6.72 \text{ mA} \times 470 \text{ }\Omega)]$$

$$= 11.8 \text{ V} \qquad \blacksquare$$

REVIEW QUESTIONS

21. If drain current were to increase in the self-bias circuit of Fig. 9-6, then V_{GS} would (increase or decrease).

22. What would be the value of V_{GS} in the circuit of Fig. 9-6 if the drain current were 5 mA?

23. What would be the gate current in the circuit of Fig. 9-6 if the drain current were 5 mA?

24. What is the voltage at the drain in the circuit of Fig. 9-6 when the drain current is 5 mA?

25. What would be the value of V_S in the circuit of Fig. 9-7 if $V_{GS} = -1$ V?

26. What would be the value of I_D in the circuit of Fig. 9-7 if $V_{GS} = -1$ V?

27. What would be the value of V_D in the circuit of Fig. 9-7 if $V_{GS} = -1$ V?

28. What would be the value of V_{DS} in the circuit of Fig. 9-7 if $V_{GS} = -1$ V?

29. What is the term used to describe a situation when a large value of voltage or current overwhelms a smaller value?

9-3 JFET AMPLIFIERS

JFETs do not amplify very well. They are useful for providing a high input impedance. The method for identifying the various configurations is the same as was used for BJTs. Look for where the input and output are taken from, and the remaining lead is considered common.

Common-Source Amplifier   The common-source amplifier is illustrated in Fig. 9-8. The input is on the gate and the output is off the drain. Self-biasing is used here with capacitive coupling. A bypass capacitor, C_3, increases the voltage gain of the circuit. Even with the bypass capacitor, the gain is still very low, as you will see.

A drain current of 6 mA is designed to occur in the circuit of Fig. 9-8. The transconductance curve from Fig. 9-5(d) was used in laying out this circuit. Remember that the voltage across R_S is equivalent to the gate–source voltage, but opposite in polarity.

$$V_{GS} = I_D \times R_S$$
$$= 6 \text{ mA} \times 330 \text{ }\Omega$$
$$= -1.98 \text{ V}$$

An input voltage of 1 V p–p causes V_{GS} to vary by 0.5 V in each direction. Using this information, we can determine upper and lower limits of V_{GS} and thereby determine the change in I_D. The values for $V_{GS(OFF)}$ and I_{DSS} used in the square law are taken from the transconductance curve of Fig. 9-5(d).

When $V_{GS} = -1.98 \text{ V} + 0.5 \text{ V pk.} = -1.48 \text{ V}$, then

$$I_D = \left(1 - \frac{-1.48 \text{ V}}{-6.6 \text{ V}}\right)^2 \times 12.8 \text{ mA}$$

$$= 7.7 \text{ mA}$$

When $V_{GS} = -1.98 \text{ V} - 0.5 \text{ V pk.} = -2.48 \text{ V}$, then

$$I_D = \left(1 - \frac{-2.48 \text{ V}}{-6.6 \text{ V}}\right)^2 \times 12.8 \text{ mA}$$

$$= 4.99 \text{ mA}$$

The transconductance for this circuit can be determined by determining the change in both V_{GS} and I_D.

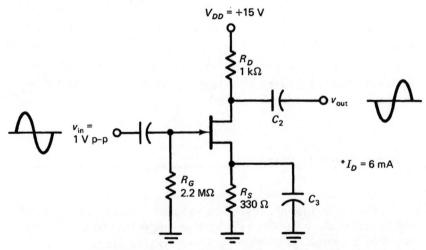

FIGURE 9-8   Common-source amplifier.

$$\Delta V_{GS} = -2.48 \text{ V} - (-1.48 \text{ V})$$
$$= -1 \text{ V}$$

and

$$\Delta I_D = 7.7 \text{ mA} - 4.99 \text{ mA}$$
$$= 2.71 \text{ mA}$$

so

$$g_m = \frac{\Delta I_D}{\Delta V_{GS}}$$
$$= \frac{2.71 \text{ mA}}{-1 \text{ V}}$$
$$= 2710 \ \mu\text{S}$$

The value of transconductance is used in the formula for finding voltage gain. From the circuit of Fig. 9-8, notice that there is a 180° phase inversion between input and output. The value of the amplified output is found by multiplying the input voltage by the voltage gain. Voltage gain for the common source is found by

$$A_v = -g_m \times R_D$$
$$= -2710 \ \mu\text{S} \times 1 \text{ k}\Omega$$
$$= -2.71$$

The minus sign indicates phase inversion. It does not indicate polarity. Using this value along with the value of the input signal, we find that

$$v_{\text{out}} = v_{\text{in}} \times A_v$$
$$= 1 \text{ V p–p} \times (-2.71)$$
$$= -2.71 \text{ V p–p}$$

Once again, the minus sign indicates phase inversion and not polarity. A greater voltage gain can be caused by experimenting with different resistor values or a FET with different drain characteristic curves.

The input impedance for a common-source circuit using self-bias is equal to the value of the gate resistor. In the circuit of Fig. 9-8, the value of R_G is 2.2 MΩ; therefore,

$$z_{\text{in}} = R_G$$
$$= 22 \text{ M}\Omega$$

EXAMPLE 9-4 _____

The following values in the circuit of Fig. 9-8 are changed: $R_S = 100 \ \Omega$, $R_D = 560 \ \Omega$, and $I_D = 9.22 \text{ mA}$. All remaining values are the same. The JFET data sheet indicates that $V_{GS(\text{OFF})} = -6.6 \text{ V}$ and $I_{DSS} = 12.8 \text{ mA}$. Determine the ac output voltage of this circuit.

Solution: The gate-to-source voltage in this circuit is equal to the voltage across the source resistor. Since drain current must flow through R_S, the voltage across this resistor, and, therefore, V_{GS}, is

$$V_{GS} = V_{R_S} = 9.22 \text{ mA} \times 100$$
$$= -922 \text{ mV}$$

The 1 V p–p input signal causes V_{GS} to vary 0.5 V in both directions. Drain current for each of these instances can be calculated by adding and subtracting this 0.5 V to and from V_{GS} and then applying the square law.

$$I_D = \left(1 - \frac{-1.42 \text{ V}}{-6.6 \text{ V}}\right)^2 \times 12.8 \text{ mA}$$
$$= 7.88 \text{ mA}$$

and

$$I_D = \left(1 - \frac{-422 \text{ mV}}{-6.6 \text{ V}}\right)^2 \times 12.8 \text{ mA}$$
$$= 11.2 \text{ mA}$$

Next, the transconductance is found by calculating the change in both V_{GS} and I_D.

$$\Delta V_{GS} = -1.42 - (-422 \text{ mV})$$
$$= 1 \text{ V}$$

and

$$\Delta I_D = 11.2 \text{ mA} - 7.88 \text{ mA}$$
$$= 3.32 \text{ mA}$$

so

$$g_m = \frac{3.32 \text{ mA}}{1 \text{ V}}$$
$$= 3320 \text{ } \mu\text{S}$$

Gain for a common-source amplifier is determined by

$$A_v = -3320 \text{ } \mu\text{S} \times 560 \text{ } \Omega$$
$$= -1.86$$

Therefore,

$$v_{\text{out}} = -1.86 \times 1 \text{ V p–p}$$
$$= -1.86 \text{ V p–p} \qquad \blacksquare$$

Practice Problem: What would be the new transconductance, voltage gain, and v_{out} for the circuit of Fig. 9-8 if the initial value of V_{GS} were -3 V? Use the values of $V_{GS(OFF)}$ and I_{DSS} from Fig. 9-5(d). *Answer:* $g_m = 2120 \text{ } \mu\text{S}$, $A_V = -2.12$, and $v_{\text{out}} = -2.12 \text{ V p–p}$.

Common-Drain Amplifier The common-drain amplifier of Fig. 9-9 is reminiscent of the common-collector amplifier using BJTs. The output is in phase with the input and the voltage gain is almost unity. A voltage-divider bias is used to provide a different approach to input impedance. Finally, the gate–source voltage is noted in the figure for convenience.

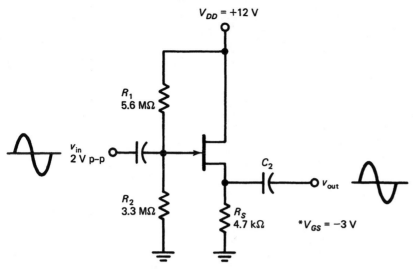

FIGURE 9-9 Common-drain amplifier.

Transconductance for this circuit is found the same way as in the previous circuit. The input signal causes the gate–source voltage to vary. An initial value of -3 V is given, so the upper and lower limits of V_{GS} can be found by adding and subtracting, respectively, the peak voltage of the input signal. The square law is then applied to these values. Once again, the values of $V_{GS(OFF)}$ and I_{DSS} are taken from the transconductance curve of Fig. 9-5(d).

When $V_{GS} = -3$ V $+ 1$ V pk. $= -2$ V, then

$$I_D = \left(1 - \frac{-2 \text{ V}}{-6.6 \text{ V}}\right)^2 \times 12.8 \text{ mA}$$

$$= 6.22 \text{ mA}$$

When $V_{GS} = -3$ V $- 1$ V pk. $= -4$ V, then

$$I_D = \left(1 - \frac{-4 \text{ V}}{-6.6 \text{ V}}\right)^2 \times 12.8 \text{ mA}$$

$$= 1.99 \text{ mA}$$

The changes in V_{GS} and I_D can be calculated as

$$\Delta V_{GS} = -4 \text{ V} - (-2 \text{ V}) = -2 \text{ V}$$

$$\Delta I_D = 6.22 \text{ mA} - 1.99 \text{ mA} = 4.23 \text{ mA}$$

and transconductance is equal to

$$g_m = \frac{4.23 \text{ mA}}{-2 \text{ V}}$$

$$= 2115 \text{ } \mu\text{S}$$

The method for finding the voltage gain is different. There is no drain resistor present, so attention is focused on R_S. It was mentioned earlier that the voltage gain for the common drain is almost unity. The formula for finding it is

$$A_v = \frac{R_S}{R_S + (1/g_m)}$$

$$= \frac{4.7 \text{ k}\Omega}{4.7 \text{ k}\Omega + (1/2115 \text{ }\mu S)}$$

$$= 0.909$$

The voltage gain for the circuit of Fig. 9-9 is slightly less than 1; therefore, the output voltage is slightly less than the input voltage. The input voltage multiplied by the voltage gain gives us the output voltage.

$$v_{\text{out}} = v_{\text{in}} \times A_V$$

$$= 2 \text{ V p–p} \times 0.909$$

$$= 1.82 \text{ V p–p}$$

The input impedance offered by the common drain is very high. In the circuit of Fig. 9-9, it is determined by R_1 in parallel with R_2.

$$z_{\text{in}} = R_1 \parallel R_2$$

$$= 5.6 \text{ M}\Omega \parallel 3.3 \text{ M}\Omega$$

$$= 2.08 \text{ M}\Omega$$

Practice Problem: What would be the new transconductance, voltage gain, and v_{out} for the circuit of Fig. 9-9 if the initial value of V_{GS} is -5 V? Use the values of $V_{GS(\text{OFF})}$ and I_{DSS} from Fig. 9-5(D). *Answer:* $g_m = 940 \text{ }\mu S$, $A_V = 0.815$, and $v_{\text{out}} = 1.63$ V p–p.

Common-Gate Amplifier
The common-gate amplifier of Fig. 9-10 is the most seldom used of the three amplifiers. One major reason is that it does not provide the high input impedance of the other amplifiers.

The transconductance of the common gate is found the same way as in the previous examples. The input signal continues to exercise its control over V_{GS}. We assume a g_m of 2000 μS.

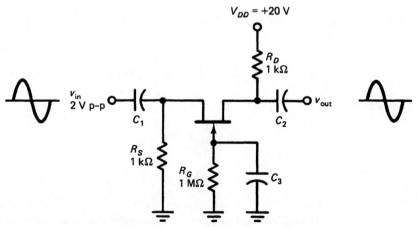

FIGURE 9-10 Common-gate amplifier.

The voltage gain for the common gate is similar to that of the common source. The only difference is the absence of the minus sign indicating no phase inversion.

$$A_v = g_m \times R_D$$
$$= 2000 \ \mu S \times 1 \ k\Omega$$
$$= 2$$

Output voltage in this case is

$$v_{out} = v_{in} \times A_v$$
$$= 2 \ V \ p\text{--}p \times 2$$
$$= 4 \ V \ p\text{--}p$$

Now we come to the disadvantage of the common gate, the input impedance. The input current is equal to the drain current in this circuit. As a result, the input impedance is greatly reduced.

$$z_{in} = \frac{1}{g_m}$$
$$= \frac{1}{2000 \ \mu S}$$
$$= 500 \ \Omega$$

REVIEW QUESTIONS

30. What biasing method is used in the common-source amplifier of Fig. 9-8?
31. The common-source amplifier offers a very low input impedance. True or false?
32. The common-source amplifier provides a 180° phase shift between input and output. True or false?
33. How is the input impedance found for self-bias circuits?
34. How is the input impedance found for voltage-divider bias circuits?
35. The common-drain amplifier offers a very high input impedance. True or false?
36. The common-drain amplifier provides a 180° phase shift between input and output. True or false?
37. What is the typical voltage gain of a common-drain amplifier?
38. What is the major disadvantage of the common-gate amplifier?
39. Does the common-gate amplifier provide a phase shift?
40. The common-gate amplifier provides a voltage gain of less than unity. True or false?

Metal-Oxide Semiconductor Field-Effect Transistor (MOSFET): A unipolar device whose gate is insulated from the channel by a layer of metal-oxide. Gate-source voltage (V_{GS}) is used to control the drain current (I_D).

D-Type MOSFET: A normally on device that can be controlled by a voltage of either polarity at its gate.

Substrate: A region of MOSFETs used to control the width of the channel.

Metal-Oxide Insulator: A thin layer of silicon dioxide that insulates the gate from the channel in MOSFETs. Other insulating materials can be used.

Depletion Mode: A condition where the channel is depleted of current carriers, causing a decrease in drain current.

Enhancement Mode: A condition where current carriers are increased, causing an increase in drain current.

The acronym **MOSFET** represents metal-oxide semiconductor field-effect transistor. MOSFETs are sometimes referred to as IGFETs because they have insulated gates. Two types of MOSFETs are discussed in the remainder of this chapter, beginning with the depletion-type, or D-type, MOSFET.

Figure 9-11 shows the block diagram of a **D-type MOSFET.** The source, drain, and gate are familiar from the discussion of JFETs. An additional region is found, the **substrate (SS),** which is used to reduce the width of the channel under certain conditions. The gate is insulated from the channel by a layer of metal oxide. Silicon dioxide (glass) is often used as the insulator.

The schematic symbol is also presented in Fig. 9-11. The arrow pointing inward indicates that the device is an N-channel MOSFET. The solid channel line is what distinguishes a D-type MOSFET from an E-type MOSFET. Also, most MOSFETs have their source and substrate internally connected. In this case, there are only three external leads. The final view of Fig. 9-11 shows one type of packaging used for MOSFETs. A dashed line indicates that a substrate lead may be found in certain cases. Consult the manufacturer's spec sheet for the identification of the leads.

Modes of Operation   Unlike the JFET, the D-type MOSFET can be biased with either a positive or negative gate voltage. The polarity of the gate voltage determines the mode of operation. Be careful not to confuse mode and type. The name of this device is D-type, yet it can be operated in either **depletion** or **enhancement mode.**

The drain curves of Fig. 9-12 will be helpful in explaining the different modes of operation. Using a V_{GS} of 0 V as the starting point, notice how a negative voltage on the gate decreases the drain current. This occurs because a negative voltage on the gate repels the electrons from the channel. The channel is depleted of current carriers and positive ions begin to form. Consequently, the channel begins to narrow and drain current is reduced. At $V_{GS(OFF)}$, the channel is blocked and the flow of drain current is cut off. All voltages below a V_{GS} of 0 V in Fig. 9-12 cause the depletion mode of operation to occur.

A positive voltage applied to the gate causes drain current to increase according to Fig. 9-12. The positive voltage has the opposite effect by drawing electrons into the channel. This relates to the first law of electrostatics, which states that like charges repel each other and unlike charges attract. The greater the positive voltage, the greater the attraction of electrons into and through the chan-

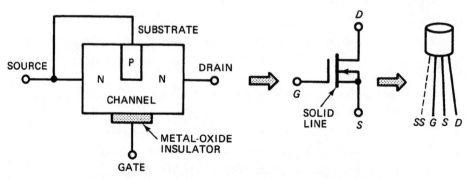

FIGURE 9-11 N-channel depletion-type MOSFET.

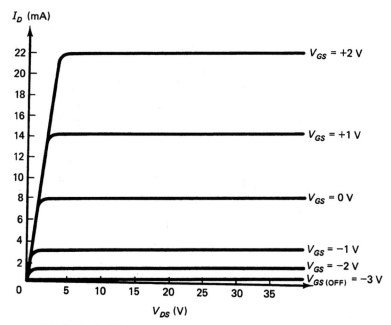

FIGURE 9-12   Drain curves for a D-type MOFSET.

nel. Since current is enhanced by voltages above a V_{GS} of 0 V in Fig. 9-12, it is referred to as enhancement mode.

The D-type MOSFET is a square-law device. Drain current can be calculated the same way that it was for JFETs. I_{DSS} is still the value of current when the source is shorted to the gate, in other words, when $V_{GS} = 0$ V. Apply the square law to the drain-curve chart of Fig. 9-12. Suppose you wished to know the drain current when $V_{GS} = -2$ V.

$$I_D = \left(1 - \frac{V_{GS}}{V_{GS(OFF)}}\right)^2 \times I_{DSS}$$

$$= \left(1 - \frac{-2\ \text{V}}{-3\ \text{V}}\right)^2 \times 8\ \text{mA}$$

$$= 889\ \mu\text{A}$$

And what if $V_{GS} = +2$ V?

$$I_D = \left(1 - \frac{+2\ \text{V}}{-3\ \text{V}}\right)^2 \times 8\ \text{mA}$$

$$= 22\ \text{mA}$$

The final point about MOSFETs is actually a warning. BE CAREFUL OF STATIC DISCHARGE! A MOSFET's insulation can be easily damaged by static electricity. When purchased, the leads are usually shorted with a piece of wire, metal foil, or conductive plastic. The wire, metal foil, or conductive plastic is removed just before inserting the component into the circuit. Always handle the MOSFET by its case.

JFET and E-Type MOSFET
(foam protects against
static).

EXAMPLE 9-5 _____

What is the value of drain current when $V_{GS} = 0$ V in Fig. 9-12?

Solution: We can arrive at a value of 8 mA by simply looking at the drain curves. We can also prove that drain current is 8 mA by applying the square law.

$$I_D = \left(1 - \frac{0 \text{ V}}{-3 \text{ V}}\right)^2 \times 8 \text{ mA}$$

$$= 8 \text{ mA} \qquad \blacksquare$$

REVIEW QUESTIONS _____

41. What does the acronym MOS represent?
42. Which lead of a MOSFET is sometimes connected internally to the source?
43. Does maximum current flow through a D-type MOSFET when $V_{GS} = 0$ V?
44. When a negative voltage is applied to the gate of an N-channel D-type MOSFET, it is operating in the _____ mode.
45. Use the square law to determine the drain current in Fig. 9-12 when $V_{GS} = +1$ V.
46. Use the square law to determine the drain current in Fig. 9-12, when $V_{GS} = -1$ V.
47. Electrons are drawn into the channel of a D-type MOSFET when operated in the _____ mode.
48. MOSFETs can be easily damaged by _____ _____ .

9-5 E-TYPE MOSFETs _____

E-Type MOSFET: A normally off device that can be made to conduct by increasing the voltage at the gate with the appropriate polarity.

The **enhancement-type MOSFET** differs both in construction and operation from the D-type. Figure 9-13 shows the substrate extending all the way across to the gate. The channel is nonexistent when $V_{GS} = 0$ V. The gate is still insulated by a layer of metal oxide.

The schematic symbol shows the source and substrate connected internally. The broken channel line is what distinguishes the E-type from the D-type

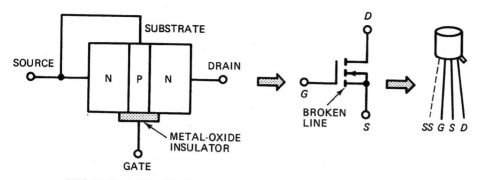

FIGURE 9-13 N-channel enhancement-type MOSFET.

MOSFET. An inward-pointing arrow continues to inform us that it is an N-channel device.

Mode of Operation Notice that the word "mode" is singular in the subtitle. Since the channel is initially cut off, there is no need to deplete it any further. The one mode of operation for an E-type MOSFET is the enhancement mode. Figure 9-14 shows the drain curves for an E-type MOSFET. The gate–source voltages have only one polarity. As the voltage at the gate becomes more positive with respect to the source, drain current increases.

When a positive voltage is applied to the gate, it attracts electrons, which, in turn, begin to fill the holes of the P-type substrate. As the holes along the insulation area are filled, it creates an **"inversion layer."** At a certain point, called the **threshold voltage** ($V_{GS(TH)}$), enough holes are filled so as to allow the remaining electrons to flow along the inversion layer as current carriers. This is the start of drain current and begins at the value of +2 V for the device of Fig. 9-14. The threshold voltage normally has a value of 1 to 5 V.

Specific values of drain current can be calculated using a modification of the square law. It requires knowing the constant (k) of the device. Finding the value

Inversion Layer: An area along the gate of an E-type MOSFET that allows drain current to flow. It is created by voltage on the gate of the same polarity as the substrate.
Threshold Voltage: The minimum voltage required for conduction in an E-type MOSFET.

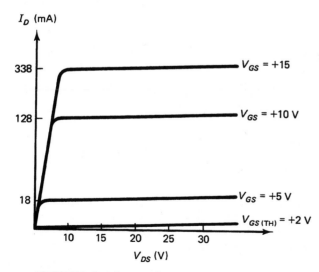

FIGURE 9-14 Drain curves for an E-type MOSFET.

of this constant can be done through measurement or by making reference to two corresponding values on the drain-curve chart. The values $V_{GS} = +5$ V, $I_D = 18$ mA, and $V_{GS(TH)} = +2$ V come from the chart.

$$k = \frac{I_D}{(V_{GS} - V_{GS(TH)})^2}$$

$$= \frac{18 \text{ mA}}{(5 \text{ V} - 2 \text{ V})^2}$$

$$= 0.002$$

By using this constant, the value of drain current for any value of V_{GS} can be calculated. The formula is

$$I_D = k \times (V_{GS} - V_{GS(TH)})^2$$

The I_D for a V_{GS} of 7 V is

$$I_D = 0.002 \times (7 \text{ V} - 2 \text{ V})^2$$

$$= 50 \text{ mA}$$

The E-type MOSFET is as susceptible to static discharge as the D-type. Caution should be used when handling these devices.

EXAMPLE 9-6

Determine the value of drain current for an E-type MOSFET in which $k = 0.003$, $V_{GS(TH)} = 2$ V, and is operating with a V_{GS} of 10 V.

Solution: Using the modified square-law formula, we find that

$$I_D = 0.003 \times (10 \text{ V} - 2 \text{ V})^2$$

$$= 192 \text{ mA} \qquad \blacksquare$$

REVIEW QUESTIONS

49. The "E" in E-type stands for _____.
50. A negative voltage on the gate of a N-channel E-type MOSFET brings about the depletion mode. True or false?
51. What is the name of the area in the substrate that is formed along the wall of insulation and allows current to flow?
52. What is the name for the minimum value of V_{GS} that allows drain current to flow?
53. What is the value of drain current for a V_{GS} of +12 V in Fig. 9-14?

9-6 MOSFET AMPLIFIERS

MOSFET amplifiers offer the advantage of an extremely high input impedance. This, in turn, allows for direct coupling. When no coupling capacitors are used, a flatter frequency response results. A flat frequency response means that the amplifier responds well to low frequencies as well as high frequencies. There is no loss of signal due to capacitive reactance.

Zero Bias   Though different biasing methods are available, the zero bias is especially convenient for D-type MOSFETs. If the drain curves for a D-type MOSFET (Fig. 9-12) are examined, we will see that a V_{GS} of 0 V allows for a reasonable signal swing. Figure 9-15 shows an example of zero bias being used in a common-source amplifier.

The source is tied directly to ground and the gate is tied to ground through the gate resistor. No current flows through R_G; therefore, the gate is held to a 0-V potential. The potential difference between the gate–source junction is 0 V and that is reason for its name, zero bias.

The shorted gate–source condition of this circuit allows a slightly different approach to be used in finding transconductance. The label g_{m0} stands for transconductance when the gate–source voltage is zero. Use the values from Fig. 9-12 for I_{DSS} and $V_{GS(OFF)}$ in the following equation.

$$g_{m0} = \frac{(-2 \times I_{DSS})}{V_{GS(OFF)}}$$

$$= \frac{(-2 \times 8 \text{ mA})}{-3 \text{ V}}$$

$$= 5333 \ \mu S$$

Common-source amplifiers provide a phase inversion as indicated in Fig. 9-15. The voltage gain for MOSFETs is found the same way as it was for JFETs.

$$A_v = -g_m \times R_D$$

$$= -5333 \ \mu S \times 2 \text{ k}\Omega$$

$$= -10.7$$

The output voltage for the circuit of Fig. 9-15 is

$$v_{out} = v_{in} \times A_v$$

$$= 100 \text{ mV p–p} \times (-10.7)$$

$$= -1.07 \text{ V p–p}$$

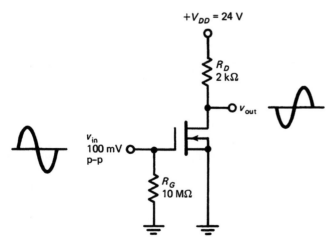

FIGURE 9-15   Common-source amplifier using zero bias.

The input impedance for the circuit of Fig. 9-15 is equal to the gate resistor.

$$z_{in} = R_G$$

$$= 10 \text{ M}\Omega$$

Drain-Feedback Bias Figure 9-16 shows another common-source amplifier except this time an E-type MOSFET is used. The drain feedback offers a different view of biasing. Examine the circuit closely and notice that $V_{GS} = V_{DS}$. This is because the source is tied to ground and the drain is tied back to the gate through R_G.

If a gate–source voltage of 5 V is desired, then the corresponding drain current must be found on the drain curves. The values of Fig. 9-14 are used in the following equation.

$$I_D = k \times (V_{GS} - V_{GS(TH)})^2$$

$$= 0.002 \times (5 \text{ V} - 2 \text{ V})^2$$

$$= 18 \text{ mA}$$

Since a gate–source voltage of 5 V has been chosen, V_{DS} must also equal 5 V. The balance of V_{DD} in Fig. 9-16 is dropped across R_D. The value of R_D in Fig. 9-16 was chosen to meet these conditions by Ohm's law.

$$R_D = \frac{V_{DD} - V_{DS}}{I_D}$$

$$= \frac{32 \text{ V} - 5 \text{ V}}{18 \text{ mA}}$$

$$= 1.5 \text{ k}\Omega$$

The transconductance of this circuit can be found by

$$g_m = \frac{2 \times k}{V_{GS} - V_{GS(TH)}}$$

$$= \frac{2 \times 0.002}{5 \text{ V} - 2 \text{ V}}$$

$$= 1333 \text{ }\mu\text{S}$$

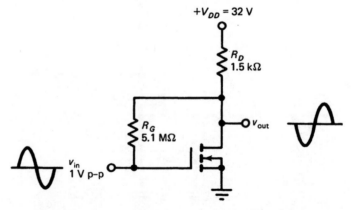

FIGURE 9-16 Common-source amplifier using drain-feedback bias.

The value of transconductance allows the voltage gain of this circuit to be solved. All common-source circuits use the same method of solving for A_v.

$$A_v = -g_m \times R_D$$
$$= -1333 \ \mu S \times 1.5 \ k\Omega$$
$$= -2$$

Therefore,

$$v_{out} = v_{in} \times A_v$$
$$= 1 \ V \ p\text{–}p \times (-2)$$
$$= -2 \ V \ p\text{–}p$$

Keep in mind that a dc supply looks like a short to ac. If V_{DD} of Fig. 9-16 is imagined as being at ground potential, then the input impedance would be equal to the sum of R_G and R_D. Because R_G is so much larger than R_D, the drain resistor is ignored.

$$z_{in} = R_G$$
$$= 5.1 \ M\Omega$$

EXAMPLE 9-7

The circuit of Fig. 9-16 is modified so that $R_G = 10 \ M\Omega$, $R_D = 15 \ k\Omega$, $V_{DD} = 32 \ V$, and $v_{in} = 10 \ mV \ p\text{–}p$. The specifications for the E-type MOSFET are $V_{GS(TH)} = +2 \ V$, $k = 0.002$, and it is operating with a V_{GS} of $+3 \ V$. Determine the ac output voltage.

Solution: The first step in determining the ac output voltage is to determine the transconductance of the MOSFET.

$$g_m = \frac{2 \times 0.002}{3 \ V - 2 \ V}$$
$$= 4000 \ \mu S$$

and

$$A_v = -4000 \ \mu S \times 15 \ k\Omega$$
$$= -60$$

Therefore,

$$v_{out} = -60 \times 10 \ mV \ p\text{–}p$$
$$= -600 \ mV \ p\text{–}p \qquad \blacksquare$$

REVIEW QUESTIONS

54. What is the value of V_{GS} when zero bias is used?
55. What is the value of the drain current when zero bias is used?
56. What is the value of V_{DS} in the circuit of Fig. 9-15 using the drain curves of Fig. 9-12?
57. The MOSFET of Fig. 9-16 is biased to operate in the _____ mode only.

58. $V_{GS} = V_{DS}$ in circuits using drain-feedback bias. True or false?

59. What size R_D would be needed for a V_{GS} of 10 V in the circuit of Fig. 9-16 if the E-type MOSFET had the characteristics of Fig. 9-14?

60. What type of amplifier provides phase inversion?

9-7 FETs AS SWITCHES

The use of BJTs as switches was discussed in the previous chapter. JFETs and MOSFETs can also be used in switch-type applications.

JFET Switch   A switch has two states: off and on. The JFET can be made to operate like a switch by controlling the gate voltage. A JFET is considered to be a normally on device because when zero volts are applied to the gate, it conducts.

Figure 9-17 shows a simple example of how a JFET can be used to switch an LED on or off. A switch at the gate is used to apply either −5 V or ground potential. It is currently in the −5-V position; therefore, the JFET is cut off and acts like an open switch. Since the drain–source path for current is cut off, current is forced to flow through the LED, which causes it to light.

If SW1 were opened, the gate would return to a 0-V potential. The channel would no longer be cut off and current would flow through the JFET. This path of lower resistance through the JFET would, in effect, short out the LED. With 0 V applied to the gate, the LED would be unlit.

MOSFET Switch   In this example, an E-type MOSFET is used in a switching application. The E-type MOSFET is considered a normally off device because it does not conduct current when $V_{GS} = 0$ V. It is in series with the LED of Fig. 9-18. A succession of pulses are shown at the gate of the device.

Remember that when zero volts are applied to the gate, the E-type MOSFET does not conduct. When it receives a positive pulse, an inversion layer is created and current begins to flow, as depicted in Fig. 9-18. The LED lights whenever a positive pulse is applied to the gate.

A frequency of 8 Hz was chosen to make the blinking of the LED visible to the human eye. Generally, a frequency of about 10 Hz or less allows for easy viewing of the blinking LED.

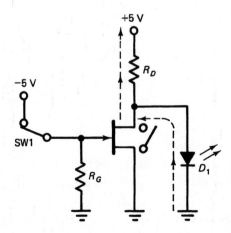

FIGURE 9-17   JFET switch.

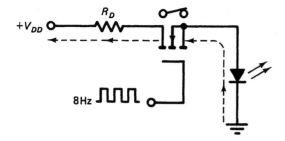

FIGURE 9-18 MOSFET switch.

61. The JFET is considered a normally (on or off) device.

62. A P-channel JFET requires a (positive or negative) voltage at its base to cause it to cut off.

63. Maximum drain current flows through a JFET when V_{GS} = _____ .

64. The LED in Fig. 9-17 lights when V_{GS} = _____ .

65. E-type MOSFETs are normally (on or off) devices.

66. D-type MOSFETs are normally (on or off) devices.

67. The LED in Fig. 9-18 blinks faster as frequency is (increased or decreased).

68. A P-channel E-type MOSFET conducts whenever the pulse at the gate is at a 0-V potential. True or false?

9-8 TROUBLESHOOTING

Troubleshooting FETs in both switch and amplifier circuits is discussed here. Opens and shorts due to solder bridges, bad resistors, and damaged traces are not mentioned in the following explanations. A close visual inspection should reveal most of these. Our discussion is limited to the topic of FETs.

FETs in Switch Applications   The applications of FETs as switches in the previous section offer a visual symptom if a problem exists. An LED that always remains lit in the circuit of Fig. 9-17 indicates a problem with the JFET. If SW1 is opened and zero volts are measured at the gate, then the JFET should be conducting and the LED should be unlit. The JFET in this condition should be removed and tested with an ohmmeter, as described later in this section.

If the LED in Fig. 9-17 is always unlit, then the two chief possibilities are the JFET and the LED itself. Piggybacking a good LED across the suspected bad one may reveal that the problem lies in the LED. Assuming the LED is good, the JFET's gate voltage should be checked for −5 V. If the proper gate voltage exists, then the depletion regions are not being properly formed around the gate area and the device is remaining on. Remove the JFET and perform an ohmmeter test.

The MOSFET in Fig. 9-18 is checked in much the same way. First, check for the proper V_{DD} and pulsing signal. If both are present, then place an oscilloscope across the drain–source region; it should be pulsing. If not, remove the MOSFET and change it for a good one. A pulsing signal across the drain–source region and an unlit LED indicate that the LED is bad.

FETs in Amplifiers   The preamplifier of Fig. 9-19 is used in our discussion. Notice that a microphone is used to produce an input signal of approximately

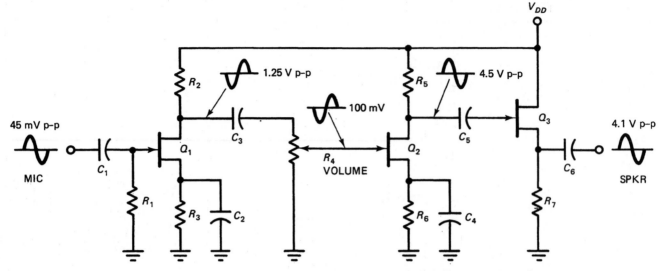

FIGURE 9-19 Microphone preamplifier.

45 mV p–p. The first stage amplifies and inverts this signal into a 1.25-V p–p signal at the drain of Q_1.

The potentiometer (R_4) is used to reduce the input signal voltage of the second stage to a desirable level. It acts as a volume control in this circuit. A 100-mV p–p signal is indicated at the wiper and/or gate of Q_2. This signal is, in turn, amplified to a 4.5-V p–p signal by the second stage.

The final stage consists of a common-drain amplifier. The common-drain amplifiers does not provide a voltage gain and so the final output is approximately 4.1 V p–p.

Once again, this circuit provides an audible symptom. There may be no sound, too much sound, or a distorted sound. In any of these cases, a frequency generator can be used to supply a test signal of 10 kHz to the input of the circuit. An oscilloscope is then used to trace the waveforms throughout the circuit.

No output on the first stage indicates that Q_1 is bad. A greatly reduced signal may mean that C_2, the bypass capacitor, is open, lowering the voltage gain. A clipped signal can mean that C_2 is shorted. An ohmmeter is used to check the capacitor or JFET for proper resistance measurements.

The signal at the gate of Q_2 should be monitored while varying the potentiometer. If voltage does not vary, check the potentiometer for a possible open or short. Varying the potentiometer should also cause the signal at the drain of Q_2 to vary. If it does not, check the gate–source resistance of the JFET. Problems with amplitude and clipping lead you to C_4, the bypass capacitor of the second stage.

The final stage should show an output signal that can also be varied by adjusting the potentiometer. If the output does not vary, then remove and check Q_3 with an ohmmeter.

A loss of the signal between any of the stages indicates a problem with the associated coupling capacitor. C_1, C_3, C_5, and C_6 are coupling capacitors. If the signal seems all right at the output, then the speaker should be inspected. Change the speaker to confirm or reject this possibility.

This whole procedure takes less than five minutes to perform when you become accustomed to the process. Remember that the common lead of the oscilloscope is always connected to ground and you are simply moving the vertical input of the scope from point to point throughout the circuit.

Testing FETs with an Ohmmeter JFETs should measure a low drain–source resistance. Both regions are made of the same type of material, so it does not matter which way the leads are connected. The forward resistance of both the gate–source and gate–drain junctions should register some resistance. The reverse resistance for each of these should measure infinity.

D-type MOSFETs should also measure a low drain–source resistance in either direction. The remaining combination of connections between the leads should measure infinity.

E-type MOSFETs are a bit more difficult to check with an ohmmeter because they are normally off devices. This means that all measurements measure infinity. A zero-ohm reading between any leads indicates a short. The surest method of troubleshooting this device is to perform the in-circuit tests discussed earlier.

REVIEW QUESTIONS

69. The measured drain–source resistance of a JFET should be (low or infinite).

70. A (positive, negative, or ground) potential between the gate–source leads of an N-channel JFET cause it to cut off.

71. The reverse gate–source resistance of a JFET should be (low or infinite).

72. A common-drain amplifier usually provides a voltage gain. True or false?

73. A (positive, negative, or ground) potential between the gate–source leads of an N-channel E-type MOSFET causes it to conduct.

74. Which type of FET is considered a normally off device?

75. A (bypass or coupling) capacitor is responsible for increasing the voltage gain of an amplifier.

SUMMARY

1. FET stands for field-effect transistor.

2. FETs are unipolar devices; they rely on majority current carriers only.

3. Junction FETs (JFETs) consist of a source, drain, and gate.

4. The channel connects the source and drain. There are N-channel and P-channel JFETs.

5. JFETs are controlled through gate voltage.

6. A reverse voltage on the gate reduces current flow through the channel of a JFET and zero volts on the gate allow maximum current to flow.

7. I_{DSS} is the drain current when $V_{GS} = 0$ V.

8. $V_{GS(OFF)}$ is the voltage at which there is no drain current.

9. Drain current for various values of V_{GS} can be calculated using the square-law formula:

$$I_D = \left(1 - \frac{V_{GS}}{V_{GS(OFF)}}\right)^2 \times I_{DSS}$$

10. The drain-source voltage at which drain current levels off is called the pinch-off voltage.

11. There is no gate current for a JFET.

12. JFETs have an extremely high input impedance.

13. Transconductance is the effect a change in V_{GS} has on I_D and can be calculated by using

$$g_m = \frac{\Delta I_D}{\Delta V_{GS}}$$

14. Important formulas for the self-biasing of a JFET are

$$V_{GS} = V_{RG} + (-V_{R_S}) \qquad V_{R_D} = I_D \times R_D$$
$$V_{R_S} = I_D \times R_S \qquad V_D = V_{DD} - V_{R_D}$$

15. Important formulas for the voltage-divider biasing of a JFET are

$$V_G = \frac{R_2}{R_1 + R_2} V_{DD} \qquad V_{RD} = I_D \times R_D$$
$$V_S = V_G - (-V_{GS}) \qquad V_D = V_{DD} - V_{R_D}$$
$$I_D = \frac{V_S}{R_S} \qquad V_{DS} = V_D - V_S$$

16. The CS (common-source) amplifier provides a 180° phase shift between the input on the gate and the output off the drain.

17. Important formulas for the CS amplifier using self-bias are

$$A_v = -g_m \times R_D$$
$$v_{\text{out}} = v_{\text{in}} \times A_v$$
$$z_{\text{in}} = R_G$$

18. The output off the source of a CD (common-drain) amplifier is in phase with the input signal on the gate.

19. Important formulas for the CD amplifier using voltage-divider bias are

$$A_v = \frac{R_S}{R_S + (1/g_m)}$$
$$v_{\text{out}} = v_{\text{in}} \times A_v$$
$$z_{\text{in}} = R_1 \parallel R_2$$

20. The output off the drain of a CG (common-gate) amplifier is in phase with the input signal on the source.

21. Important formulas for the CG amplifier are

$$A_v = g_m \times R_D$$
$$v_{\text{out}} = v_{\text{in}} \times A_v$$
$$z_{\text{in}} = \frac{1}{g_m}$$

22. The acronym MOSFET represents metal-oxide semiconductor field-effect transistor. The MOSFET is also known as an IGFET (insulated-gate field-effect transistor).

23. MOSFETs contain a source, drain, gate, and substrate.

24. MOSFETs are extremely susceptible to static discharge.

25. Forward biasing the gate of a D-type (depletion-type) MOSFET causes it to operate in the enhancement mode (increases current).

26. Reverse biasing the gate of a D-type MOSFET causes it to operate in the depletion mode (decreases current).

27. E-type (enhancement-type) MOSFETs are only operated in the enhancement mode since they are initially in a cut-off state.

28. MOSFET amplifiers provide extremely high input impedance along with a flat frequency response.

29. g_{m0} is the transconductance when $V_{GS} = 0$ V.

30. Important formulas for a D-type MOSFET amplifier using zero bias are

$$g_{m0} = \frac{-2 \times I_{DSS}}{V_{GS(OFF)}}$$

$$A_v = -g_m \times R_D$$

$$v_{out} = v_{in} \times A_v$$

$$z_{in} = R_G$$

31. $V_{GS(TH)}$ is minimum value of voltage necessary to cause an E-type MOSFET to conduct.

32. Important formulas for an E-type MOSFET amplifier using drain-feedback bias are

$$I_D = k \times (V_{GS} - V_{GS(TH)})^2$$

$$R_D = \frac{V_{DD} - V_{DS}}{I_D}$$

$$g_m = \frac{2 \times k}{V_{GS} - V_{GS(TH)}}$$

$$A_v = -g_m \times R_D$$

$$v_{out} = v_{in} \times A_v$$

$$z_{in} = R_G$$

33. A JFET is switched on when $V_{GS} = 0$ V and off when a value of $V_{GS(OFF)}$ is applied.

34. An E-type MOSFET is switched on when $V_{GS} > V_{GS(TH)}$ and off when $V_{GS} = 0$ V.

GLOSSARY

Bipolar: A device that consists of both majority and minority carriers.

Bipolar Junction Transistor (BJT): A semiconductor device that relies on both majority and minority carriers for conduction.

Channel: The area joining the source to the drain.

Depletion Mode: A condition where the channel is depleted of current carriers, causing a decrease in drain current.

Drain: An area of the FET through which electron current normally exits. It is symmetrical with the source.

Drain Curves: A pictorial representation of the relationship between I_D, V_{GS}, and V_{DS}.

D-Type MOSFET: A normally on device that can be controlled by a voltage of either polarity at its gate.

Enhancement Mode: A condition where current carriers are increased, causing an increase in drain current.

E-Type MOSFET: A normally off device that can be made to conduct by increasing the voltage at the gate with the appropriate polarity.

Field-Effect Transistor (FET): A transistor whose field around the gate is affected by the gate voltage.

Gate: An area of a FET that is insulated from the channel in MOSFETs and made of the opposite type of material from the channel for JFETs.

I_{DSS}: The drain current that flows when the source is shorted to the gate.

Inversion Layer: An area along the gate of an E-type MOSFET that allows drain current to flow. It is created by voltage on the gate of the same polarity as the substrate.

Junction Field-Effect Transistor (JFET): A unipolar device consisting of a source, drain, and gate. Its drain current (I_D) is controlled by the gate-to-source voltage (V_{GS}).

Majority Carrier: The current carrier that is in the majority: electrons for N-type material and holes for P-type material.

Metal-Oxide Insulator: A thin layer of silicon dioxide that insulates the gate from the channel in MOSFETs. Other insulating materials can be used.

Metal-Oxide Semiconductor Field-Effect Transistor (MOSFET): A unipolar device whose gate is insulated from the channel by a layer of metal-oxide. Gate-source voltage (V_{GS}) is used to control the drain current (I_D).

Minority Carrier: The current carrier that is in the minority: electrons for P-type material and holes for N-type material.

Pinch-Off Voltage: The value of V_{DS} right at the knee of the drain curves. A V_{DS} above the pinch-off voltage maintains a steady I_D.

Source: An area of a FET through which electron current normally enters the device. It is symmetrical with the drain.

Substrate: A region of MOSFETs used to control the width of the channel.

Symmetrical: Same in size and shape.

Threshold Voltage: The minimum voltage required for conduction in an E-type MOSFET.

Transconductance: The control that V_{GS} exercises over the drain current. A g_m of 2000 μS predicts an average change of 2000 μA in drain current for every volt of gate–source voltage.

Unipolar: A device that relies only on majority carriers for current flow.

V_{GS}: Gate–source voltage is used to control I_D in FETs.

$V_{GS(OFF)}$: The value of V_{GS} that cuts off JFETs and D-type MOSFETs.

PROBLEMS

SECTION 9-1

9-1. Draw and label the schematic symbols for both the N-channel and P-channel JFET.

9-2. Sketch a drain-curve chart for a P-channel JFET.

9-3. Show the proper polarity voltages necessary to correctly bias a P-channel and N-channel JFET.

9-4. Explain the formation and effects of the depletion region on drain current.

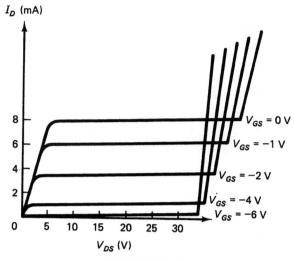

I_D (mA)

8

6

4

2

0   5   10   15   20   25   30

V_{DS} (V)

$V_{GS} = 0$ V

$V_{GS} = -1$ V

$V_{GS} = -2$ V

$V_{GS} = -4$ V

$V_{GS} = -6$ V

FIGURE 9-20

9-5. Why does the JFET have such a high input impedance?

9-6. What type of channel JFET is represented in Fig. 9-20?

9-7. What is the value of I_{DSS} in Fig. 9-20?

9-8. What is the value of $V_{GS(OFF)}$ in Fig. 9-20?

9-9. What value of gate current flows through the device depicted in Fig. 9-20?

9-10. What is the pinch-off voltage when $V_{GS} = 0$ V in Fig. 9-20?

9-11. What is the name given to the pictorial representation of the operation of the device in Fig. 9-20?

9-12. Create a transconductance curve for a JFET whose $I_{DSS} = 15$ mA and $V_{GS(OFF)} = -9$ V.

9-13. Calculate the transconductance of at least three different operating ranges from the transconductance curve of Problem 12.

SECTION 9-2

The following problems refer to the circuit of Fig. 9-21.

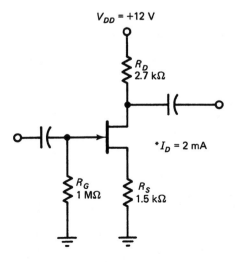

$V_{DD} = +12$ V

R_D
2.7 kΩ

$*I_D = 2$ mA

R_G
1 MΩ

R_S
1.5 kΩ

FIGURE 9-21

9-14. How much gate current is flowing?

9-15. What is the name of this circuit?

9-16. What is the value of V_{GS}?

9-17. Explain the effect of an increase in I_D on V_{GS}.

9-18. Determine the voltage at the drain.

9-19. Calculate the value of V_{DS}.

9-20. What is the voltage at the source?

9-21. What value of R_S is needed to produce an I_D of 3 mA if $V_{GS(OFF)} = -7$ V and $I_{DSS} = 12$ mA? It may be helpful to create a transconductance curve.

The following problems refer to the circuit of Fig. 9-22.

9-22. Determine the gate voltage.

9-23. What is the voltage at the source?

9-24. What is the value of drain current?

9-25. Find the value of V_D.

9-26. Calculate the drain-to-source voltage.

9-27. Explain the effect of an increase in I_D on V_{GS}.

9-28. Explain how swamping applies to this circuit.

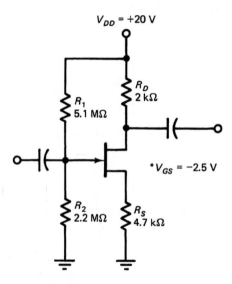

FIGURE 9-22

SECTION 9-3

The following problems refer to the circuit of Fig. 9-23. Assume $V_{GS(OFF)} = -7$ V and $I_{DSS} = 9$ mA for the JFET in this circuit. The initial value of $V_{GS} = -3$ V.

9-29. What are the upper and lower limits of V_{GS}?

9-30. Calculate the change in drain current.

9-31. Determine the transconductance.

9-32. What is the voltage gain?

9-33. Find the peak–peak output voltage.

9-34. What is the input impedance of this circuit?

9-35. If the initial value of V_{GS} were made more negative, would the transconductance of the circuit increase or decrease?

9-36. What type of amplifier is this?

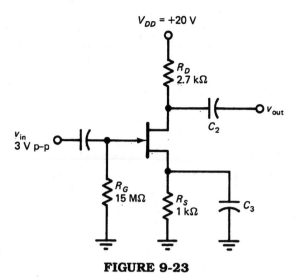

FIGURE 9-23

The following problems refer to the circuit of Fig. 9-24. Assume $V_{GS(OFF)} = -5$ V and $I_{DSS} = 7$ mA for the JFET in this circuit. The initial value of $V_{GS} = -2$ V.

9-37. What type of amplifier is this?

9-38. What are the upper and lower limits of V_{GS}?

9-39. Calculate the change in drain current.

9-40. Determine the transconductance.

9-41. What is the voltage gain?

9-42. Find the peak–peak output voltage.

9-43. What is the input impedance of this circuit?

9-44. What is the disadvantage of the common-gate amplifier?

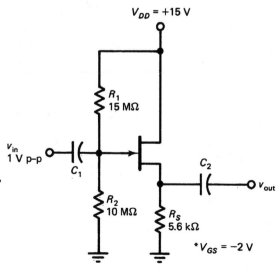

FIGURE 9-24

9-45. Draw and label the schematic symbols for both the N-channel and P-channel D-type MOSFETs.

9-46. Create a characteristic drain-curve chart for a D-type MOSFET whose $V_{GS(OFF)} = -3$ V and $I_{DSS} = 10$ mA.

9-47. What is the I_D at +2 V for a D-type MOSFET whose $V_{GS(OFF)} = +4$ V and $I_{DSS} = 7$ mA?

9-48. Draw a block diagram of a P-channel D-type MOSFET.

9-49. Describe enhancement and depletion modes for a P-channel D-type MOSFET.

SECTION 9-5

9-50. Draw and label the schematic symbols for both the N-channel and P-channel E-type MOSFETs.

9-51. Create a characteristic drain-curve chart for an E-type MOSFET whose $V_{GS(TH)} = +2$ V and $k = 0.003$.

9-52. What is I_D when $V_{GS} = +7.5$ V for an E-type MOSFET whose $V_{GS(TH)} = +4$ V and $k = 0.002$?

9-53. Draw a block diagram of a P-channel E-type MOSFET.

9-54. Describe the enhancement mode for a P-channel E-type MOSFET and explain why there is not a depletion mode.

SECTION 9-6

The following problems refer to the circuit of Fig. 9-25. Assume $V_{GS(TH)} = +3$ V and $k = 0.003$.

9-55. Determine the drain current for a V_{GS} of +5 V.

9-56. What is the drain–source voltage when $V_{GS} = +5$ V?

9-57. Calculate what size drain resistor is necessary to bias this circuit for a V_{GS} of +5 V.

9-58. What is the value of transconductance when $V_{GS} = +5$ V?

9-59. Find the voltage gain using the g_m of Problem 53.

9-60. What is the input impedance of this circuit?

9-61. Explain how drain-feedback bias offsets an attempted increase in drain current.

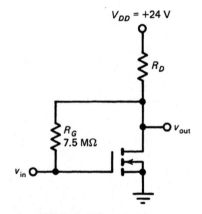

$V_{DD} = +24$ V

R_D

R_G
7.5 MΩ

v_{out}

v_{in}

FIGURE 9-25

SECTION 9-7

9-62. List the gate voltages required to produce the on and off states for the JFET, E-type MOSFET, and D-type MOSFET.

9-63. Draw a circuit using a P-channel JFET as a switch. Refer to the circuit of Fig. 9-17.

9-64. Redraw the circuit of Fig. 9-18 using a JFET and explain its operation.

9-65. Draw a circuit using a P-channel E-type MOSFET as a switch. Refer to the circuit of Fig. 9-18.

9-66. Redraw the circuit of Fig. 9-17 using a MOSFET and explain its operation.

SECTION 9-8

The following problems refer to the circuit of Fig. 9-26.

9-67. What type of amplifier is found in the first stage?

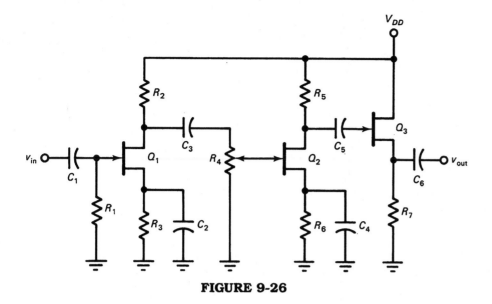

FIGURE 9-26

9-68. What would be the effect on the output if C_2 were to open?

9-69. Determine the possible causes for a complete loss of signal at the gate of Q_2. Note: a good signal is measured at the drain of Q_1.

9-70. What type of amplifier is shown in the final stage?

9-71. What is the purpose of the potentiometer?

9-72. Describe the troubleshooting procedure you would follow if a signal were measured at the gate of Q_2 but not at the output.

9-73. Create a chart listing the proper resistance measurements of each type of FET along the V_{GS} necessary to turn the device on and off.

ANSWERS TO REVIEW QUESTIONS

1. unipolar	**2.** majority	**3.** polarity
4. voltage-	**5.** Source, gate, and drain	**6.** N
7. P	**8.** Symmetrical	**9.** Negative
10. I_{DSS}	**11.** Depletion region	**12.** $V_{GS(OFF)}$
13. source	**14.** decreases	**15.** 0 V
16. large	**17.** pinch-off voltage	**18.** Transconductance
19. siemens	**20.** False	**21.** increase
22. $V_{GS} = -1.2$ V	**23.** $I_G = 0$ A	**24.** $V_D = 7.45$ V
25. $V_S = 4.72$ V	**26.** $I_D = 4.72$ mA	**27.** $V_D = 12.8$ V
28. $V_{DS} = 8.06$ V	**29.** Swamping	**30.** Self-bias
31. False	**32.** True	**33.** $z_{in} = R_G$
34. $z_{in} = R_1 \parallel R_2$	**35.** True	**36.** False
37. Almost unity	**38.** Low input impedance	**39.** No
40. False	**41.** Metal-oxide semiconductor	**42.** The substrate
43. No	**44.** depletion	**45.** 14.2 mA
46. 3.56 mA	**47.** enhancement	**48.** static electricity
49. enhancement	**50.** False	**51.** The inversion layer
52. The threshold voltage	**53.** $I_D = 200$ mA	**54.** $V_{GS} = 0$ V
55. It equals I_{DSS}	**56.** $V_{DS} = 8$ V	**57.** enhancement

58. True	**59.** $R_D = 172\ \Omega$	**60.** Common source
61. on	**62.** positive	**63.** 0 V
64. −5 V	**65.** off	**66.** on
67. increased	**68.** False	**69.** low
70. negative	**71.** infinite	**72.** False
73. positive	**74.** E-Type MOSFET	**75.** bypass

Thyristors and Unijunction Devices

When you have completed this chapter, you should be able to:

- Analyze the unique characteristics of thyristors and unijunction transistors.
- Comprehend the different methods of switching dc and ac current offered by these devices.
- Investigate signal generators.
- Determine whether a thyristor or UJT is faulty.

INTRODUCTION

Thyristors: A group of PNPN junction devices.

The diodes, BJTs, and FETs of the previous chapters are considered general-purpose devices. They have a wide range of applications. **Thyristors** and unijunction devices are special-purpose components. Their applications are solely in the areas of switching and triggering. Some are used in purely dc applications and others are designed for ac applications. All rely on some value of voltage or current to produce either an "on state" or "off state." Focus your attention on the requirements for the switching action of the following components.

10-1 SHOCKLEY DIODES

Shockley Diode: Also called a PNPN junction diode, it conducts at the breakover voltage.

The **Shockley diode,** invented by William Shockley, is also referred to as a four-layer diode or PNPN diode. Each of these names represents the device shown in Fig. 10-1.

Operation The block diagram explains how the names PNPN and four-layer diode evolved. There are three PN junctions all together and only two external leads. A forward **breakover voltage** (V_{BO}) applied to this device must overcome the PN junctions in order to cause conduction. A Shockley diode is turned on by V_{BO}.

Breakover Voltage: The voltage used by certain thyristors to cause conduction.

Figure 10-1 also shows the schematic symbol for a Shockley diode. It has an anode and cathode, just like a conventional PN junction diode. The open switch signifies that the diode is not conducting.

Once the Shockley diode is turned on by the breakover voltage, it must receive a certain level of current in order to maintain conduction. The amount of current necessary to hold the switch closed is called the **holding current** (I_H). This value is indicated on the spec sheet of the diode along with V_{BO}. If the current through the diode drops below the value of I_H, the diode will cease to conduct.

Holding Current: The minimum value of current necessary to maintain conduction.

The I–V curve of the Shockley diode in Fig. 10-2 shows the relationship between current and voltage. Locate V_{BO}, the value of forward breakover voltage necessary to cause the diode to conduct. It has a value of 12 V. Shockley diodes come in a variety of voltage ratings.

As soon as V_{BO} is reached, an initial forward switching current begins to flow (labeled I_S). The circuit in which the diode is used must provide this initial value of current for the switch to successfully close. In this case, the current is 0.5 mA.

When the switch is closed, the forward current through the diode increases rapidly, whereas the voltage across the diode decreases. Follow the curve from I_S back toward the vertical axis until you reach V_F. This is the forward voltage that is maintained across the conducting diode. Figure 10-2 shows this value as 1 V. In

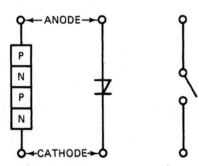

FIGURE 10-1 Shockley diode.

the meantime, the forward current continues to rise to a level determined by the circuit in which the diode is used.

The diode is now conducting and continues to conduct as long as the current through it is kept at a value above I_H. As soon as the current drops to 5 mA or less, the diode ceases to conduct.

The lower left corner of Fig. 10-2 shows the effect of reverse biasing a Shockley diode. It does not conduct until a value of reverse voltage is reached that forces it to conduct. At this point, the device breaks down. The Shockley diode should be operated in forward bias only.

Relaxation-Oscillator Application

The "relaxation oscillator" of Fig. 10-3, powered by a steady dc input, produces a pulsating dc output. The Shockley diode can be viewed as a switch that closes each time the capacitor charges to 12 V. Upon closing, it discharges the capacitor to the 1-V forward-voltage value of the diode. Conduction stops shortly thereafter because current falls below the value of I_H. The current at this point can be found by realizing that 1 V is dropped across the diode and the remainder of V_{IN} is dropped across R. The current that the circuit offers the diode at this point is

$$I_D = \frac{V_{IN} - V_F}{R}$$

$$= \frac{18\ V - 1\ V}{10\ k\Omega}$$

$$= 1.7\ mA$$

This value of current is far below the value of I_H, which is 5 mA. The diode acts like an open switch and the capacitor begins to charge again. Once it reaches 12 V, the whole process is repeated.

As the diode closes and opens, it causes the lamp to blink, and a pulsating waveform can be measured at the output, as shown in Fig. 10-3. The peak voltage of this waveform is equal to the breakover voltage of the diode. The frequency at which the lamp blinks is determined in part by the RC **time constant** of the circuit.

The lamp in this circuit blinks at a rate of approximately 1 Hz, or once every second. The specific frequency can be found by applying the following formula:

$$f = \frac{1}{(R \times C) \times \ln[V_{IN}/(V_{IN} - V_{BO})]}$$

$$= \frac{1}{(10\ k\Omega \times 100\ \mu F) \times \ln[18\ V/(18\ V - 12\ V)]}$$

$$= 0.91\ Hz$$

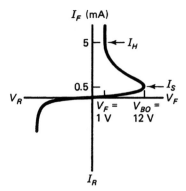

FIGURE 10-2 *I–V* curve of a Shockley diode.

Time Constant: The length of time it takes for a 63% change of current or voltage in a *RC* or *RL* circuit.

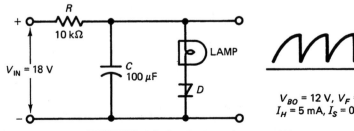

FIGURE 10-3 Relaxation oscillator.

The denominator of the formula solves for the charge time of the capacitor. It uses a natural logarithm (ln) to derive the number of time constants it takes for the capacitor to charge to the value of the Shockley diode. The value of one RC time constant is then multiplied by the number of time constants to arrive at the overall charge time. The reciprocal of the charge time is equal to the frequency.

The discharge time of this circuit is not included in the calculations for two reasons. First, the discharge time through the low-resistance path of the diode is negligible when compared to the charge time. Second, approximate values are used. Capacitors alone typically have a tolerance of 20%.

A final point about the design of this circuit should be made. The value of the resistor must be chosen to deliver the appropriate current when needed. The maximum value of resistance is governed by the switching current needed to successfully turn the diode on. An I_S of at least 0.5 mA is needed in the circuit of Fig. 10-3. So the largest R that can be used is

$$R_{MAX} = \frac{V_{IN} - V_{BO}}{I_S}$$

$$= \frac{18\text{ V} - 12\text{ V}}{0.5\text{ mA}}$$

$$= 12\text{ k}\Omega$$

The diode should stop conducting immediately after discharging the capacitor. This means that the current available after the discharge must be held to a value below I_H. If the value of the resistor is too small, the current will remain above I_H and the diode will continue to conduct. The minimum value of resistance is, therefore,

$$R_{MIN} = \frac{V_{IN} - V_F}{I_H}$$

$$= \frac{18\text{ V} - 1\text{ V}}{5\text{ mA}}$$

$$= 3.4\text{ k}\Omega$$

Oscillate: To fluctuate back and forth between two voltage levels.

Choosing a resistor outside of these values prevents the circuit from **oscillating**. Finally, the value of capacitance can be changed to provide a different frequency.

EXAMPLE 10-1 _____

The values in the circuit of Fig. 10-3 are changed so that $V_{IN} = 15$ V, $R = 7.5$ kΩ, and $C = 10$ μF; the Shockley diode's specifications are $V_{BO} = 10$ V, $V_F = 1$ V, $I_H = 3$ mA, and $I_S = 0.5$ mA. Determine (a) the frequency at which the lamp blinks and (b) the minimum and maximum resistance values for R.

Solution:

(a) Remember that the frequency of the blinking light is dependent upon the charge time of the capacitor. The reciprocal of the charge time is equal to the frequency of the output.

$$f = \frac{1}{(7.5\text{ k}\Omega \times 10\text{ }\mu\text{F}) \times \ln[15\text{ V}/(15\text{ V} - 10\text{ V})]}$$

$$= 12.1\text{ Hz}$$

(b) The value of resistance affects both the RC time constant and the levels of current within the circuit. Certain values of current must be maintained for the diode to operate properly. If the resistor is too large, the diode will not switch on, and if the resistor is too small, the diode will not switch off.

$$R_{MAX} = \frac{15\text{ V} - 10\text{ V}}{0.5\text{ mA}}$$

$$= 10\text{ k}\Omega$$

and

$$R_{MIN} = \frac{15\text{ V} - 1\text{ V}}{3\text{ mA}}$$

$$= 4.67\text{ k}\Omega \qquad \blacksquare$$

REVIEW QUESTIONS

1. What are the main applications of thyristors and unijunction devices?
2. Another name for the Shockley diode is _____ .
3. There are _____ PN junctions in a Shockley diode.
4. What is the name of the voltage used to overcome the PN junctions within the Shockley diode?
5. What are the names associated with the two leads of a Shockley diode?
6. _____ current is the minimum value of current required to maintain conduction for the Shockley diode.
7. The forward voltage of the Shockley diode is approximately 12 V. True or false?
8. The output of a relaxation oscillator is a _____ _____ .
9. The diode in the circuit of Fig. 10-3 switches on when the voltage across the capacitor reaches _____ volts.
10. Decreasing the resistance in the circuit of Fig. 10-3 causes the lamp to blink (more, less, or just as) frequently.
11. What is the voltage across the diode when it is conducting?
12. A capacitor charges to 63% of the difference between its voltage and the applied voltage in one _____ _____ .
13. What would be the appropriate frequency of the circuit of Fig. 10-3 if the capacitance were changed to 4.7 μF?
14. The discharge time is equal to the charge time in the circuit of Fig. 10-3. True or false?
15. What would be the minimum value of resistance allowable if the I_H in the circuit of Fig. 10-3 were 3 mA?

10-2 SCR

The **silicon-controlled rectifier (SCR)** also belongs to the thyristor family. It differs from the Shockley diode in that it is turned on by a trigger signal on its gate rather than by a breakover voltage. Large SCRs can handle up to several hundred amps of current.

Silicon-Controlled Rectifier (SCR): A thyristor that conducts when a positive trigger voltage is received at the gate.

SCR 200 V, 6 A.

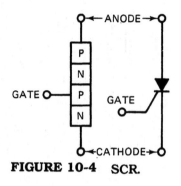

FIGURE 10-4   SCR.

Figure 10-4 shows the block diagram of an SCR. It looks very much like the PNPN diode except that it has a gate lead. The SCR acts like an open switch until a positive pulse is received on its gate. A trigger voltage of 1 or 2 V is sufficient to turn it on. The maximum value of gate voltage needed is listed on the spec sheet for each SCR.

Once the SCR is turned on, it remains on until current falls below the holding current. The value of holding current, I_H, varies from one SCR to another. The maximum values range anywhere from 30 to 75 mA.

From the schematic symbol for the SCR, we see the symbol for a diode within it. The SCR acts like a rectifier diode in that conduction occurs in one direction only. It must be forward biased to conduct. A voltage of 1 or 2 V develops across it while it is conducting. When used in ac applications, the SCR turns off near the zero crossing point of the signal as the forward current drops below I_H.

DC Overvoltage-Protection Application   The circuit of Fig. 10-5 shows a typical use for the SCR. This circuit might be used to protect a microprocessor circuit from an accidental increase in the supply voltage. Under normal circumstances, the SCR and zener diode do not conduct. But if V_{IN} should accidentally rise to 6 V or more, the circuit kicks in to protect against the overvoltage.

The zener diode begins to conduct at a voltage slightly above its zener voltage of 5.1 V. Any voltage above 5.1 V develops across the resistor. When the voltage across the resistor reaches a value of 1 V, it turns on the SCR. The supply voltage at this point is 6.1 V.

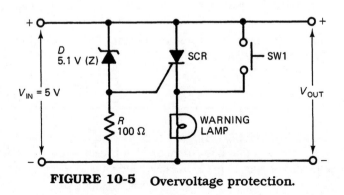

FIGURE 10-5   Overvoltage protection.

256

Immediately upon the conduction of the SCR, the warning lamp lights to indicate an overvoltage problem. The SCR and lamp shunt most of the current away from the circuit connected to V_{OUT}.

The SCR continues to conduct until its current is reduced below I_H. SW1 redirects current around the SCR when pressed. It provides a means for resetting the circuit back to a nonconducting state. Of course, the problem with V_{IN} should be corrected before resetting the circuit.

AC Half-Wave Control Application

The SCR can also be used in ac applications. An example is found in Fig. 10-6, wherein the SCR conducts on only one alternation. The alternation that causes conduction and the path of current flow are illustrated. On this alternation, the SCR starts in an off condition. As current flows through the resistor voltage divider and diode, R_2 develops the necessary trigger voltage. The SCR is triggered on and drops approximately 1 V. Once the SCR is triggered on, it conducts until the current through it falls below the value of the holding current. Meanwhile, current is flowing through the lamp during this alternation.

The lamp remains lit until the polarities at the input change. On the next alternation, the SCR is reverse biased and does not conduct. The diode is also reverse biased, so current cannot flow through the voltage divider. The lamp does not receive current during this alternation and is unlit.

Current flows through the lamp on only one alternation under the control of the SCR. This simple circuit illustrates the operation of the SCR in an ac environment. Keep in mind that a positive voltage of about 1 V at the gate is needed to close the SCR switch and a forward current of less than I_H opens it.

EXAMPLE 10-2

How is the SCR used in the circuit of Fig. 10-6 to control the brightness of the lamp?

Solution: The SCR and the voltage divider formed by R_1 and R_2 allow us to vary the percentage of the half wave that is fed to the lamp. If R_1 is larger than R_2, then a greater portion of the input signal is needed to develop the 1 V across R_1 in order to turn on the SCR. Only the remaining portion of the input half wave is passed on to the lamp through the SCR. The less voltage that is passed on to the lamp by the SCR, the dimmer it lights. ∎

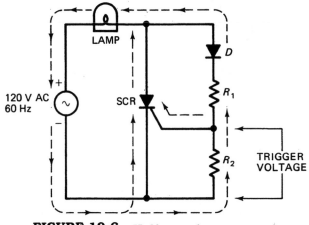

FIGURE 10-6 Half-wave lamp control.

16. What does SCR represent?

17. How is the SCR turned on?

18. Name the three leads of an SCR.

19. An SCR can be turned off by a negative trigger voltage at its gate. True or false?

20. Can an SCR conduct in both directions?

21. How much voltage is dropped across an SCR once it is turned on?

22. What is the name given to the minimum amount of current required to maintain conduction?

23. What is the normal range of trigger voltages for SCRs?

24. How does SW1 in the circuit of Fig. 10-5 turn off the SCR?

25. Why does the SCR in the circuit of Fig. 10-6 turn off on the negative alternation?

10-3 SCR TYPES

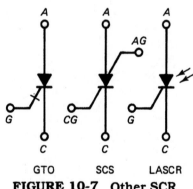

FIGURE 10-7 Other SCR types.

The SCR that was just discussed is the most widely used type. However, the idea of controlling the switching action has spawned some ingenious methods of controlling the gate. Each of the following SCR types has its own way to fill that unique situation when such characteristics are required.

GTO or GCS   The gate turn-off (GTO) switch is also called a gate-controlled switch (GCS). It is similar to the SCR in that a positive pulse at the gate turns it on. The difference is that it can be turned off by a negative pulse. It allows for complete voltage control of the switching action. Figure 10-7 shows the schematic symbol for a GTO.

SCS   A circumstance may arise where a switching device needs to be operated by two separate sources. In that case, the silicon-controlled switch (SCS) fits the bill. Figure 10-7 shows the schematic symbol for the SCS, which sometimes goes by the name of "tetrode thyristor."

The schematic symbol depicts the usual anode and cathode leads, but now there are two gates instead of one. *CG* is the cathode gate and *AG* is the anode gate. The gates are named because of their proximity to the cathode and anode leads, respectively.

A forward-bias trigger on either *CG* or *AG* causes the SCS to conduct. The *CG* requires a positive trigger and the *AG* a negative trigger pulse. It conducts in one direction only.

Conduction continues until either of the gate inputs receives a reverse trigger pulse. The polarity of the reverse trigger is the opposite of that required to close the switch.

LASCR   Finally, light itself can be used to trigger the gate of a switching device, as in the case of the light-activated silicon-controlled rectifier (LASCR). The schematic symbol in Fig. 10-7 shows the inward-pointing arrows indicating light. When arrows point toward a device it means that the device is affected by light.

In this case, if a sufficient amount of light enters the lens of the device, it will cause the LASCR to conduct. The sensitivity to the light can be controlled by connecting a resistor between the gate and cathode. The lower the value of resistance, the more light required to close the switch. Leaving the gate open results in maximum sensitivity to light.

To open the switch, the current must be reduced below the necessary holding-current value. Other forms of the LASCR may allow for a negative trigger to stop conduction.

REVIEW QUESTIONS

26. What is another name for the GTO?
27. What does GTO represent?
28. How is the GTO turned on?
29. How is the GTO turned off?
30. Name the leads of an SCS.
31. What does SCS represent?
32. How is the SCS turned on?
33. How is the SCS turned off?
34. What does LASCR represent?
35. How is the LASCR turned on?
36. How is the LASCR turned off?
37. How is the sensitivity of the LASCR adjusted?

10-4 DIAC

The **diac** is made to conduct in both directions. Its name is a contraction of *diode ac* switch. It is used in high ac voltage and current applications.

The schematic symbol of a diac is shown in Fig. 10-8 along with a block diagram of its internal construction. Several PN junctions must be overcome in order for it to conduct. There is no anode and cathode, so the leads are labeled as terminal 1 (T_1) and terminal 2 (T_2).

Figure 10-9 shows operational features of the diac by the use of an *I–V* curve. In the upper right quadrant, the forward voltage increases until it reaches the value of the positive breakover voltage ($+V_{BO}$). This value usually runs between 25 and 40 V. When this point is reached, the voltage rapidly decreases to about 10 or 15 V and levels off. Current increases according to the load conditions.

The same scenario exists in the lower left quadrant, only the polarities are opposite. Summarizing, a diac is made to conduct with either a positive or negative breakover voltage and drops a voltage of between 10 and 15 V while conducting.

Looking at the vertical axis of Fig. 10-9, we see the conditions of forward and reverse current. The positive and negative holding currents are marked. The diac is turned off when the current falls below the value of I_H. Although there is a switching action with the diac, the voltages involved are considerably higher than those discussed with previous devices.

Diac: A diode ac switch. An ac thyristor that has two leads and conducts in either direction at a positive or negative breakover voltage.

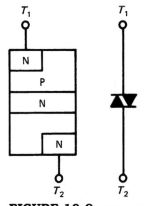

FIGURE 10-8 The diac.

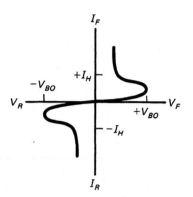

FIGURE 10-9 The diac *I–V* curve.

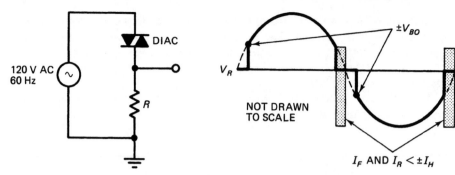

FIGURE 10-10 The diac as an ac switch.

AC Switch The circuit of Fig. 10-10 describes the operation of the diac. A much more practical circuit is shown in the next section when the diac and triac are combined. For now, let us examine the switching action of the diac alone.

An ac line signal is applied to the input. The output is measured across the resistor. The dashed portions of the output signal denote the voltage that would have appeared across the resistor if the diac were conducting at that time. The diac does not turn on until the input signal reaches a positive or negative 30 V. At that point, the diac turns on and drops approximately 15 V. The remaining 15 V at this instant are dropped across the resistor. So the voltage across the resistor rises almost instantly from a 0-V condition to 15 V. The point at which V_{BO} occurs is noted in Fig. 10-10 and a small portion of the positive alternation is shown missing.

The diac continues to conduct throughout the majority of the positive alternation. Eventually, however, the current through the diac drops below the value of the holding current. At this point, the diac turns off and current ceases to flow. This period of time is indicated by the shaded area in Fig. 10-10. The figure states that the forward current at this point is less than the holding current ($I_F < I_H$).

When the negative alternation of the input signal is applied in Fig. 10-10 and a voltage of −30 V is reached, the diac switches on once again. The same procedure just outlined occurs again except that the polarity has changed. Notice that the beginning portion of the negative alternation is missing due to the initial nonconduction of the diac. Once the diac turns on, the waveform picks up as it did for the positive alternation.

Finally, at the extreme right of the waveform in Fig. 10-10, the current through the diac drops below the value of $-I_H$. The shaded area is used to point out this period of time.

EXAMPLE 10-3 _____

How does the value of the resistor affect the circuit in Fig. 10-10?

Solution: The diac attempts to turn on at its rated breakover voltage regardless of the size of the resistor. However, a certain value of holding current is needed to enable the diac to conduct. If the resistor is too large, the diac will not receive the necessary current. ∎

REVIEW QUESTIONS _____

38. How was the name diac formed?
39. What is the range of the breakover voltages for a diac?

CHAP. 10 / THYRISTORS AND UNIJUNCTION DEVICES

40. A positive breakover voltage is used to initiate conduction and negative trigger voltage is used to stop it in a diac. True or false?

41. What happens to the voltage across the diac at the point that it begins to conduct?

42. Does current flow through the resistor alternately in both directions in Fig. 10-10?

43. What causes the diac to stop conducting during each alternation in the circuit in Fig. 10-10?

10-5 TRIAC

The **triac** is another ac switching device. Its name is a conjunction of *tri*ode *ac* switch. The triac differs from the diac by the addition of a gate. Figure 10-11 shows the block diagram and schematic symbol for a triac.

The use of a gate allows greater control over the operation of the device. Figure 10-12 illustrates how the gate can be used to change the $I–V$ characteristics of the device. With no gate current or voltage available to the triac, a maximum value of $+V_{BO}$ across the terminals is required to turn it on. As the value of gate current (I_G) and voltage are increased, the triac conducts at a lower value of breakover voltage. Once the device is conducting, the gate no longer influences it. This explanation is true for the reverse characteristics in the lower left quadrant of the figure.

Turning off the triac is accomplished by reducing the current below the necessary holding current (I_H).

AC Phase Control   The speed of a motor or the intensity of a light can be controlled by an ac phase-control circuit. A general label of "load" is used in Fig. 10-13 to represent any device whose operation varies with a change in the amount of ac current or voltage that it receives.

The operation of the circuit of Fig. 10-13 relies on the action of phase shifting. Two values are said to be out of phase if their voltage and current do not increase and decrease at the same time. The voltage across the capacitor is out of phase with the input voltage. The extent of this phase shift is determined by the resistance of the phase-control rheostat in the circuit of Fig. 10-13.

Triac: The triode ac switch conducts in either direction when a sufficient level of breakover voltage and current is supplied.

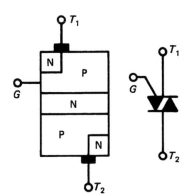

FIGURE 10-11   The triac.

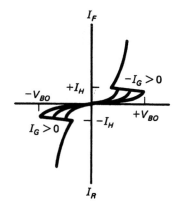

FIGURE 10-12   The triac $I–V$ curves.

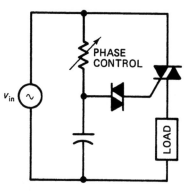

FIGURE 10-13   AC phase control.

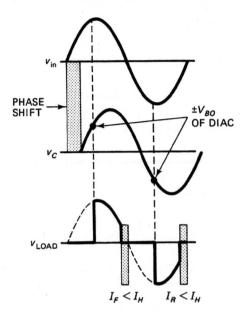

FIGURE 10-14 Phase control waveforms.

As the phase-control resistance is increased, the waveform of the capacitor moves toward the right and out of phase with the input voltage, as shown in Fig. 10-14. If the phase-control resistance is increased to its maximum value, the phase shift between v_C and v_{in} will be almost 90°. The phase shift is about 45° in Fig. 10-14.

A decrease to zero ohms in the resistance of the phase control places the voltage across the capacitor in phase with the input voltage. Therefore, the v_C waveform can be shifted in either direction by adjusting the resistance of the phase control.

The voltage across the capacitor is used to trigger the diac in the circuit of Fig. 10-13. Both the positive and negative breakover voltages are noted on the waveform of the capacitor in Fig. 10-14. Once these levels of voltage are reached, the diac is turned on. The diac in turn triggers the gate of the triac.

Up to this point, the triac has been turned off, so current could not flow through the load. When the triac is finally turned on, a good portion of v_{in} has already occurred. Figure 10-14 shows the load conducting at the present value of v_{in}. In other words, the beginning portion of the input waveform is prevented from reaching the load as a result of the phase shift.

The triac continues to conduct through the positive alternation until the current falls below the value of the holding current. At that point, the triac turns off and current ceases to flow through the load. This sudden drop in current and voltage across the load is also indicated by the shaded area in Fig. 10-14.

The negative alternation repeats the process just outlined, starting with turning on the diac. The portion of v_{in} between the $-V_{BO}$ and the $-I_H$ is passed on to the load.

Adjusting the phase control allows regulation of the amount of ac current and voltage that reaches the load. Less current and voltage to the load results in a dimmer light or slower motor speed.

REVIEW QUESTIONS _____

44. As the value of gate current and voltage increases for a triac, the value of V_{BO} _____ .

CHAP. 10 / THYRISTORS AND UNIJUNCTION DEVICES

45. How is the triac turned off?

46. Can a triac conduct in both directions?

47. An increase in the resistance of the phase control in Fig. 10-13 causes a (decrease or increase) in the phase shift between the input voltage and that of the capacitor.

48. A decrease in the resistance of the phase control of Fig. 10-13 causes (more or less) ac load voltage.

49. What is the maximum phase shift that can occur between the voltage across a capacitor and the input voltage?

50. The resistance of the phase control should be (increased or decreased) to brighten the light (if it were the load).

10-6 UJT

The **unijunction transistor (UJT)** receives its name because it has only one (uni) PN junction. It is composed of an emitter and two bases. Notice that the emitter is closer to B_2 in Fig. 10-15. The resistance of this junction (R_{B2}) is less than that of the emitter-to-B_1 junction (R_{B1}). Together they make up a parameter called the interbase resistance (R_{BB}).

$$R_{BB} = R_{B1} + R_{B2}$$

When a voltage is applied to the bases of the UJT, the voltage is divided across the internal base resistances. By determining the ratio of R_{B1} to R_{BB}, another parameter, called the intrinsic standoff ratio, is found. This quantity is expressed on the spec sheet next to the Greek letter η (eta). If $R_{B2} = 1$ kΩ and $R_{B1} = 3$ kΩ, then η is equivalent to

$$\eta = \frac{R_{B1}}{R_{BB}}$$

$$= \frac{3 \text{ k}\Omega}{4 \text{ k}\Omega}$$

$$= 0.75$$

Unijunction Transistor (UJT): The UJT has only one PN junction. It conducts when $V_E = V_P$.

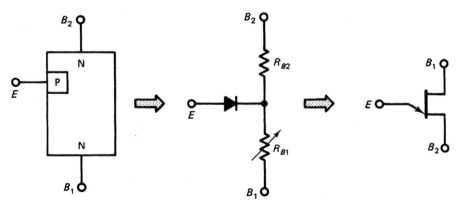

FIGURE 10-15 UJT.

This formula is essentially the voltage-divider formula used throughout this book. The value of η is usually between 0.5 and 0.8.

If 12 V were applied to the bases, then the voltage across R_{B1} would be

$$V_{RB1} = \eta \times 12 \text{ V}$$
$$= 0.75 \times 12 \text{ V}$$
$$= 9 \text{ V}$$

The voltage across R_{B1} is equivalent to the voltage at the cathode of the emitter diode in Fig. 10-15. In order to forward bias the emitter diode, the anode would have to be 0.7 V higher in potential than the cathode. This means that the voltage at the emitter must equal a minimum of 9.7 V in order to cause conduction between B_1 and E. This value is called **peak voltage** (V_P) and is used to turn on the UJT. All of the preceding facts can be summarized in the following formula:

$$V_P = (\eta \times V_{BB}) + V_{PN}$$
$$= (0.75 \times 12 \text{ V}) + 0.7 \text{ V}$$
$$= 9.7 \text{ V}$$

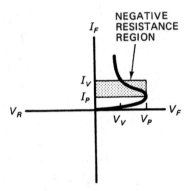

FIGURE 10-16 UJT I–V curve.

Turning On the UJT   Figure 10-16 explains the action within the diode upon conduction. First, a voltage that is at least equal to V_P must be applied to the emitter. Upon receiving the necessary peak voltage, the emitter begins to conduct and the resistance of R_{B1} decreases. Decreasing the resistance of R_{B1} causes a further increase in I_E. At the same time, the voltage at the emitter begins to decrease. The effect of increasing current while voltage is decreasing is called "negative resistance."

Eventually, the UJT is saturated and V_E levels off to a steady value. The device is now on and both its emitter and base currents increase.

Turning Off the UJT   Another value found on the spec sheets of UJTs is that of **valley current** (I_V). If the emitter current is reduced below the value of I_V, the UJT will cut off the emitter current. The relationship between I_V and valley voltage, V_V, is also depicted in Fig. 10-16.

Sound Generator   The sound generator circuit of Fig. 10-17 is basically a relaxation oscillator with a speaker connected as a load. The RC time constant of R_1 and C_1 determine the frequency of the tone or oscillation.

The peak voltage to turn on the UJT is

$$V_P = (\eta \times V_{BB}) + V_{PN}$$
$$= (0.6 \text{ V} + 10 \text{ V}) + 0.7 \text{ V}$$
$$= 6.7 \text{ V}$$

Initially, when the V_{BB} supply is turned on, the UJT is off and the capacitor is at a 0-V potential. The capacitor then begins to charge, and when it reaches V_P, it turns on the UJT. The UJT conducts, and current through the speaker rapidly increases as the resistance of R_{B2} decreases. The low resistance provided by the UJT allows the capacitor to discharge through R_{B2} and the speaker. Soon the current falls below I_V and the UJT turns off. This process is repeated over and over.

CHAP. 10 / THYRISTORS AND UNIJUNCTION DEVICES

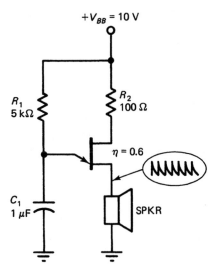

FIGURE 10-17 Sound generator.

The pulsing current through the speaker emits a tone and a sawtooth waveform appears at B_2, as shown in Fig. 10-17. Decreasing the resistance of R_1 increases the frequency and tone of the circuit.

EXAMPLE 10-4

If η of the UJT in Fig. 10-17 were 0.8, what value of peak voltage would be needed to turn it on?

Solution: The capacitor must charge to a peak voltage of

$$V_P = (0.8 \times 10 \text{ V}) + 0.7 \text{ V}$$
$$= 8.7 \text{ V}$$
∎

REVIEW QUESTIONS

51. What does UJT represent?
52. How many PN junctions does the UJT have?
53. What are the names of the leads on a UJT?
54. What is the name given to the overall resistance between B_1 and B_2 in a UJT?
55. What parameter does the Greek letter η represent?
56. How is a UJT turned on?
57. How is a UJT turned off?
58. An increase in current accompanied by a decrease in voltage is referred to as

_____ _____ .

59. What would be the value for V_P if $\eta = 0.5$ in the circuit of Fig. 10-17?
60. If R_1 were changed to 10 kΩ in the circuit of Fig. 10-17, would the frequency at B_2 increase or decrease?

10-7 TROUBLESHOOTING

All the devices discussed in this chapter exhibit switchlike characteristics. In-circuit troubleshooting involves checking for the appropriate breakover voltage or gate voltage for turning on the device. It may be necessary to force this condition to occur. Observation of the voltage across the device should show a decrease when the device is turned on.

Shockley Diode   An extremely high forward and reverse resistance is measured. This is due to the internal construction of the device. There are three PN junctions between the anode and cathode.

A forward breakover voltage is used to cause conduction. It must be a forward voltage since the diode does not conduct in reverse bias. A current below I_H stops conduction in this device.

When used in an oscillator circuit, a pulsing waveform should be observed with an oscilloscope. If this waveform is not present, then the device can be measured for a short with an ohmmeter. An open is difficult to detect because of the normally high resistance. A good diode could be piggybacked across the suspected open diode. This can save unnecessary desoldering and soldering.

SCR   A very high forward and reverse resistance is measured between the anode and cathode of this device. The gate–cathode forward resistance is low. This junction is similar to a conventional PN junction diode (see Fig. 10-4). The gate–cathode reverse resistance is very high. Finally, the gate–anode forward and reverse resistances are high. An additional PN junction lies between these two leads, as shown in Fig. 10-4.

A positive gate voltage of between 0.8 and 2 V is used to turn on this device. Lowering the current below I_H turns off the device. The SCR should be forward biased. An oscilloscope or VOM connected across the anode–cathode junction reveals if the SCR is switching properly. A trigger pulse should be applied to the gate while checking for the switching action.

Diac   A very high resistance is measured in both directions for this device. The diac is found in high ac voltage and current applications.

It is turned on by exceeding the breakover voltage in either direction. The diac is turned off when current drops below I_H. Normally, a distorted waveform is observed across this device. If the diac is shorted, 0 V will be measured across it, and when opened, an undistorted input signal will appear across it. An ohmmeter test only reveals a short because of the normally high resistance.

Triac   The resistance from T_1 to T_2 is very high in both directions. The gate-to-T_1 resistance is low in both directions. Figure 10-11 illustrates the reason. On the other hand, the gate-to-T_2 resistance is very high in both directions because of the extra PN junctions.

A positive or negative signal applied to the gate reduces the V_{BO} for the TRIAC. Reducing the current below I_H turns it off. Once again, it may be necessary to force the necessary signal at the gate for in-circuit measurements of voltage.

UJT   The resistance of B_1–B_2 should measure between 5 and 10 kΩ in either direction. The forward resistance of either base–emitter junction shows a low resistance. Figure 10-15 illustrates the polarities of the emitter and bases. Finally, the reverse resistance of the base–emitter junction is very high.

The UJT is turned on by the appropriate peak voltage at the emitter. It is turned off when the emitter current falls slightly below the value of the valley current, I_V.

A circuit in which this device is typically found is the relaxation oscillator. It provides a pulsed output. An oscilloscope is used to view the output while the emitter voltage is varied by means of a charging and discharging capacitor.

REVIEW QUESTIONS

61. Is the forward resistance of a Shockley diode low or high?
62. How is a Shockley diode turned on?
63. Which measurement of an SCR provides a low resistance?
64. How is an SCR turned on?
65. Diacs and Triacs are used in _____ circuits.
66. How are diacs and triacs turned off?
67. What is the T_1-to-T_2 resistance for diacs and triacs?
68. How is a UJT turned on?
69. The B_1–B_2 resistance of a UJT is infinite. True or false?
70. How is a UJT turned off?

SUMMARY

1. The Shockley, or PNPN, diode has three PN junctions and only two external leads.
2. A forward breakover voltage (V_{BO}) is required to turn on the Shockley diode.
3. The holding current I_H is the value of current necessary to maintain conduction.
4. A current below I_H turns off the Shockley diode.
5. Important formulas associated with the Shockley diode are

$$I_D = \frac{V_{IN} - V_F}{R}$$

$$f = \frac{1}{(R \times C) \times \ln[V_{IN}/(V_{IN} - V_{BO})]}$$

$$R_{MAX} = \frac{V_{IN} - V_{BO}}{I_S}$$

$$R_{MIN} = \frac{V_{IN} - V_F}{I_H}$$

6. The SCR (silicon-controlled rectifier) is turned on by a trigger signal at its gate.
7. The SCR is turned off when current falls below I_H.
8. A GTO (gate turn-off) switch or GCS (gate-controlled switch) is turned on by a positive pulse at its gate and turned off by a negative pulse at its gate.
9. The SCS (silicon-controlled switch) has two gates, either of which turns on the device when a forward-biased trigger pulse is applied and turns off when a reverse-biased trigger pulse is applied.
10. The LASCR (light-activated silicon-controlled rectifier) is turned on by light and turned off when current falls below I_H.

11. The diac (diode ac switch) can conduct in either direction.

12. The diac turns on when its positive or negative breakover voltage is reached.

13. The diac turns off when current falls below the positive or negative I_H.

14. The triac (triode ac switch) is turned on by a breakover voltage whose value is determined in part by the value of gate current. As I_G increases, $+V_{BO}$ decreases.

15. The UJT (unijunction transistor) contains an emitter and two bases.

16. The UJT is turned on by a peak voltage (V_P) applied to its emitter.

17. The UJT is turned off by reducing the emitter current below its rated valley current (I_V).

18. Important formulas associated with the UJT are

$$R_{BB} = R_{B1} + R_{B2}$$
$$\eta = \frac{R_{B1}}{R_{BB}}$$
$$V_{RB1} = \eta \times V_{BB}$$
$$V_P = (\eta \times V_{BB}) + V_{PN}$$

19. In-circuit troubleshooting involves checking for the correct values of turn on/off voltage and current.

GLOSSARY

Breakover Voltage: The voltage used by certain thyristors to cause conduction.

Diac: A diode ac switch. An ac thyristor that has two leads and conducts in either direction at a positive or negative breakover voltage.

Holding Current: The minimum value of current necessary to maintain conduction.

Oscillate: To fluctuate back and forth between two voltage levels.

Peak Voltage: A certain value of voltage that when applied to the emitter of a UJT enables conduction.

Silicon-Controlled Rectifier (SCR): A thyristor that conducts when a positive trigger voltage is received at the gate.

Shockley Diode: Also called a PNPN junction diode, it conducts at the breakover voltage.

Thyristors: A group of PNPN junction devices.

Time Constant: The length of time it takes for a 63% change of current or voltage in an RC or RL circuit.

Triac: The triode ac switch conducts in either direction when a sufficient level of breakover voltage and current is supplied.

Unijunction Transistor (UJT): The UJT has only one PN junction. It conducts when $V_E = V_P$.

Valley Current: That value of current below which I_E stops conduction in a UJT.

PROBLEMS

SECTION 10-1

The following problems refer to the circuit of Fig. 10-18.

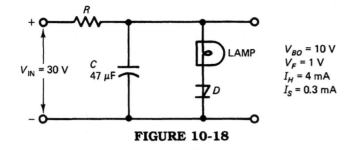

$V_{BO} = 10$ V
$V_F = 1$ V
$I_H = 4$ mA
$I_S = 0.3$ mA

FIGURE 10-18

10-1. What type of circuit is this and what is its purpose?

10-2. Find the minimum value that R can assume in this circuit.

10-3. Find the maximum value that R can assume in this circuit.

10-4. Calculate the value of one time constant when $R = 15$ kΩ.

10-5. Determine the frequency if $R = 15$ kΩ.

10-6. Explain the effect on frequency when R and C are either increased or decreased individually.

10-7. Provide a brief explanation of V_{BO}, V_F, I_H, and I_S.

10-8. Design a circuit with a frequency of 2 Hz by substituting alternate values for R and C.

SECTION 10-2

The following problems refer to the circuit of Fig. 10-19. The SCR in this circuit requires a trigger voltage of 1 V and $I_H = 20$ mA. It drops 1.5 V when conducting and the lamp drops 5 V when lit. The rheostat, R_1, is set at 100 Ω.

10-9. How is an SCR turned on and turned off?

10-10. Determine the value of current through R_3 when the SCR is not conducting.

10-11. What value of resistance from wiper to ground causes the SCR to turn on?

10-12. Once the SCR is turned on, what is the voltage across the branch containing the lamp and the SCR?

10-13. What is the voltage across R_1?

10-14. Find the current through R_1. Remember it is adjusted to 100 Ω.

10-15. Calculate the minimum value of R_1 necessary to turn off the SCR.

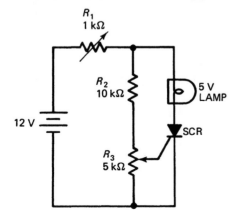

FIGURE 10-19

SECTION 10-3

10-16. List each SCR type along with its requirements for turning on and turning off.

10-17. Design a simple circuit using the LASCR that turns on a porch light when the sun sets. You might use a rheostat to adjust the sensitivity.

10-18. Name an application where a GTO or SCS can be used.

SECTIONS 10-4 AND 10-5

The following problems refer to the circuit of Fig. 10-20.

10-19. How is the diac turned on?

10-20. Explain the purpose of phase control.

10-21. What is the relationship between v_{in} and v_C?

10-22. How is the triac turned on?

10-23. What is the relationship between the resistance of the phase control and the current through the load?

10-24. How is the triac turned off?

10-25. How is the diac turned off?

10-26. Draw two waveforms that are in phase and two that are out of phase by 90°.

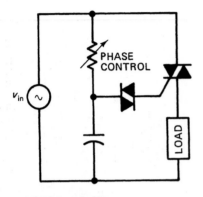

FIGURE 10-20

FIGURE 10-21

SECTION 10-6

The following problems refers to the circuit of Fig. 10-21.

10-27. Describe the operation of this circuit.

10-28. Label the legs of the UJT.

10-29. What is the peak voltage?

10-30. Draw the waveforms that are found at each of the legs.

10-31. Explain the meaning of intrinsic standoff ratio, interbase resistance, valley current, and negative resistance.

SECTION 10-7

10-32. Draw and label the schematic symbols for a Shockley diode, SCR, DIAC, TRIAC, and UJT.

10-33. List the proper resistance measurements for the devices in Problem 32.

10-34. List the methods of turning each of the devices in Problem 32 on and off.

10-35. Describe the type of waveform that is expected across each of the devices in Problem 32.

10-36. Compare the oscillator circuits of Figs. 10-3 and 10-17. Explain the difference in operation.

ANSWERS TO REVIEW QUESTIONS

1. Switching and triggering	**2.** PNPN, or four-layer diode	**3.** three
4. Breakover voltage	**5.** Anode and cathode	**6.** Holding
7. False	**8.** pulsating waveform	**9.** 12
10. more	**11.** 1 V	**12.** time constant
13. 19.4 Hz	**14.** False	**15.** 5.67 kΩ

16. Silicon-controlled rectifier

17. By a positive trigger at the gate

18. Anode, cathode, and gate

19. False

20. No

21. 1 or 2 V

22. Holding current

23. 1 or 2 V

24. Reduces the current below I_H

25. Forward current is removed

26. Gate-controlled switch (GCS)

27. Gate turn-off switch

28. By a positive trigger

29. By a negative trigger

30. Anode, cathode, anode gate, and cathode gate

31. Silicon-controlled switch

32. By a forward trigger on the anode gate or the cathode gate

33. By a reverse trigger on the anode gate or the cathode gate

34. Light-activated silicon-controlled rectifier (LASCR)

35. By adding light

36. By a low I or reverse-trigger V

37. By a gate resistor

38. By the contraction of $d$iode ac switch

39. 25 to 40 V

40. False

41. It is reduced

42. Yes

43. I_F or $I_R < I_H$

44. decreases

45. The current $< I_H$

46. Yes

47. increase

48. more

49. 90°

50. decreased

51. Unijunction transistor

52. One

53. Emitter, base 1, and base 2

54. Interbase resistance

55. Intrinsic standoff ratio

56. $V_E = > V_P$

57. $I_E < I_V$

58. negative resistance

59. $V_P = 5.7$ V

60. Decrease

61. High

62. By the breakover voltage

63. The forward gate–cathode resistance

64. By a positive trigger on the gate

65. ac

66. By low current

67. Very high

68. By V_P at the emitter

69. False

70. $I_E < I_V$

Operational Amplifiers

When you have completed this chapter, you should be able to:

- Investigate the important characteristics of the op amp.
- Analyze the operation of the inverting amplifier.
- Analyze the operation of the noninverting amplifier.
- Measure and observe open-loop and closed-loop characteristics.
- Recognize the faulty operation of the op amp.

INTRODUCTION

The operational amplifier is an integrated circuit that provides excellent gain over a broad range of frequencies. Convenience and low cost make it a fine choice for designing circuits. It is often simply called an op amp.

Our discussion centers around the 741 op amp, which is one of the more popular general-purpose op amps. There are dozens of other op amps that provide characteristics that vary from those of the 741. However, the theory discussed in this chapter applies equally well to any op amp.

11-1 DIFFERENTIAL AMPLIFIER

The first stage within an op amp consists of a differential amplifier. It amplifies the difference between the two input signals. Figure 11-1 shows an example of a differential amplifier.

DC Analysis   The differential amplifier of Fig. 11-1 is biased by both positive and negative dc supply voltages. The emitter supply (V_{EE}) provides 0.7 V for the

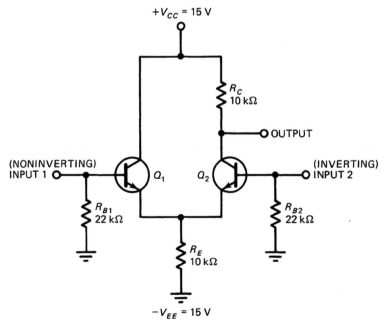

FIGURE 11-1   Differential amplifier.

base–emitter junctions of the transistors. A relatively small base current for each transistor results in nearly zero volts dropped across the base resistors. Therefore, if zero volts are dropped across the base resistors and 0.7 V is dropped across the base–emitter junction, the remainder of V_{EE} must be dropped across R_E.

$$V_{R_E} = V_{EE} - V_{BE}$$
$$= 15\ V - 0.7\ V$$
$$= 14.3\ V$$

Apply Ohm's law to determine the current through R_E.

$$I_{R_E} = \frac{V_{R_E}}{R_E}$$
$$= \frac{14.3\ V}{10\ k\Omega}$$
$$= 1.43\ mA$$

The current through R_E is split between the two transistors. Considering the transistors to be evenly matched allows for evenly dividing the emitter resistors' current. The result is the emitter current for each transistor.

$$I_E = \frac{I_{R_E}}{2}$$
$$= \frac{1.43\ mA}{2}$$
$$= 715\ \mu A$$

Since the collector current is approximately equal to the emitter current, it can be stated that $I_C = I_E$. The voltage drop across R_C can now be determined.

$$V_{R_C} = I_C \times R_C$$
$$= 715\ \mu A \times 10\ k\Omega$$
$$= 7.15\ V$$

Finally, the dc voltage at the output is determined as follows:

$$V_{OUT} = V_{CC} - V_{R_C}$$
$$= 15\ V - 7.15\ V$$
$$= 7.85\ V$$

AC Analysis   The goal of the ac analysis is to determine the gain of the differential amplifier of Fig. 11-1. First, the ac emitter resistance (r'_e) must be determined. The value of dc emitter current is taken from the previous dc analysis.

$$r'_e = \frac{25\ mV}{I_E}$$
$$= \frac{25\ mV}{715\ \mu A}$$
$$= 35\ \Omega$$

Since two identical transistors are used in this circuit, the value of r_e' is multiplied by a factor of 2 in the following gain formula.

$$Av = \frac{R_C}{2r_e'}$$

$$= \frac{10 \text{ k}\Omega}{2 \times 35\Omega}$$

$$= 143$$

A signal applied to either input of the differential amplifier is amplified by a factor of 143 at the output.

Noninverting Operation   The differential amplifier in Fig. 11-2 is set up for the noninverting mode of operation. An ac signal is applied to the base of Q_1, which acts as an emitter follower. The signal at the emitter of Q_1 is of the same value and phase as the input signal. This signal is then applied to the common-base amplifier of Q_2. Recall that a signal applied to the emitter of a common-base circuit is amplified and in phase at the collector. Looking at the process as a whole, you find that a signal applied to the noninverting input is amplified and in phase at the output.

Earlier it was stated that a differential amplifier amplifies the difference between its two input signals. In the case of the circuit of Fig. 11-2, the signal at input 1 is 10 mV p–p and Input 2 is tied to ground (0 V). It is the difference between these two signals that is amplied by the gain found during the ac analysis.

$$v_{in} = \text{Input 1} - \text{Input 2}$$

$$= 10 \text{ mV p–p} - 0 \text{ V}$$

$$= 10 \text{ mV p–p}$$

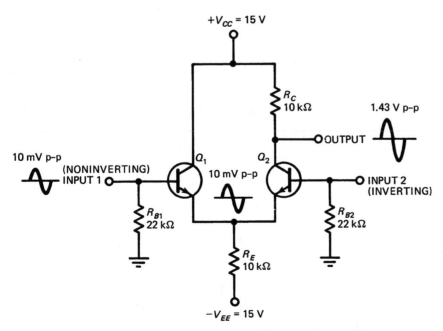

FIGURE 11-2   Noninverting differential amplifier.

and

$$v_{out} = v_{in} \times A_v$$
$$= 10 \text{ mV p–p} \times 143$$
$$= 1.43 \text{ V p–p}$$

You may be wondering what would happen at the output if both inputs were of the same value and phase. The answer is that if the transistors were perfectly matched, the output would be zero. It is the differential amplifier's ability to reject signals that are common to both of its inputs that is defined as common-mode rejection. This is a very useful feature for rejecting noise that may be present at the inputs. The differential amplifier is able to reject noise while at the same time amplifying the desired input signal. A ratio of these two values is called the **common-mode rejection ratio (CMRR)**. It can be determined by comparing the gain of the desired input signal to that of the noise. The gain of the noise is normally less than 1 and ideally zero. Therefore, the value of CMRR is usually anywhere from a few hundred to infinity. The larger the CMRR, the better the amplifier is able to reject noise.

Common-Mode Rejection Ratio (CMRR): Indicates how well an op amp is able to reject noise; the larger the value, the better.

$$\text{CMRR} = \frac{A_v}{A_{\text{NOISE}}}.$$

Inverting Operation Figure 11–3 shows the inverting mode of operation for the differential amplifier. Q_2 acts as a common-emitter amplifier in this circuit. The signal applied to its base is amplified and inverted (180° phase shift) at the collector.

Keep in mind that it is actually the difference between the two inputs that is amplified. In this example, 10 mV p–p is applied to Input 2 and Input 1 is tied to ground (0 V). The difference between the two inputs is 10 mV p–p.

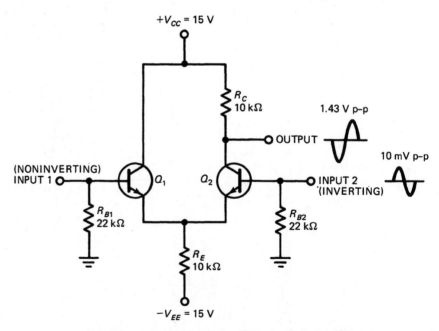

FIGURE 11-3 Inverting differential amplifier.

$$v_{in} = \text{Input 1} - \text{Input 2}$$
$$= 0 \text{ V} - 10 \text{ mV p–p}$$
$$= 10 \text{ mV p–p}$$

and

$$v_{out} = v_{in} \times -A_V$$
$$= 10 \text{ mV p–p} \times (-143)$$
$$= -1.43 \text{ V p–p}$$

The polarity is dropped in the case of v_{in} because it is an ac signal. A minus sign was added to the value of gain, not to denote polarity, but to indicate the phase shift. The input signal is amplified by a factor of 143 and inverted. Figure 11-3 illustrates the 180° phase shift between the input and output signals.

EXAMPLE 11-1

The resistor values of the circuit of Fig. 11-3 are changed so that $R_E = 15 \text{ k}\Omega$, $R_C = 15 \text{ k}\Omega$, $R_{B1} = 33 \text{ k}\Omega$, and $R_{B2} = 33 \text{ k}\Omega$. The supply voltages are the same. Determine (a) I_E and (b) A_v.

Solution:

(a) First, find the current of the emitter resistor and then split it in half for the emitter current of each transistor.

$$I_{R_E} = \frac{14.3 \text{ V}}{15 \text{ k}\Omega}$$
$$= 953 \text{ } \mu A$$

and

$$I_E = \frac{953 \text{ } \mu V}{2}$$
$$= 477 \text{ } \mu A$$

(b) Start by finding the ac emitter resistance and then apply it to the gain formula. Notice that a minus sign is used to denote the phase inversion.

$$r_e' = \frac{25 \text{ mV}}{477 \text{ } \mu A}$$
$$= 52.4 \text{ } \Omega$$

$$A_v = \frac{-15 \text{ k}\Omega}{2 \times 52.4 \text{ } \Omega}$$
$$= -143 \qquad \blacksquare$$

REVIEW QUESTIONS

1. The differential amplifier amplifies the difference in voltage between its two inputs. True or false?
2. What is the name given to a differential amplifier's ability to reject signals common to both inputs?

3. The emitter current of Q_2 is equal to the current flowing through R_E in the circuit of Fig. 11-3. True or false?

4. What would be the output of the circuit of Fig. 11-3 if Input 1 = 10 mV p–p and Input 2 = 30 mV p–p?

5. Determine the collector current of Q_2 in the circuit of Fig. 11-3 if the supply voltages were changed to +12 V.

11-2 OP-AMP CHARACTERISTICS

The op amp is designed to provide good amplification over a wide range of frequency and impedance conditions. The following characteristics accomplish this goal. The actual spec sheet for the 741 op amp is also shown for reference (see page 432).

TYPES uA741M, uA741C
GENERAL-PURPOSE OPERATIONAL AMPLIFIERS

electrical characteristics at specified free-air temperature, V_{CC+} = 15 V, V_{CC-} = −15 V

PARAMETER		TEST CONDITIONS†		uA741M			uA741C			UNIT
				MIN	TYP	MAX	MIN	TYP	MAX	
V_{IO}	Input offset voltage	V_O = 0	25°C		1	5		1	6	mV
			Full range			6			7.5	
$\Delta V_{IO(adj)}$	Offset voltage adjust range	V_O = 0	25°C		±15			±15		mV
I_{IO}	Input offset current	V_O = 0	25°C		20	200		20	200	nA
			Full range			500			300	
I_{IB}	Input bias current	V_O = 0	25°C		80	500		80	500	nA
			Full range			1500			800	
V_{ICR}	Common-mode input voltage range		25°C	±12	±13		±12	±13		V
			Full range	±12			±12			
V_{OM}	Maximum peak output voltage swing	R_L = 10 kΩ	25°C	±12	±14		±12	±14		V
		$R_L \geq$ 10 kΩ	Full range	±12			±12			
		R_L = 2 kΩ	25°C	±10	±13		±10	±13		
		$R_L \geq$ 2 kΩ	Full range	±10			±10			
A_{VD}	Large-signal differential voltage amplification	$R_L \geq$ 2 kΩ	25°C	50	200		20	200		V/mV
		V_O = ±10 V	Full range	25			15			
r_i	Input resistance		25°C	0.3	2		0.3	2		MΩ
r_o	Output resistance	V_O = 0, See Note 6	25°C		75			75		Ω
C_i	Input capacitance		25°C		1.4			1.4		pF
CMRR	Common-mode rejection ratio	V_{IC} = V_{ICR} min	25°C	70	90		70	90		dB
			Full range	70			70			
k_{SVS}	Supply voltage sensitivity ($\Delta V_{IO}/\Delta V_{CC}$)	V_{CC} = ±9 V to ±15 V	25°C		30	150		30	150	μV/V
			Full range			150			150	
I_{OS}	Short-circuit output current		25°C		±25	±40		±25	±40	mA
I_{CC}	Supply current	No load, V_O = 0	25°C		1.7	2.8		1.7	2.8	mA
			Full range			3.3			3.3	
P_D	Total power dissipation	No load, V_O = 0	25°C		50	85		50	85	mW
			Full range			100			100	

†All characteristics are measured under open-loop conditions with zero common-mode input voltage unless otherwise specified. Full range for uA741M is −55°C to 125°C and for uA741C is 0°C to 70°C.
NOTE 6: This typical value applies only at frequencies above a few hundred hertz because of the effects of drift and thermal feedback.

Internal Amplifiers   Inside the op amp are dozens of transistors, resistors, and capacitors arranged to create several different circuits. The two inputs of the op amp connect to a differential amplifier. The differential-amplifier stage amplifies the difference in voltage between the two inputs.

The output of the differential-amplifier stage is passed onto a second stage called the voltage amplifier. As its name implies, it provides a tremendous voltage

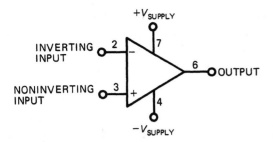

FIGURE 11-4   Op amp.

Operational amplifier.

gain. It also serves to match impedances between the first and third stages within the op amp.

The third stage is comprised of a power amplifier. It is designed to provide the necessary power for low-impedance loads. Together, these stages provide the overall internal circuitry of the op amp.

The internal structure of the op amp is not accessible. Therefore, attention is focused on the IC form of op amp. Figure 11-4 shows a typical schematic symbol for the op amp. The numbers next to the leads refer to the pin numbers of the IC itself.

Pinout   The eight-pin DIP op amp is shown in Fig. 11-5. DIP represents **dual in-line package** and is a standard method of packaging ICs. The name arises because the pairs (dual) of pins are directly across (in line) from each other.

Normally, only five out of eight pins are used, as shown in Fig. 11-4. But, this discussion includes all eight pins in numerical order. Notice that the pins run from 1 to 8 in a counter clockwise direction when viewed from above. This is true of all ICs. Pin 1 can be located by looking for the side containing a notch or circular indentation, as shown in Fig. 11-5.

Dual In-Line Package (DIP): A common form of packaging ICs. The pins are connected to the chip in pairs in line with one another.

- **Pin 1.** The offset null is used to balance the op amp. A potentiometer may be connected from pin 1 to pin 5 with the wiper connected to the negative voltage supply, as shown in Fig. 11-6. The potentiometer is adjusted so as to zero the inputs of the op amp for critical applications.

- **Pin 2.** The inverting input is indicated by a minus sign on the schematic symbol. It provides a 180° phase shift between the input and output. If a +1 V is placed at this input and a gain of 5 is provided, then the output voltage would be −5 V. Note the change in polarity.

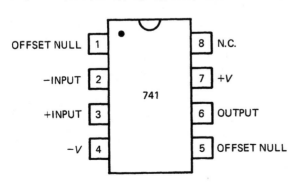

FIGURE 11-5   Pinout of the 741 op amp.

FIGURE 11-6   Potentiometer used to adjust the offset null.

- **Pin 3.** This is the noninverting input. As its name implies, it does not invert the polarity of the input while amplifying it.

- **Pin 4.** The negative supply voltage is used to bias the op amp. Its voltage usually is the same value as pin 7, but with opposite polarity. The recommended voltage for the 741 op amp is between -3 and -15 V.

- **Pin 5.** See the description under pin 1 for the use of the offset-null inputs.

- **Pin 6.** A single output is provided by the 741 op amp. The output is measured with respect to the ground of the connected circuit. Its voltage cannot exceed that of the power supplies in either a negative or positive direction.

- **Pin 7.** A positive supply voltage is needed to properly bias the 741 op amp. All together, two supplies are used. The recommended value of this supply is between $+3$ and $+15$ V (see Pin 4).

- **Pin 8.** N.C. means no connection. This pin is provided for standardization only. The industry has decided to limit the variety of packages for the sake of convenience. All DIP ICs have an even number of pins.

Open Loop: When no feedback is used.

Feedback: Returning all or a portion of the output signal back to the input. Positive feedback enhances, or adds, to the input signal, and negative feedback subtracts from the input signal.

Closed Loop: Indicates that some form of feedback is used.

Voltage Gain   The op amp provides a voltage gain of between 20,000 and 200,000 in the open-loop configuration. **Open loop** means that **feedback** resistors are not connected from the output of the op amp back to either input. This is a tremendous amount of voltage gain. As a matter of fact, it is too much. And that is why closed-loop operation is discussed most of the time. **Closed loop** means that feedback resistors are connected from the output to the inputs. By applying the concept of negative feedback, we can reduce the gain to a more usable level. This topic is discussed in the next section.

Input and Output Impedances   The input impedance of the op amp is very high. It is rated at 2 MΩ for the open-loop configuration. This allows it to be connected to a high-impedance source. Because of this high input impedance, the op amp draws an extremely small amount of input current. This fact applies to both inputs.

The output impedance, on the other hand, is very low. The spec sheet lists it as 75 Ω for the open-loop configuration. This allows the output to be connected to low-impedance loads.

Common-Mode Rejection   A specification called the common-mode rejection ratio (CMRR) describes how well the op amp rejects noise (unwanted signals). Common-mode signals are those unwanted signals that are present at both inputs at the same time. The tremendous voltage gain of the op amp would cause major problems if it indiscriminately amplified noise along with the desired signal. Since there is no difference between the noise on the two inputs, ideally, the differential amplifier should eliminate it. The op amp comes close to this ideal; however, there is a certain amount that is amplified. The ratio of the overall gain as compared to the gain of the noise is what determines the CMRR. The spec sheet for the 741 lists the CMRR at 90 dB, which means it will amplify the desired signal over 30 thousand times more than the unwanted noise.

Decibel (dB): Used to express gain.

The **decibel** (dB) is another way of expressing gain. It can be converted to the ordinary value of gain by using the antilogarithm, or 10^x, function.

$$A_v = 10^{dB/20}$$

$$= inv \; log(dB/20)$$

To go from ordinary gain to decibels,

$$dB = 20 \log(A_v)$$

EXAMPLE 11-2

Convert (a) +40 dB from decibels to a value for A_v, and (b) an A_v of 1000 to decibel form.

Solution:

(a) The number 20 in the formula is a constant.

$$A_v = \text{inv} \log(+40 \text{ dB}/20)$$
$$= 100$$

(b)

$$dB = 20 \log(1000)$$
$$= +60 \text{ dB} \quad \blacksquare$$

Frequency vs. Voltage Gain   The op amp has a flat frequency response that extends from 0 Hz (dc) up to 1 MHz ac. A relationship between frequency and voltage gain does exist, as shown in Fig. 11-7.

As the voltage gain is lowered, the range of operating frequencies increases. At a voltage gain of 1, the range of frequencies extends from 0 Hz to 1 MHz. When the voltage gain is 10^3, the frequency range is reduced to 0–1 KHz.

The ability of the op amp to amplify dc as well as ac is one of its unique characteristics. The range of frequencies is referred to as the **bandwidth**. It can be said that voltage gain and bandwidth are inversely proportional.

Bandwidth: A range of frequencies.

REVIEW QUESTIONS

6. The inputs of an op amp are connected internally to a _____ amplifier.
7. The final amplifier stage within the op amp is the _____ amplifier.
8. What type of IC package is used in Fig. 11-5?

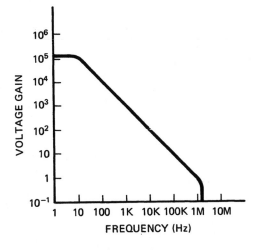

FIGURE 11-7   Frequency range vs. voltage gain.

9. What pair of pins are used to balance the input of the op amp?

10. Which input provides a 180° phase shift?

11. What range of supply voltages is normally used to bias the 741 op amp?

12. What do the letters N.C. refer to on IC schematic symbols?

13. A voltage gain of 20,000 to 200,000 is provided by the op amp when used in an _____ _____ configuration.

14. What is the approximate input impedance of the 741 op amp in an open-loop configuration?

15. The higher the CMRR, the better. True or False?

16. As voltage gain is decreased, the bandwidth is _____.

17. What is the equivalent value of decibels for a voltage gain of 1000?

18. Express 25 dB in terms of ordinary voltage gain.

11-3 INVERTING AMPLIFIER

The inverting amplifier takes its name because the inverting input serves as the input to the amplifier. Figure 11-8 shows an example of the inverting amplifier. A +2-V dc voltage is the input signal for the circuit.

Feedback: Returning all or a portion of the output signal back to the input. Positive feedback enhances, or adds, to the input signal, and negative feedback subtracts from the input signal.

Negative Feedback A resistor tied from the output to the input of any circuit provides what is called **feedback**. This is because the output signal is fed back to the input through the resistor. When the feedback signal opposes or negates all or part of the original input signal it is called negative feedback.

The positive input voltage of the circuit of Fig. 11-8 is amplified and inverted. A feedback resistor, R_F, feeds back a portion of this negative output voltage to the input. The net result is that almost all of the original input signal is cancelled. A condition called "virtual ground" occurs at the inverting input of the op amp itself.

Virtual Ground: A point in the circuit that appears at ground potential but is not directly connected to ground.

It is called **virtual ground** because it acts like ground although it is not connected to ground. If the virtual ground's voltage were measured with reference to ground, it would register 0 V. It is not really zero; it is just too small to easily measure with conventional meters.

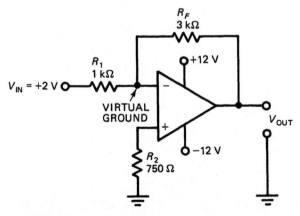

FIGURE 11-8 Inverting amplifier using negative feedback.

Voltage Gain Using Negative Feedback The voltage gain of the op amp can be controlled through external resistors. Remember that open-loop gain can be as high as 200,000. Reducing the voltage gain to a usable level is accomplished through negative feedback. The exact value of voltage gain of the inverting amplifier is determined by

$$A_v = \frac{-R_F}{R_1}$$

$$= \frac{-3 \text{ k}\Omega}{1 \text{ k}\Omega}$$

$$= -3$$

The minus sign is included to denote the phase inversion that occurs as a result of using the inverting input. One can readily see that voltage gain can be adjusted by changing the value of one or both of these resistors. Since the input voltage is +2 V in the circuit of Fig. 11-8, the output voltage is

$$V_{OUT} = V_{IN} \times A_v$$

$$= +2 \text{ V} \times (-3)$$

$$= -6 \text{ V}$$

The output voltage is measured with reference to ground. Earlier, we mentioned a point called virtual ground. If the output is measured with reference to virtual ground, -6 V will still be read. The voltage between virtual ground and the output in the circuit of Fig. 11-8 is equivalent to the voltage drop across R_F. The voltage across R_F is actually positive with reference to the actual ground.

$$V_{R_F} = V_{OUT}$$

$$= 6 \text{ V}$$

Knowing the voltage across R_F allows you to calculate the current through R_F.

$$I_{R_F} = \frac{V_{R_F}}{R_F}$$

$$= \frac{6 \text{ V}}{3 \text{ k}\Omega}$$

$$= 2 \text{ mA}$$

This is the same value of current that flows through R_1 in the circuit of Fig. 11-8. It can be proven by considering that the voltage across R_1 with reference to virtual ground is equal to the +2-V input voltage. Therefore,

$$V_{R1} = V_{IN}$$

$$I_{R1} = \frac{V_{R1}}{R_1}$$

$$= \frac{2V}{1 \text{ k}\Omega}$$

$$= 2 \text{ mA}$$

The current flowing into the op amp is so small (nanoamps) that it is considered to be zero. High input impedance is the reason for this extremely small current. The output impedance of the op amp is very low. It provides the 2 mA that flows back through R_F and R_1 to the positive input voltage.

The maximum output voltage in this circuit is theoretically ± 12 V, as determined by the values of the supply voltages. Actually, the maximum would be slightly less (about ± 10 V or ± 11 V) due to the voltage drop within the op amp itself.

Practice Problem: What would be the output voltage in the circuit of Fig. 11-8 if $R_1 = 2.7$ kΩ and $R_F = 10$ kΩ? *Answer:* $V_{OUT} = -7.41$ V.

Input and Output Impedances
Although the input impedance of the op amp may be large, the overall input impedance of the amplifier in Fig. 11-8 is reduced due to the use of negative feedback. By using virtual ground as the starting point, it is obvious that R_1 is the only resistance between virtual ground and the input voltage. This resistance is in parallel with the very large input impedance of the op amp. Therefore, the overall input impedance of the amplifier is approximately equal to R_1.

$$z_{in} = R_1$$
$$= 1 \text{ k}\Omega$$

The output impedance is approached with the assumption that the open-loop gain of this particular op amp is 100,000. We also use the open-loop output impedance value of 75 Ω from the spec sheet. This information, along with the present closed-loop gain of the circuit, allows us to approximate the output impedance.

$$z_{out} = \frac{75}{100,000/A_v}$$
$$= \frac{75}{100,000/3}$$
$$= 0.002 \ \Omega$$

Notice that we are using open-loop values as our starting point. The open-loop gain of the op amp was stated as being somewhere between 20,000 and 200,000. Even if the low-end value of gain were used, the output impedance would still be well under 1 Ω.

EXAMPLE 11-3 _____

The resistor values of the circuit of Fig. 11-8 are changed so that $R_1 = 2$ kΩ, $R_F = 10$ kΩ, and $R_2 = 1.6$ kΩ. Determine the value of (a) A_v, (b) V_{OUT}, (c) z_{in}, and (d) z_{out}.

Solution:
(a) The gain of the amplifier is determined by the values of feedback and input resistors.

$$A_v = \frac{-10 \text{ k}\Omega}{2 \text{ k}\Omega}$$
$$= -5$$

(b) The output voltage is solved for by using the gain.

$$V_{OUT} = +2 \text{ V} \times (-5)$$
$$= -10 \text{ V}$$

(c) The input impedance is equal to the value of the input resistor.

$$z_{in} = 2 \text{ k}\Omega$$

(d) We use the open-loop output impedance and gain of the op amp to aid us in finding the output impedance of this circuit.

$$z_{out} = \frac{75}{100,000/5}$$
$$= 3.75 \text{ m}\Omega \qquad \blacksquare$$

Practice Problem: What are the input and output impedances of the circuit of Fig. 11-9? Use the open-loop values of 75Ω for the output impedance and 100,000 for the gain.   *Answer:* $z_{in} = 5 \text{ k}\Omega$ and $z_{out} = 0.015 \text{ }\Omega$.

Offset Resistor   The resistor that is tied from the noninverting input to ground in the circuit of Fig. 11-9 is called the offset resistor. Its function is to balance the resistance at the inputs of the op amp. When the op amp "looks" out of its inverting input, it "sees" R_F in parallel with R_1. Therefore, in designing the op amp circuit, it is a good idea to place a similar amount of resistance at the noninverting input. Therefore,

$$R_2 = R_F \| R_1$$
$$= 4.76 \text{ k}\Omega$$

The closest standard value of resistance is 4.7 kΩ. Using the offset resistor reduces the amount of **offset voltage** developed between the inputs.

Bandwidth   The inverting amplifier of Fig. 11-9 operates over a wide range of frequencies. It was demonstrated earlier that a relationship exists between voltage

Offset Voltage: An unwanted voltage that may exist between the inputs of the op amp. The offset-null inputs of the op amp are used to eliminate it.

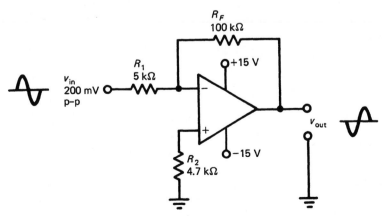

FIGURE 11-9   Inverting amplifier with ac input.

gain and bandwidth. The gain of this circuit is 20, and using that value, we can find the bandwidth of frequencies at which this circuit operates.

But, first, you must find a quantity called the gain–bandwidth product. It is found by taking any value of gain from the vertical axis in Fig. 11-7 and multiplying it by the frequency read at the adjacent point of the sloping line. For example, a voltage gain of 10^3 meets the slope at the point where frequency is 1 kHz. If these two values are multiplied, the result is a constant called the gain–bandwidth product (GBP).

$$\text{GBP} = A \times f$$
$$= 10^3 \times 1 \text{ kHz}$$
$$= 1 \text{ MHz}$$

The same GBP value will be found regardless of which values of gain and frequency are used. A GBP of 1 MHz is standard for all 741 op amps. The bandwidth can be found by

$$\text{BW} = \frac{\text{GBP}}{A}$$
$$= \frac{1 \text{ MHz}}{20}$$
$$= 50 \text{ kHz}$$

This means that the amplifier of Fig. 11-9 provides a gain of 20 for any input signal from 0 Hz to 50 kHz. Of course, the output must remain within the limits of the positive and negative supply voltages. The concept of bandwidth takes on greater significance when filters are discussed in the next chapter.

Practice Problem: What is the bandwidth available to the amplifier of Fig. 11-8? *Answer:* BW = 333 kHz.

Positive Feedback   Just as negative feedback reduces the input signal, positive feedback enhances, or reinforces, the input signal. An example of positive feedback is the high-pitched screeching noise that results when a microphone is held in front of a loudspeaker. What happens is that the output from the loudspeaker enters the mic and is amplified. The newly amplified sound enters the mic and is amplified again. As this process is repeated over and over, the output of the amplifier reaches an extremely loud and annoying pitch. Positive feedback causes the signal to eventually saturate the amplifier.

Voltage Gain Using Positive Feedback   Theoretically, the voltage gain for the circuit of Fig. 11-10 is infinite. But practically, this circuit is limited by the value of its supply voltages. When positive feedback is used, the op amp is forced into saturation. The output reaches its maximum voltage, which is usually about 1 or 2 V below the value of the supply voltages. The 1 or 2 V are dropped internally by the op amp.

Let us start by assuming that the output of the circuit of Fig. 11-10 is saturated. If the output were saturated positively, then the output voltage would be equal to +8 V. This +8 V is split evenly across the two 5-kΩ resistors and the voltage at the noninverting input of the op amp is equal to the voltage across R_2.

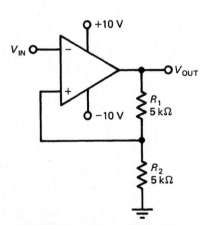

FIGURE 11-10 Inverting amplifier using positive feedback.

CHAP. 11 / OPERATIONAL AMPLIFIERS

The output remains in this saturated condition until V_{IN} is made greater than +4 V. As soon as V_{IN} is increased beyond +4 V, the op amp amplifies the positive difference between its inputs by the infinite gain. The output saturates negatively due to the phase inversion. A -4 V are developed across R_2 and fed back to the noninverting input. This increases the positive difference between the inputs, which, in turn, reinforces the output. Once the output saturates, there is no room for further amplification.

At this point, the output remains negatively saturated until the voltage at the inverting input exceeds -4 V. When V_{IN} is made more negative than -4 V, the negative difference between the inputs is inverted, amplified, and reinforced by the positive feedback. The output returns to a positively saturated condition.

Another name for this circuit is the Schmitt trigger. Useful applications of this circuit are discussed in the section bearing that name.

REVIEW QUESTIONS

19. When the output of a circuit is returned to the input it is called _____ .
20. A point in a circuit that is at a potential of zero volts but is not connected to ground is called _____ _____ .
21. How is the gain of the op amp reduced?
22. Why is a minus sign used in the gain formula for an inverting amplifier?
23. The current through the feedback resistor is equal to the current through R_1 in the circuit of Fig. 11-8. True or false?
24. Negative feedback (increases or decreases) the input impedance of the inverting amplifier.
25. Op amps provide a very (small or large) output impedance.
26. What is the name given to the resistor that is used to balance the inputs of the op amp?
27. What is the value of the gain–bandwidth product for the 741 op amp?
28. The bandwidth is (increased or decreased) as voltage gain is increased.
29. What type of feedback can cause an op amp to saturate?
30. If the output of the circuit of Fig. 11-10 were saturated to -8 V, what value of V_{IN} would cause it to saturate positively?

11-4 NONINVERTING AMPLIFIER

The noninverting amplifier of Fig. 11-11 uses the noninverting input to receive the input signal. It offers several advantages over the inverting amplifier such as increased input impedance and voltage gain. There is no phase shift between input and output signals. This may also be advantageous in certain circumstances.

Negative Feedback   Negative feedback is used in the circuit of Fig. 11-11. The positive input voltage is amplified and appears at the output . The output is fed back, in turn, through R_F to the inverting input. A voltage develops across R_1 that is equivalent to the value of V_{IN}. This reduces the difference between the inverting and noninverting inputs, thus having a negative effect on the difference to be amplified.

Keep in mind that if negative feedback were not used, the open-loop gain of the circuit would saturate the output. There is, of course, a small difference

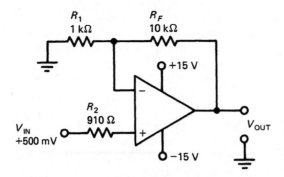

FIGURE 11-11 Noninverting amplifier using negative feedback.

between the inverting and noninverting inputs otherwise there would be nothing to amplify. Negative feedback reduces this to a level acceptable by the op amp.

Voltage Gain Using Negative Feedback The voltage gain of the noninverting amplifier is slightly better than that of the inverting amplifier. When the output is measured with reference to ground in the circuit of Fig. 11-11, both R_1 and R_F are included.

The output voltage of the inverting amplifier consists of only the voltage across R_F. Now the output voltage consists of the voltage across R_F and the voltage across R_1. Therefore, the modified gain formula applied to the circuit of Fig. 11-11 is

$$A_v = \frac{R_F}{R_1} + 1$$

$$= \frac{10 \text{ k}\Omega}{1 \text{ k}\Omega} + 1$$

$$= 11$$

The output voltage is, therefore,

$$V_{OUT} = A_V \times V_{IN}$$

$$= 11 \times (+500 \text{ mV})$$

$$= 5.5 \text{ V}$$

Notice that there is no phase inversion. The noninverting amplifier, as its name implies, does not invert the input signal.

It was mentioned earlier that the voltage across R_1 is equal to the input voltage. Using this value, we can find the amount of feedback current that is flowing. The same current flows through R_1 and R_F.

$$V_{R1} = V_{IN}$$

$$= +500 \text{ mV}$$

so

$$I_{R1} = \frac{V_{R1}}{R_1}$$

$$= \frac{+500 \text{ mV}}{1 \text{ k}\Omega}$$

$$= 500 \text{ }\mu\text{A}$$

and

$$I_{R_F} = I_{R1}$$
$$= 500 \ \mu A$$

Therefore,

$$V_{R_F} = I_{R_F} \times R_F$$
$$= 500 \ \mu A \times 10 \ k\Omega$$
$$= 5 \ V$$

The sum of adding the voltages of R_F and R_1 equals V_{OUT}. So a slightly better gain without the phase shift is offered by the noninverting amplifier.

Practice Problem: What would be the output voltage and feedback current in the circuit of Fig. 11-11 if $R_F = 15 \ k\Omega$ and $R_1 = 2.7 \ k\Omega$? *Answer:* $V_{OUT} = +3.28 \ V$ and $I_{R_F} = 185 \ \mu A$.

Input and Output Impedances The absence of virtual ground at the input of this circuit greatly increases the input impedance. We continue to use the open-loop values of 100,000 for gain and 2 MΩ for input impedance in our formulas for the closed-loop circuit of Fig. 11-11. The input impedance is found by

$$z_{in} = 2 \ M\Omega \times \frac{100,000}{A_V}$$
$$= 2 \ M\Omega \times \frac{100,000}{11}$$
$$= 18.2 \ G\Omega$$

That's right, Gigaohms! The input impedance of the noninverting amplifier is tremendous when compared to that of the inverting amplifier. This value of input impedance will not load down signal sources having even the largest of resistances.

Since the output impedance is extremely small, the op amp can drive very small load resistances. Output impedance is found the same way as it was for the inverting amplifier. The open-loop values of 100,000 for gain and 75 Ω for output impedance are used.

$$z_{out} = \frac{75\Omega}{100,000/A_V}$$
$$= \frac{75\Omega}{100,000/11}$$
$$= 0.008 \ \Omega$$

Practice Problem: What would be the z_{in} and z_{out} for a circuit whose gain is 150? Use the open-loop values stated before. *Answer:* $z_{in} = 1.33 \ G\Omega$ and $z_{out} = 0.113\Omega$.

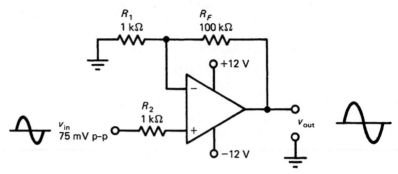

FIGURE 11-12 Noninverting amplifier with ac input.

Offset Resistor   The offset resistor in the circuit of Fig. 11-11 is R_2. It offsets any difference in resistance between the inputs of the op amp. The resistance at the inverting input is the parallel sum of R_1 and R_F. So

$$R_2 = R_1 \parallel R_F$$
$$= 1 \text{ k}\Omega \parallel 10 \text{ k}\Omega$$
$$= 909 \ \Omega$$

The closest standard value of resistance available is 910 Ω.

Practice Problem: What is the exact value of offset resistance for the circuit of Fig. 11-12?   *Answer:*   Offset resistance = 990 Ω.

EXAMPLE 11-4

The resistor values in the circuit of Fig. 11-11 are changed so that $R_1 = 2.7$ kΩ, $R_F = 33$ kΩ, and R_2 is unknown. Determine the value of (a) A_V, (b) V_{OUT}, (c) z_{in}, and (d) R_2.

Solution:

(a) The gain of the noninverting amplifier is determined by the value of the feedback resistor and R_1.

$$A_V = \frac{33 \text{k}\Omega}{2.7 \ \Omega} + 1$$
$$= 13.2$$

(b) We use the gain to determine the output voltage.

$$V_{OUT} = 13.2 \times 500 \text{ mV}$$
$$= 6.6 \text{ V}$$

(c) The input impedance of the noninverting amplifier is much greater than that of the inverting amplifier. The spec sheet of the op amp shows the open-loop input impedance as 2 MΩ and we are using an open-loop gain of 100,000.

$$z_{in} = 2 \text{ M}\Omega \times \frac{100,000}{13.2}$$
$$= 15.2 \text{ G}\Omega$$

(d) The value of the offset resistor is equal to the parallel sum of R_1 and R_F.

$$R_2 = 2.7 \text{ k}\Omega \parallel 33 \text{ k}\Omega$$
$$= 2.5 \text{ k}\Omega \qquad \blacksquare$$

Bandwidth The procedure for finding bandwidth is the same for both inverting and noninverting amplifiers. First, recall that the gain–bandwidth product is equal to 1 MHz for the 741 op amp. Second, keep in mind that a slightly different gain formula exists for noninverting amplifiers. Putting it all together, we find that

$$BW = \frac{GBP}{A_V}$$
$$= \frac{1 \text{ MHz}}{101}$$
$$= 9.9 \text{ kHz}$$

The circuit of Fig. 11-12 provides a gain of 101 to a range of frequencies extending from 0 Hz to 9.9 kHz. Notice that there is no phase inversion between input and output in the circuit.

Practice Problem: What is the bandwidth provided by the amplifier in the circuit of Fig. 11-11? *Answer:* BW = 90.9 kHz.

Positive Feedback Positive feedback always enhances or adds to the input signal. The end result in the circuit of Fig. 11-13 is saturation. If the output were to saturate positively, then the positive signal would be fed back to the noninverting input and the saturated condition would be maintained. A $-V_{\text{IN}}$ must overcome the positive voltage across R_1 in order to force the circuit to saturate negatively.

The easiest way to determine the input voltage necessary for switching the output state is to begin with V_{IN} at 0 V and the output saturated positively. If the negative saturation voltage for the circuit of Fig. 11-13 were -10 V, then the "trip voltage" necessary to cause negative saturation would be

$$-V_T = -V_{\text{SAT}} \times \frac{R_1}{R_1 + R_F}$$
$$= -10 \text{ V} \times \frac{1 \text{ k}\Omega}{1 \text{ k}\Omega + 10 \text{ k}\Omega}$$
$$= -0.91 \text{ V}$$

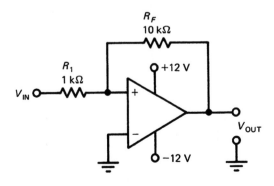

FIGURE 11-13 Noninverting amplifier using positive feedback.

This formula is simply a voltage-divider formula using $-V_{SAT}$ as the supply voltage. The trip voltage is actually the voltage across R_1. As V_{IN} is gradually made more and more negative, it eventually reaches a -0.91 V. At this point, V_{IN} and V_{R1} cancel each other and the voltage at the noninverting input is 0 V. Just a slight increase in the negative voltage of V_{IN} is enough to drive the amplifier into negative saturation. The positive feedback kicks in and the op amp remains in a negatively saturated condition.

Once the op amp is saturated negatively, it takes a trip voltage of $+0.91$ V to force the op amp back into positive saturation. The trip point can be adjusted by changing the values of the resistors.

Practice Problem: What would be the trip voltage in the circuit of Fig. 11-13 if R_1 were changed to a 2-kΩ resistor and the saturation voltage were 10 V? *Answer:* $V_T = 1.67$ V.

Voltage Gain Using Positive Feedback The voltage gain of a circuit using positive feedback approaches infinity unless impeded by some outside factor. In this case, there is the practical limitation of the supply voltages. So, theoretically, we may say that the voltage gain is infinite, yet, in actuality, it is limited by the supply voltages of the particular circuit.

REVIEW QUESTIONS

31. The noninverting amplifier offers a larger input impedance than the inverting amplifier. True or false?
32. Negative feedback is used to (increase or decrease) the overall gain of the circuit.
33. Through which resistor(s) does feedback current flow in the circuit of Fig. 11-11?
34. What would be the voltage gain in the circuit of Fig. 11-11 if $R_1 = 3.3$ kΩ and $R_F = 47$ kΩ?
35. The output polarity of noninverting amplifiers is always positive regardless of the polarity of the input. True or false?
36. What is the output impedance of a 741 op amp whose closed-loop gain is 50?
37. What is the input impedance of a 741 op amp whose closed-loop gain is 95?
38. What is the input impedance of the circuit of Fig. 11-12?
39. What is the output impedance of the circuit of Fig. 11-12?
40. What is the output voltage of the circuit of Fig. 11-12?
41. Positive feedback causes the circuit of Fig. 11-13 to saturate. True or false?
42. What would be the V_T in the circuit of Fig. 11-13 if $R_1 = 2.7$ kΩ, $R_F = 47$ kΩ, and $V_{SAT} = 10$ V?

11-5 VOLTAGE FOLLOWER

The voltage follower is sometimes referred to as a "unity gain amplifier." The latter name indicates that there is a gain of 1 with no phase shift between input and output. It is normally used to isolate a high-impedance source from a low-impedance load.

$$A_v = 1$$

Figure 11-14 shows the voltage follower. The output is tied directly to the inverting input, providing maximum negative feedback. As the input signal increases, the output signal increases. Remember that the op amp amplifies the difference between its inputs. The signal on the inverting input comes very close to equaling the signal on the noninverting input. The reason that they can never be equal is because there would be no difference between the inputs and the op amp would cease to function. Voltage at the output is, therefore, very close to equaling the input. It is so close that they are treated as being the same.

$$v_{out} = v_{in}$$
$$= 2V \text{ p-p}$$

Input impedance is found the same way as before. The open-loop values are 100,000 for the gain and 2 MΩ for the input impedance. Since the closed-loop gain of this circuit is unity, or 1:

$$z_{in} = 2 \text{ M}\Omega \times \frac{100,000}{A_V}$$
$$= 2 \text{ M}\Omega \times \frac{100,000}{1}$$
$$= 200 \text{ G}\Omega$$

Output impedance is also found using the same procedure as before. The open-loop output impedance is listed as 75 Ω for the 741 op amp. So

$$z_{out} = \frac{75\Omega}{100,000/A_V}$$
$$= \frac{75\Omega}{100,000/1}$$
$$= 0.00075 \ \Omega$$

The voltage follower supplies a tremendously large input impedance and a small output impedance.

Practice Problem: What would be the input and output impedances for a voltage follower whose op amp had an open-loop gain of 20,000, an input impedance of 2 MΩ, and an output impedance of 75Ω? *Answer:* z_{in} = 40 GΩ and z_{out} = 0.00375Ω.

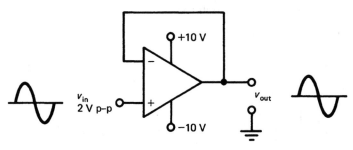

FIGURE 11-14 Voltage follower.

43. What is another name for the voltage follower?

44. The voltage follower has a very large _____ impedance.

45. The voltage follower does not provide a phase shift. True or false?

46. Find z_{in} for a voltage follower with an open-loop gain of 200,000, an input impedance of 1 MΩ, and an output impedance of 50 Ω?

47. Find z_{out} for a voltage follower with an open-loop gain of 200,000, an input impedance of 1 MΩ, and an output impedance of 50 Ω?

11-6 SUMMING AMPLIFIER

The summing amplifier adds the amplified inputs to produce an output. Figure 11-15 shows a summing amplifier used as a mixer. Voice and instrument signals are combined and amplified to produce a unified sound at the output.

Rheostats R_1 to R_4 control the level of each input. It may be desirable to make the guitar louder than the keyboard or the microphones louder than the instruments. Each of the input signals can be made louder (amplified more) by reducing the resistance to the input signal. As the rheostat is decreased, that particular input is amplified to a greater extent. The level of resistance for each rheostat is noted in Fig. 11-15.

The feedback rheostat acts as an overall volume control. The larger its resistance, the greater the overall amplification and, therefore, the louder (larger) the output signal. The volume in the circuit of Fig. 11-15 is at its maximum when R_F is adjusted to its maximum resistance of 1 MΩ.

The individual gain of each input is to be determined. This is accomplished by dividing R_F by the resistance of each rheostat. The gain is then used to determine the output voltage or strength of that particular signal. Once the values of the four output signals are calculated, these signals are added to provide the sum, or total output voltage.

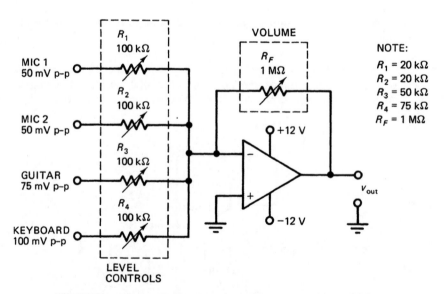

FIGURE 11-15 Summing amplifier used as a mixer.

$$A_v \text{ of MIC1} = \frac{1 \text{ M}\Omega}{20 \text{ k}\Omega} = 50$$

$$v_{\text{out}} \text{ of MIC1} = 50 \text{ mV p–p} \times 50 = 2.5 \text{ V p–p}$$

$$A_v \text{ of MIC2} = \frac{1 \text{ M}\Omega}{20 \text{ k}\Omega} = 50$$

$$v_{\text{out}} \text{ of MIC2} = 50 \text{ mV p–p} \times 50 = 2.5 \text{ V p–p}$$

$$A_v \text{ of GUITAR} = \frac{1 \text{ M}\Omega}{50 \text{ k}\Omega} = 20$$

$$v_{\text{out}} \text{ of GUITAR} = 75 \text{ mV p–p} \times 20 = 1.5 \text{ V p–p}$$

$$A_v \text{ of KEYBOARD} = \frac{1 \text{ M}\Omega}{75 \text{ k}\Omega} = 13.3$$

$$v_{\text{out}} \text{ of KEYBOARD} = 100 \text{ mV p–p} \times 13.3 = 1.33 \text{ V p–p}$$

$$v_{\text{out}} = 2.5 \text{ V p–p} + 2.5 \text{ V p–p} + 1.5 \text{ V p–p} + 1.33 \text{ V p–p}$$

$$= 7.83 \text{ V p–p}$$

The microphones are louder than the instruments in this example since the output signals for the mics are larger. A maximum limit of about 20 V p–p for the output signal is imposed by the power supplies. Finally, a virtual ground condition still exists at the inverting input of the op amp.

EXAMPLE 11-5 _____

The level control rheostats in the circuit of Fig. 11-15 are adjusted to 30 kΩ each and the feedback resistor is kept at 1 MΩ. Determine the output voltage.

Solution:
Since all of the level-control rheostats are adjusted to 30 kΩ, each of the input signals is amplified by a gain of 33.3.

$$A_v = \frac{1 \text{ M}\Omega}{30 \text{ k}\Omega}$$
$$= 33.3$$

The output is equal to the sum of the amplified input signals.

$$v_{\text{out}} \text{ of MIC1} = 50 \text{ mV p–p} \times 33.3 = 1.67 \text{ V p–p}$$
$$v_{\text{out}} \text{ of MIC2} = 50 \text{ mV p–p} \times 33.3 = 1.67 \text{ V p–p}$$
$$v_{\text{out}} \text{ of GUITAR} = 75 \text{ mV p–p} \times 33.3 = 2.5 \text{ V p–p}$$
$$v_{\text{out}} \text{ of KEYBOARD} = 100 \text{ mV p–p} \times 33.3 = 3.33 \text{ V p–p}$$
$$v_{\text{out}} = 1.67 \text{ V p–p} + 1.67 \text{ V p–p} + 2.5 \text{ V p–p} + 3.33 \text{ V p–p}$$
$$= 9.17 \text{ V p–p} \qquad \blacksquare$$

Practice Problem: Determine the overall output signal if rheostats R_1 to R_4 were adjusted to 50 kΩ and R_F were adjusted to 200 kΩ. *Answer:* $v_{\text{out}} = 1.1$ V p–p.

48. A summing amplifier (adds, subtracts, multiplies, or divides) the amplified results of the inputs.

49. MIC1 can be made louder than MIC2 by (increasing or decreasing) the resistance of R_1 in the circuit of Fig. 11-15.

50. The overall volume of the circuit can be decreased by (increasing or decreasing) the resistance of R_F in the circuit of Fig. 11-15.

51. All input signals are inverted by the amplifier in Fig. 11-15. True or false?

52. What would be the output signal level of MIC1 if R_1 were increased to 50 kΩ in the circuit of Fig. 11-15?

53. What would be the output signal level of the GUITAR if R_3 were decreased to 25 kΩ in the circuit of Fig. 11-15?

54. What would be the overall output voltage if R_F were decreased to 500 kΩ in the circuit of Fig. 11-15?

55. The current through R_F is equal to the sum of the individual rheostat currents of R_1 to R_4. True or false?

11-7 DIFFERENCE AMPLIFIER

The amplifiers discussed up to this point used either the inverting or noninverting input. Both inputs are used in the difference amplifier of Fig. 11-16. This circuit is also known as a "subtractor" because it performs the arithmetic operation of subtraction.

The operation of this circuit can be expressed algebraically as

$$V_{OUT} = B - A$$

All of the resistors are the same size to keep the subtraction as straightforward as possible. If different-sized resistors are used, then different multipliers would have to be used with the variables in the formula.

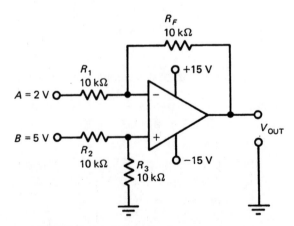

FIGURE 11-16 Difference amplifier.

The easiest way to understand how this circuit works is to approach each input individually and then combine the amplified results at the output.

The subtraction takes place as a result of the phase inversion provided by the inverting input of the op amp. A gain of 1 is provided for the inverting input by R_F and R_1. Therefore, the +2 V of the A input is inverted to a −2 V at the output.

A voltage divider is used to split the +5 V at input B across R_2 and R_3. The voltage at the noninverting input is equal to the voltage across R_3, which is +2.5 V. Recall the voltage-divider formula:

$$V_{R3} = \frac{R_3}{R_2 + R_3} V_B$$

$$= \frac{10\ k\Omega}{10\ k\Omega + 10\ k\Omega} (+5\ V)$$

$$= +2.5\ V$$

The gain for the noninverting input is found by using the same resistors that were used for the gain of the inverting input.

$$A_V = \frac{R_F}{R_1} + 1$$

$$= \frac{10\ k\Omega}{10\ k\Omega} + 1$$

$$= 2$$

A gain of 2 along with +2.5 V at the noninverting input results in +5-V output for the noninverting input. If this voltage is combined with the −2-V output of the inverting input, then the result is

$$V_{OUT} = +5\ V - 2\ V$$

$$= +3\ V$$

One last point: the voltage at the inverting input always seeks to balance itself with the voltage at the noninverting input. In the circuit of Fig. 11-16, the voltage at the inverting input is equal to +2.5 V. Remember that the difference between these two inputs appears as 0 V.

EXAMPLE 11-6

The input voltages of the circuit of Fig. 11-16 are adjusted so that $A = 7$ V and $B = 3$ V. Determine the value of the output voltage.

Solution: It was determined earlier that the difference amplifier subtracts the voltage applied at the inverting input from the voltage applied at the noninverting input.

$$V_{OUT} = 3\ V - 7\ V$$

$$= -4\ V \qquad \blacksquare$$

Practice Problem: What would be the output of the circuit of Fig. 11-13 if input A were −2 V? *Answer:* $V_{OUT} = +7$ V.

56. What is another name for the difference amplifier?
57. What algebraic function does the difference amplifier perform?
58. Which input provides the phase inversion in the circuit of Fig. 11-16?
59. What would be the value of the output voltage if both inputs A and B received +2 V in the circuit of Fig. 11-16?
60. How would the multipliers of the algebraic formula for the circuit of Fig. 11-16 change if R_F were substituted with a value of 20 kΩ?

11-8 AVERAGING AMPLIFIER

Although the averaging amplifier may look similar to the summing amplifier discussed earlier, it has a much different purpose. The averaging amplifier of Fig. 11-17 is designed specifically to produce an output voltage that is the average of the input voltages.

$$V_{OUT} = \frac{V_A + V_B + V_C}{3}$$

The design of the circuit is rather simple. All of the input resistors are the same size and the value of R_F is equal to the value of an input resistor divided by the number of inputs. In our example, there are three inputs and value of the input resistor is 10 kΩ. So

$$R_F = \frac{R}{\text{number of inputs}}$$

$$= \frac{10 \text{ k}\Omega}{3}$$

$$= 3.33 \text{ k}\Omega$$

A standard value of 3.3 kΩ was chosen in the circuit of Fig. 11-17 for the feedback resistor. We assume its measured value is 3.33 kΩ since this falls within the 5% tolerance of the resistor.

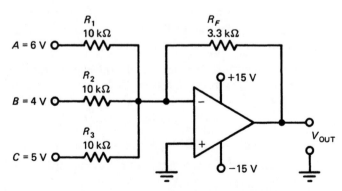

FIGURE 11-17 Averaging amplifier.

The feedback resistor is smaller than the input resistors and, therefore, causes a gain of less than 1. The gain is $\frac{1}{3}$ since three inputs are used and you are seeking their average.

$$A_V = \frac{R_F}{R}$$

$$= \frac{3.33 \text{ k}\Omega}{10 \text{ k}\Omega}$$

$$= 0.333$$

This is the gain for each of the three inputs. The respective outputs are found by multiplying the gain by the input voltage. These combine to create a single output voltage.

$$A_{out} = 6 \text{ V} \times 0.333 = 2 \text{ V}$$

$$B_{out} = 4 \text{ V} \times 0.333 = 1.33 \text{ V}$$

$$C_{out} = 5 \text{ V} \times 0.333 = 1.67 \text{ V}$$

$$V_{OUT} = 2 \text{ V} + 1.33 \text{ V} + 1.67 \text{ V} = 5 \text{ V}$$

The average of the three input voltages is 5 V. You can double check this by adding the inputs and dividing by the number of inputs.

$$V_{OUT} = \frac{6 \text{ V} + 4 \text{ V} + 5 \text{ V}}{3}$$

$$= 5 \text{ V}$$

Of course, the number of inputs can be increased or decreased. If another input were added using a 10-kΩ resistor, then R_F would be changed to 2.5 kΩ. And if one of the inputs were eliminated, the feedback resistor in this circumstance would be 5 kΩ.

EXAMPLE 11-7

In the circuit of Fig. 11-17, the values of R_1, R_2, and R_3 are changed to 68 kΩ, and $V_A = 3$ V, $V_B = 5$ V, and $V_C = 4$ V. Determine (a) the value of the feedback resistor necessary to average the inputs and (b) the value of the output voltage.

Solution:
(a) When three inputs are used, a gain of $\frac{1}{3}$, or 0.33, is needed to provide an average. The feedback resistor must therefore be one-third the size of the input resistors.

$$R_F = \frac{68 \text{ k}\Omega}{3}$$

$$= 22.7 \text{ k}\Omega$$

(b) The output voltage is equal to the average of the input voltages.

$$V_{OUT} = \frac{3 \text{ V} + 5 \text{ V} + 4 \text{ V}}{3}$$

$$= 4 \text{ V} \qquad \blacksquare$$

61. The gain of the averaging amplifier will always be less than 1 when two or more inputs are used. True or false?

62. What value for the feedback resistor would be used in the circuit of Fig. 11-17 if the input resistors were all 47 kΩ?

63. What would be the output voltage in the circuit of Fig. 11-17 if $V_A = 4$ V, $V_B = 3$ V, and $V_C = 2$ V?

64. What would be the output voltage in the circuit of Fig. 11-17 if two more inputs were added such that $V_D = 3$ V and $V_E = 2$ V?

65. What type of feedback is used in the circuit of Fig. 11-17?

11-9 SCHMITT TRIGGER

The positive feedback of the Schmitt trigger circuit of Fig. 11-18 provides two functions: it forces the op amp into saturation and it allows setting the values of the trip points. Adjusting the trip points above and below zero can eliminate unwanted noise triggering.

In Fig. 11-18, the Schmitt trigger is used as a wave shaper. A sine wave is applied to the inverting input. As the positive and negative trip points are reached, the output saturates in the opposite directions. The result is a square wave with a peak-to-peak voltage equal to the saturation voltages.

The value of the trip point is determined by the voltage across R_2. If it is assumed that the saturation voltages of the op amp are $+10$ V and -10 V, then the formula for finding the trip point is

$$+V_T = \frac{R_2}{R_1 + R_2}(+V_{SAT})$$

$$= \frac{10 \text{ k}\Omega}{100 \text{ k}\Omega + 10 \text{ k}\Omega}(+10 \text{ V})$$

$$= +0.91 \text{ V}$$

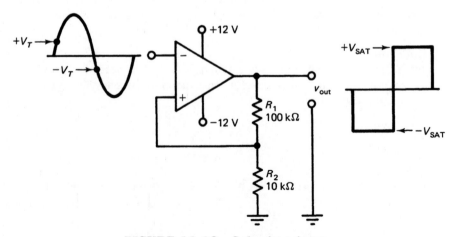

FIGURE 11-18   Schmitt trigger.

and

$$-V_T = \frac{R_2}{R_1 + R_2}(-V_{SAT})$$

$$= \frac{10 \text{ k}\Omega}{100 \text{ k}\Omega + 10 \text{ k}\Omega}(-10 \text{ V})$$

$$= -0.91 \text{ V}$$

As the incoming sine wave reaches +0.91 V, it causes the output to saturate negatively to −10 V. The change in polarity is caused by the inverting input. Likewise, when the sine wave reaches −0.91 V on its negative alternation, the output saturates to a +10 V.

Practice Problem: What would be the trip points if $R_1 = 56$ kΩ and $R_2 = 2.7$ kΩ, assuming saturation voltages of +10 V and −10 V? *Answer:* $+V_T = +460$ mV and $-V_T = -460$ mV.

REVIEW QUESTIONS

66. The positive feedback of the Schmitt trigger forces the op amp into _____.

67. Adjusting the trip point above and below zero prevents accidental _____ triggering.

68. What causes the square-wave output of the Schmitt trigger?

69. What value of resistance must R_2 be in the circuit of Fig. 11-18 in order to create trip points of +1.5 V and −1.5 V?

11-10 COMPARATOR

As the name implies, the comparator compares the input voltage to some reference voltage. In Fig. 11-19, the reference voltage is equivalent to the voltage across R_2. The circuit is designed to answer the question, "Is V_{IN} more positive than V_{R2}?" A "yes" answer is represented by positive saturation and a "no" by negative saturation. Saturation is caused by the open-loop gain of the circuit.

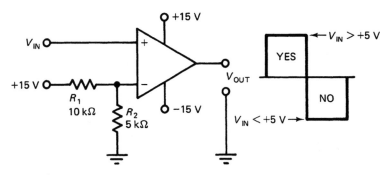

FIGURE 11-19 Comparator.

The positive supply in the circuit of Fig. 11-19 is used to service the voltage divider at the inverting input. The reference voltage across R_2 is solved by using the voltage-divider formula.

$$V_{R2} = \frac{5 \text{ k}\Omega}{10 \text{ k}\Omega + 5\text{k }\Omega} (+15 \text{ V})$$

$$= +5 \text{ V}$$

Figure 11-19 shows that if V_{IN} is greater than +5 V, the output saturates positively. If V_{IN} is less than +5 V, then the output saturates negatively.

Practice Problem: How would this circuit operate if the negative supply voltage were used to service the voltage divider? *Answer:* If V_{IN} is more negative than the reference voltage, then the output saturates negatively. If V_{IN} is less negative, then the output saturates positively.

REVIEW QUESTIONS

70. What type of configuration is used in the circuit of Fig. 11-19?
71. Calculate the reference voltage for the circuit of Fig. 11-19 if $R_2 = 3.3$ kΩ.
72. What type of output waveform would appear if a 20-V p–p sine wave were applied to the input?

11-11   INTEGRATOR

Ramp Voltage: A voltage that either increases or decreases at a linear rate.

The integrator of Fig. 11-20 is shown producing a triangle wave from a square-wave input. The wave shaping is due in part to the charging of the feedback capacitor. Initially, the capacitor acts as a low resistance, thereby causing the gain to be very low. As the capacitor begins to charge, its reactance (resistance) increases and so does the gain. The increase is linear and produces a ramping output voltage. Gain in this circuit is limited to a maximum of 10 by R_F. Limiting the gain of the integrator prevents the reactance of the capacitor at lower frequencies from saturating the op amp.

Eventually, the polarity of the square wave at the input changes. It cause an opposite ramping effect at the output. The continuously changing input produces a

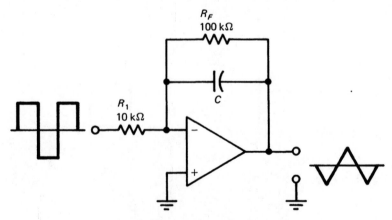

FIGURE 11-20   Integrator.

triangular output waveform. The frequency of the triangle wave is the same as the square wave.

The voltage of the output is dependent on the frequency of the input signal and the RC time constant provided by R_1 and C. A slower input frequency allows the capacitor more time to charge, which increases its reactance. The larger the reactance, the larger the gain. If the RC time constant is made shorter by lowering the value of R_1 or C, the capacitor takes less time to charge. This also increases the reactance of the capacitor and the gain of the circuit.

REVIEW QUESTIONS

73. What type of waveform does the integrator produce?
74. What is the maximum gain of the circuit of Fig. 11-20?
75. The higher the input frequency, the _____ the value of output voltage.
76. The shorter the RC time constant, the _____ the value of output voltage.

11-12   TROUBLESHOOTING

Troubleshooting will be discussed in the context of the circuit of Fig. 11-21. The first step in troubleshooting is to understand the function of the circuit. In the two-step process indicator, the inputs increase to +3 V when a particular step in some process is completed. Upon completion of both steps, the indicator LED lights.

The operation of this circuit begins with the +3-V levels. They are provided by some other circuits upon completion of two separate processes. These +3-V signals are then passed through the voltage followers, IC1 and IC2. Remember that the voltage followers do not change the polarity or voltage level of the signal; they simply provide impedance matching.

As the +3-V signals exit the voltage followers, they enter a summing amplifier, represented by IC3 and R_1 to R_3. A gain of 1 is provided for each of these signals. The two +3-V signals are added and the result at the output of IC3 is −6 V. The polarity changes due to the use of the inverting input.

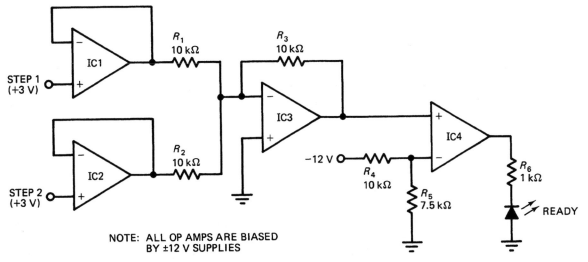

FIGURE 11-21   Two-step process indicator.

PC (printed circuit) board.

The -6 V reaches the comparator, IC4. A voltage divider at the inverting input is used to set the reference voltage at the inverting input. Using the voltage-divider formula, you find it to be -5.14 V. If the sum of the voltages exiting IC3 is more negative than -5.14 V, then the output of IC4 will saturate negatively. Otherwise, the output of IC4 is saturated positively and the LED remains off.

So both steps of the process must be completed before the LED goes on. Armed with a working knowledge of the circuit, you can intelligently approach any symptom and localize the problem. The op amps themselves cannot be repaired. If a problem exists in a particular op amp stage, then swap it for a good op amp.

LED Does Not Light

Is there a negative voltage at the output of IC4? A negative voltage is necessary to forward bias the LED. If a negative voltage is present, then the LED may be at fault.

A positive voltage at the output of IC4 means that a positive voltage at one of the earlier stages is missing. Return to the beginning of the circuit and check for $+3$ V at the inputs of IC1 and IC2. If they are present, then continue to check for the appropriate voltage levels as described before.

LED Is Always Lit

Make sure that at least one of the step voltages is not active. Force one of the voltage followers low if necessary. Next, check for the possible saturation of IC3. A faulty op amp or open feedback resistor may be the cause.

Inspect the inverting input of IC4 for a correct reference voltage. If it is too low, then only one step voltage would be necessary to turn on the LED.

Final Comments

There are many possibilities for these faults. A key to effective troubleshooting is knowing what to expect in a circuit. Each op-amp circuit of this chapter should be understood individually. Upon comprehending these, examine the operation of the circuit of Fig. 11-21 once more. Create some "what if" scenarios by opening feedback paths or shorting inputs to ground and see if you can predict the outcome.

77. What type of circuit is formed with IC1?
78. What type of circuit is formed with IC3?
79. What type of circuit is formed with IC4?
80. Determine the output voltage of IC3 when Step 1 is +3 V and Step 2 is 0 V.
81. Determine the output voltage of IC4 when Step 1 is +3 V and Step 2 is 0 V.
82. What would be the condition of the LED if IC1 were to open internally and its output were constantly low?
83. What would be the condition of the LED if only Step 1 were active and R_3 were open?
84. Determine the output voltage of IC3 when Step 1 and Step 2 are 0 V.
85. Determine the condition of the LED if Step 2 were to malfunction, resulting in a constant +12-V input voltage.

SUMMARY

1. A differential amplifier amplifies the difference between two signals.
2. The differential amplifier rejects noise that is present at both of its inputs. This is called common-mode rejection.
3. Noninverting operation of the differential amplifier results in an amplified output that is in phase with the input signal.
4. Inverting operation of the differential amplifier results in an amplified signal that is 180° out of phase with the input signal.
5. The op amp (operational amplifier) is composed of a differential amplifier, a voltage amplifier, and a power amplifier.
6. Op amps have extremely high input impedance, very low output impedance, and tremendous voltage-gain capabilities.
7. Open-loop configuration refers to the absence of feedback resistors.
8. Frequency and voltage gain for an op amp are inversely related (e.g., lower frequencies allow for higher gains).
9. Negative feedback is used to control gain.
10. Positive feedback is used to reinforce the input signal, resulting in saturation.
11. The inverting amplifier amplifies and inverts the input signal.
12. Important formulas for the inverting amplifier are

$$A_V = \frac{-R_F}{R_1} \qquad z_{in} = R_1$$

$$V_{OUT} = V_{IN} \times A_V \qquad z_{out} = \frac{R_{out(OL)}}{A_{V(OL)}/A_V}$$

$$V_{R_F} = V_{OUT} \qquad R_{OFFSET} = R_F \parallel R_1$$

$$I_{R_F} = \frac{V_{R_F}}{R_F} \qquad GBP = A \times f$$

$$V_{R1} = V_{IN} \qquad BW = \frac{GBP}{A}$$

13. The noninverting amplifier amplifies the input signal without providing a phase shift.

14. Important formulas for the noninverting amplifier are

$$A_V = \frac{R_F}{R_1} + 1 \qquad z_{in} = R_{in(OL)}\frac{A_{v(OL)}}{A_V}$$

$$V_{OUT} = V_{IN} \times A_V$$

$$V_{R1} = V_{IN} \qquad z_{out} = \frac{R_{out(OL)}}{A_{V(OL)}/A_V}$$

$$I_{R1} = \frac{V_{R1}}{R_1} \qquad R_{OFFSET} = R_F \parallel R_1$$

$$I_{R_F} = I_{R1} \qquad \text{GBP} = A \times f$$

$$V_{R_F} = I_{R_F} \times R_F \qquad \text{BW} = \frac{\text{GBP}}{A}$$

15. The voltage follower, or unity gain amplifier, provides a gain of 1 and is used to isolate high impedances from low impedances.

16. The summing amplifier produces an output signal whose voltage is equal to the sum of the input voltages.

17. The difference amplifier produces an output signal whose voltage is equal to the difference of the input voltages.

18. The averaging amplifier produces an output signal whose voltage is equal to the average of the input voltages.

19. A Schmitt trigger circuit uses positive feedback to force positive or negative saturation at the output when the trip-point voltages are exceeded.

20. A comparator compares two input voltages and saturates the output voltage accordingly. It uses an open-loop configuration.

21. An integrator circuit produces a triangle wave from a square-wave input.

GLOSSARY

Bandwidth: A range of frequencies.

Closed Loop: Indicates that some form of feedback is used.

Common-Mode Rejection Ratio (CMRR): Indicates how well an op amp is able to reject noise; the larger the value, the better.

Decibel (dB): Used to express gain.

Dual In-Line Package (DIP): A common form of packaging ICs. The pins are connected to the chip in pairs in line with one another.

Feedback: Returning all or a portion of the output signal back to the input. Positive feedback enhances, or adds, to the input signal, and negative feedback subtracts from the input signal.

Offset Voltage: An unwanted voltage that may exist between the inputs of the op amp. The offset-null inputs of the op amp are used to eliminate it.

Open Loop: When no feedback is used.

Ramp Voltage: A voltage that either increases or decreases at a linear rate.

Virtual Ground: A point in the circuit that appears at ground potential but is not directly connected to ground.

PROBLEMS

The following problems refer to the circuit of Fig. 11-22.

11-1. What type of amplifier is illustrated?

11-2. Calculate the dc emitter current of Q_1.

11-3. Determine the dc voltage at the collector of Q_2.

11-4. What is the ac voltage gain for this circuit?

11-5. What value of ac voltage is measured at the output?

11-6. Is the output signal in phase or out of phase with the input signal?

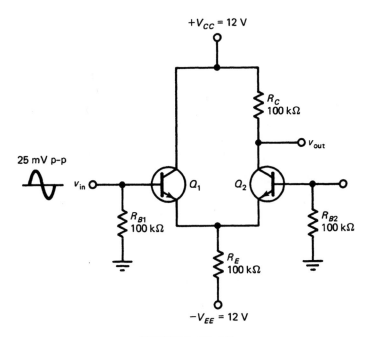

FIGURE 11-22

SECTION 11-2

11-7. Define an op amp.

11-8. Draw and label the schematic symbol of an op amp (including the pin numbers).

11-9. List the characteristics of an op amp including voltage gain, input impedance, output impedance, and frequency range.

11-10. Calculate the decibel value for each voltage gain listed in Fig. 11-7.

11-11. Explain the concepts of open loop and closed loop.

11-12. What are the offset-null inputs used for?

SECTION 11-3

The following problems refer to the circuit of Fig. 11-23.

11-13. What type of amplifier is this circuit?

11-14. Which resistor is the feedback resistor?

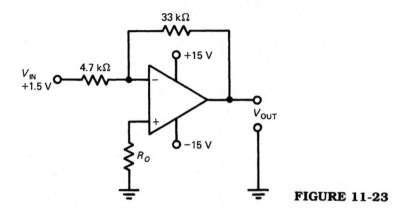

FIGURE 11-23

11-15. Determine the voltage gain.
11-16. Calculate the value of the offset resistor that is needed to balance the inputs of this circuit.
11-17. Find the value of the output voltage, including its polarity.
11-18. What bandwidth does this circuit provide?
11-19. Determine the input and output impedances.
11-20. How can this circuit be modified for a gain of 25?
11-21. Define negative feedback.
11-22. What is virtual ground?
11-23. How is the gain–bandwidth product determined?
11-24. Compare positive and negative feedbacks and explain their differences.

SECTION 11-4

The following problems refer to the circuit of Fig. 11-24.

11-25. What type of amplifier is this circuit?
11-26. Calculate the value of feedback current.
11-27. Determine the voltage gain.
11-28. What is the value of the output voltage?
11-29. Find the bandwidth.
11-30. Calculate the input and output impedances.
11-31. Determine the value of the offset resistor, R_2.
11-32. Modify this circuit for a gain of 50.
11-33. Explain the concept of using positive feedback with a noninverting amplifier.

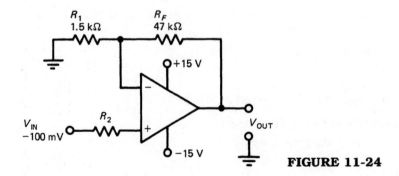

FIGURE 11-24

11-34. Draw the schematic for a voltage follower.

11-35. What is the voltage gain of this circuit?

11-36. Why can't the voltage at the inverting input exceed the voltage at the noninverting input?

11-37. What is the input impedance of a voltage follower whose open-loop gain = 150,000, input impedance = 10 MΩ, and output impedance = 60 Ω?

11-38. What is the output impedance of a voltage follower whose open-loop gain = 150,000, input impedance = 10 MΩ, and output impedance = 60 Ω?

SECTION 11-6

The following problems refer to the circuit of Fig. 11-25.

11-39. What type of circuit is this circuit?

11-40. Determine the gains for inputs A to D.

11-41. Calculate the output voltage level of each input.

11-42. What is the overall output voltage of this circuit?

11-43. How can the resistance of R_3 be modified to provide a gain of 3 for input C?

11-44. Calculate the effect of changing R_F to 150 kΩ on the output voltage.

11-45. Explain how virtual ground occurs in this circuit.

11-46. What voltage is measured across R_F in this circuit?

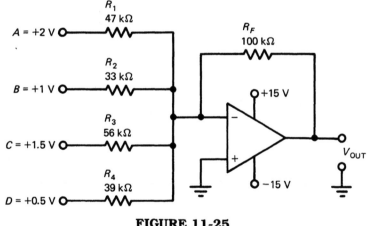

FIGURE 11-25

SECTION 11-7

The following problems refer to the circuit of Fig. 11-26.

11-47. What type of circuit is this circuit?

11-48. Write the algebraic expression for this circuit.

11-49. Determine the gains for each input.

11-50. What is the value of output voltage?

11-51. What value of voltage is found at the inverting input with reference to ground?

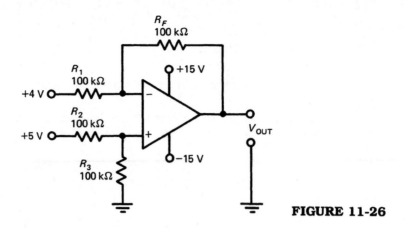

FIGURE 11-26

11-52. Write an algebraic expression for the operation of the circuit if R_F were change to 150 kΩ.

SECTION 11-8

The following problems refer to the averaging amplifier of Fig. 11-27.

11-53. What is the function of the averaging amplifier?
11-54. Calculate the value of the feedback resistor.
11-55. Solve for the individual gains and output voltages.
11-56. Determine the overall output voltage.
11-57. Write the algebraic expression for this circuit.
11-58. Design an amplifier that averages five inputs.

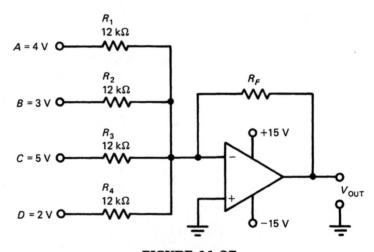

FIGURE 11-27

SECTION 11-9

11-59. Draw a schematic diagram for a Schmitt trigger. Make the supply voltages equal to +15 V and −15 V.
11-60. Choose a pair of resistors that provide a positive and negative trip point of 5 V. Assume that the saturation voltages are +13 V and −13 V.
11-61. Draw a sine-wave input and explain the resulting output waveform.
11-62. Why is positive feedback used?

11-63. Draw a schematic diagram for a comparator. Make the supply voltages equal to +15 V and −15 V.

11-64. Choose the correct supply and resistors to provide a positive reference voltage of 3 V.

11-65. Explain the concept of an open loop.

11-66. How does the comparator differ from the Schmitt trigger?

11-67. Draw an output waveform for the comparator and explain how it is generated.

11-68. Draw the schematic diagram for an integrator along with its output waveform.

11-69. Explain how input frequency and the *RC* time constant affect the output voltage of the integrator.

SECTION 11-12

The following problems refer to the circuit of Fig. 11-28. The LED lights only when a Request Access (+5 V) is made and a Clear to Access (0 V) is present. All other combinations of +5 V and 0 V at the inputs of this circuit cause the LED to remain unlit. Keep in mind that 0 V is more positive than any negative voltage when dealing with IC4.

11-70. What type of circuit is formed with IC1?

11-71. What type of circuit is formed with IC3?

11-72. What type of circuit is formed with IC4?

11-73. Calculate the gain for both the inverting and noninverting inputs of IC3.

11-74. Find the reference voltage at the inverting input of IC4.

11-75. Determine the output condition of IC1 through IC4 for all four possible combinations of Request and Clear accesses.

11-76. Explain the steps you would take to troubleshoot this circuit if faced with the symptom of an unlit LED.

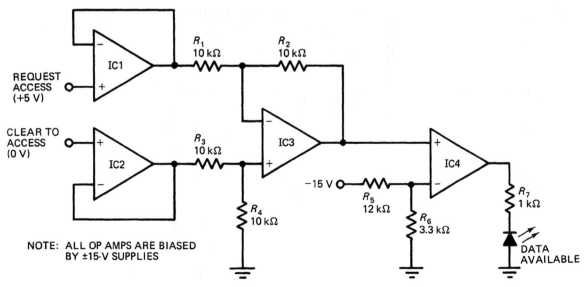

FIGURE 11-28

11-77. Explain the steps you would take to troubleshoot this circuit if faced with the symptom of an LED that is always lit.

ANSWERS TO REVIEW QUESTIONS

1. True
2. Common-mode rejection
3. False
4. 2.86 V p–p
5. 565 μA
6. differential
7. power
8. DIP
9. Offset null pins 1 and 5
10. Inverting input
11. 3 to 15 V
12. No connection
13. open-loop
14. 2 MΩ
15. True
16. increased
17. 60 dB
18. $A_v = 17.8$
19. feedback
20. virtual ground
21. By negative feedback
22. To indicate phase inversion
23. True
24. decreases
25. small
26. Offset resistor
27. 1 MHz
28. decreased
29. Positive feedback
30. $V_{IN} < -4$ V
31. True
32. decrease
33. R_1 and R_F
34. $A_V = 15.2$ V
35. False
36. $z_{out} = 0.0375$ Ω
37. $z_{in} = 2.11$ GΩ
38. $z_{in} = 1.98$ GΩ
39. $z_{out} = 0.076$ Ω
40. $v_{out} = 7.58$ V
41. True
42. $V_T = 543$ mV
43. Unity gain amplifier
44. input
45. True
46. $z_{in} = 200$ GΩ
47. $z_{out} = 0.000250$ Ω
48. adds
49. decreasing
50. decreasing
51. True
52. 1 V p–p
53. 3 V p–p
54. 3.92 V p–p
55. True
56. Subtractor
57. Subtraction
58. Inverting
59. $V_{OUT} = 0$ V
60. $V_{OUT} = 3B - 2A$
61. True
62. $R_F = 15.7$ kΩ
63. $V_{OUT} = 3$ V
64. $V_{OUT} = 4$ V
65. Negative
66. saturation
67. noise
68. Saturation
69. $R_2 = 17.65$ kΩ
70. Open loop
71. $V_{R2} = 3.72$ V
72. Square wave
73. Triangle wave
74. 10
75. lower
76. higher
77. Voltage follower
78. Summing amplifier
79. Comparator
80. -3 V
81. $+V_{SAT} = +10$ V
82. Unlit
83. Lit
84. 0 V
85. Always lit

Op Amp Filters

When you have completed this chapter, you should be able to:

- Compare active and passive filters.
- Investigate the purpose and operation of the low-pass and high-pass filters.
- Realize how cut-off frequencies are determined.
- Investigate the purpose and operation of band-pass and band-stop filters.
- Define the roll-off characteristics of the different filters.

INTRODUCTION

Filter: A circuit that passes certain frequencies that make up its passband and rejects others that make up its stop band.

The op amp is essentially an amplifier. It can be utilized in **filter circuits** by surrounding it with frequency-discriminating components. Resistors, capacitors, and coils are arranged to pass certain frequencies and block others. The op amp provides impedance matching and any necessary gain.

The filter networks of this chapter are composed exclusively of resistors and capacitors. Coils tend to be more bulky and expensive.

12-1 PASSIVE FILTERS

Components that do not depend on an external source of power are called passive components. A filter constructed from these components is called a passive filter. The circuit of Fig. 12-1 is an example of a passive filter. It is composed of two passive components: a resistor and a capacitor.

Let us begin by refreshing your memory with some basic ac theory as it applies to resistors and capacitors. A capacitor has a quality called **reactance** that can be defined as the opposition offered to alternating current. Capacitive reactance is symbolized by X_C and is measured in ohms. The formula for capacitive reactance is

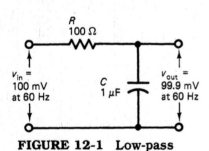

FIGURE 12-1 Low-pass passive filter below f_{co}.

Reactance: The opposition to alternating current offered by inductors or capacitors.

$$X_C = \frac{1}{2\pi f C}$$

where X_C = capacitive reactance in ohms

π = Greek symbol pi = 3.141592654 . . .

f = frequency in Hz

C = capacitance in farads

Examine the formula and note the important role that frequency plays. The lower the value of frequency, the higher the value of capacitive reactance, and vice versa. It is important to see this relationship between frequency and capacitive reactance because the entire theory of filters rests on it.

The filter of Fig. 12-1 is called a low-pass filter because a larger signal is available across the capacitor at lower frequencies, which can then be passed on to the load. Any time the output is taken across the capacitor, the circuit is considered a low-pass filter.

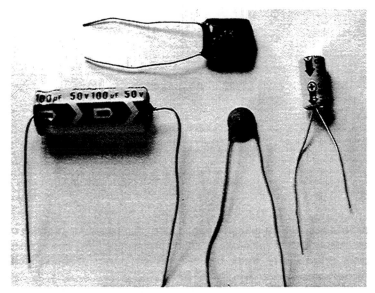

Assorted capacitors.

Below Cutoff Frequency   When we say a filter passes low frequencies, we imply frequencies lower than a certain frequency. The frequency that divides low frequencies from high frequencies is called the **cutoff frequency.** Its value depends on components used in the filter circuit. The cutoff frequency for the circuit of Fig. 12-1 is 1.59 kHz. We examine how cutoff frequency is determined in the next section.

Cutoff Frequency: The frequency at which one-half of the total power appears at the output. It defines the limit of usable frequencies.

Consider the condition of the circuit of Fig. 12-1, beginning with the capacitive reactance.

$$X_C = \frac{1}{2\pi(60 \text{ Hz})(1 \ \mu\text{F})}$$

$$= 2.65 \text{ k}\Omega$$

The total opposition to alternating-current flow is called **impedance** and is symbolized by the letter Z. Since X_C is so much larger than R, the impedance is equal to the value of X_C when rounded.

Impedance: The overall opposition offered to alternating current.

$$Z = \sqrt{R^2 + X_C^2}$$

$$= \sqrt{100 \ \Omega^2 + 2.65 \text{ k}\Omega^2}$$

$$= 2.65 \text{ k}\Omega$$

Using the value of total impedance, we can find the total current and the individual voltage drops. The sum of the voltage drops appears to exceed the total voltage, but that is because of the phase angle of the circuit. Voltage across the capacitor and voltage across the resistor are 90° out of phase with each other.

$$I_T = \frac{v_{\text{in}}}{Z}$$

$$= \frac{100 \text{ mV}}{2.65 \text{ k}\Omega}$$

$$= 37.7 \ \mu\text{A}$$

so

$$V_R = I_T \times R$$
$$= 37.7 \ \mu A \times 100 \ \Omega$$
$$= 3.77 \ mV$$

and

$$V_C = I_T \times X_C$$
$$= 37.7 \ \mu A \times 2.65 \ k\Omega$$
$$= 99.9 \ mV$$

Most of the input signal is dropped across the capacitor because of the low frequency. Also, most of the power is found at the capacitor. Power dissipated by a resistor is expressed in watts and the reactive power of the capacitor is expressed in **volt-ampere reactive,** or VARs.

Volt-Ampere Reactive (VAR): The power attributed to reactive components such as capacitors or inductors.

$$P_R = I_T \times V_R$$
$$= 37.7 \ \mu A \times 3.77 \ mV$$
$$= 142 \ nW$$

and

$$P_C = I_T \times V_C$$
$$= 37.7 \ \mu A \times 99.9 \ mV$$
$$= 3.77 \ \mu VAR$$

It is important to realize that when the frequency is below cutoff, the majority of the voltage and power of the input signal is found at the capacitor. As the frequency of the signal is increased in the following circuit sections, the voltage and the percentage of total power at the capacitor decreases. The capacitor reacts differently to different frequencies.

Cutoff Frequency   The point at which the voltage and power are split evenly between the capacitor and resistor is called the cutoff frequency. For a low-pass filter, it is the maximum usable frequency. The power at this point is exactly one-half of the total power. Anything less than half power is considered an unusable level by the output.

Figure 12-2 shows the same filter circuit with the cutoff frequency applied. The cutoff frequency is determined by

$$f_{CO} = \frac{1}{2\pi RC}$$
$$= \frac{1}{2\pi (100 \ \Omega)(1 \ \mu F)}$$
$$= 1.59 \ kHz$$

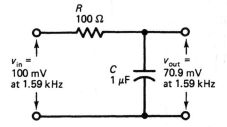

FIGURE 12-2 Low-pass passive filter at f_{CO}.

Solving the values of the circuit at this new frequency shows that the values are split evenly between the components. All of the values in this circuit change as a result of the change in capacitive reactance.

$$X_C = \frac{1}{2\pi(1.59 \text{ kHz})(1 \ \mu\text{F})}$$

$$= 100 \ \Omega$$

$$Z = \sqrt{100 \ \Omega^2 + 100 \ \Omega^2}$$

$$= 141 \ \Omega$$

$$I_T = \frac{100 \text{ mV}}{141 \ \Omega}$$

$$= 709 \ \mu\text{A}$$

$$V_R = 709 \ \mu\text{A} \times 100 \ \Omega$$

$$= 70.9 \text{ mV}$$

$$V_C = 709 \ \mu\text{A} \times 100 \ \Omega$$

$$= 70.9 \text{ mV}$$

$$P_R = 709 \ \mu\text{A} \times 70.9 \text{ mV}$$

$$= 50.3 \ \mu\text{W}$$

$$P_C = 709 \ \mu\text{A} \times 70.9 \text{ mV}$$

$$= 50.3 \ \mu\text{VAR}$$

There are several unique values that always occurs at the cutoff frequency for every circuit. The capacitive reactance always equals the value of resistance. Therefore, the voltage and power of the capacitor and resistor are equal. 70.7% of the source voltage is found across the capacitor and the resistor. The previous calculations show a slightly higher voltage due to rounding.

The cutoff frequency is sometimes referred to as the half-power point. This is because the power is evenly split between the two components at f_{CO}.

EXAMPLE 12-1 _____

What would be the cutoff frequency in the circuit of Fig. 12-2 if $R = 1 \text{ k}\Omega$ and $C = 0.22 \ \mu\text{F}$?

Solution: The cutoff frequency for an *RC* filter is found by substituting the values of the components into the f_{CO} formula.

$$f_{CO} = \frac{1}{2\pi(1 \text{ k}\Omega)(0.22 \ \mu\text{F})}$$

$$= 723 \text{ Hz} \qquad \blacksquare$$

Above Cutoff Frequency
As the frequency rises above the value of f_{CO}, the capacitor begins to perform its function as a filter. Just as the filter in a swimming pool filters out, or removes, leaves, so does the low-pass filter remove, or reject, frequencies above f_{CO}. It may not completely remove the signal, but it reduces the strength of the signal (**attenuates** it) below the usable level.

A frequency of 25 kHz is chosen in the circuit of Fig. 12-3. This is well above the cutoff frequency of 1.59 kHz. Watch as the capacitor lowers its reactance and almost shorts the signal away from the output.

Attenuate: To reduce the strength of a signal.

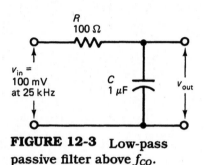

FIGURE 12-3 Low-pass passive filter above f_{CO}.

$$X_C = \frac{1}{2\pi(25 \text{ kHz})(1 \ \mu\text{F})}$$

$$= 6.37 \ \Omega$$

$$Z = \sqrt{100 \ \Omega^2 + 6.37 \ \Omega^2}$$

$$= 100 \ \Omega$$

$$I_T = \frac{100 \text{ mV}}{100 \ \Omega}$$

$$= 1 \text{ mA}$$

$$V_R = 1 \text{ mA} \times 100 \ \Omega$$

$$= 100 \text{ mV}$$

$$V_C = 1 \text{ mA} \times 6.37 \ \Omega$$

$$= 6.37 \text{ mV}$$

$$P_R = 1 \text{ mA} \times 100 \text{ mV}$$

$$= 100 \text{ mW}$$

$$P_C = 1 \text{ mA} \times 6.37 \text{ mV}$$

$$= 6.37 \text{ mVAR}$$

The signal strength at the output is greatly reduced as a result of the filtering action of the capacitor.

Overview of the Low-Pass Filter
The following table compares the effects of frequency on various values within the circuit. Capacitive reactance decreases as frequency increases, providing a low-resistance path for ac, thus redirecting the input signal away from the load. The strength of the voltage and the power of the output signal also decrease.

	BELOW f_{CO} AT 60 Hz	f_{CO} AT 1.59 kHz	ABOVE f_{CO} AT 25 kHz
X_C	2.65 kΩ	100 Ω	6.37 Ω
V_C	99.9 mV	70.9 mV	6.37 mV
P_C	3.77 mVAR	50.3 μVAR	6.37 μVAR
P_R	142 nW	50.3 μW	100 mW

High-Pass Filter

The high-pass filter is designed to pass frequencies above f_{CO} and filter out the frequencies below it. The major difference in appearance between it and the low-pass filter is that the high-pass filter provides the output off the resistor. Figure 12-4 uses the same components as the previous circuits, but notice where the output is taken from now.

Since the same components are used, all of the previous values hold true for this circuit. The cutoff frequency is still 1.59 kHz, but now it represents the minimum useful frequency.

Viewing the following table with emphasis on the resistor values shows that the signal strength at the output increases as frequency increases.

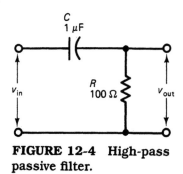

FIGURE 12-4 High-pass passive filter.

	BELOW f_{CO} AT 60 Hz	f_{CO} AT 1.59 kHz	ABOVE f_{CO} AT 25 kHz
V_R	3.77 mV	70.9 mV	100 mV
P_R	142 nW	50.3 μW	100 mW
P_C	3.77 mVAR	50.3 μVAR	6.37 μVAR

Practice Problem: What is the cutoff frequency of a filter composed of a 220-Ω resistor and a 0.047-μF capacitor? *Answer:* f_{CO} = 15.4 kHz.

REVIEW QUESTIONS

1. A filter that does not rely on an external power source is called a _____ filter.
2. What is the opposition offered to ac by a capacitor called?
3. As frequency increases, X_C _____ .
4. What is the X_C of a 22-μF capacitor at 1 kHz?
5. Determine the impedance of the circuit of Fig. 12-2 if R = 330 Ω.
6. As frequency increases, the voltage across the capacitor _____ .
7. What is f_{CO}?
8. What type of filter passes frequencies above f_{CO}?
9. The f_{CO} is sometimes referred to as the _____-power point.
10. The (resistor or capacitor) is directly affected by frequency.
11. The reactive power in a circuit is expressed in _____ .
12. What value of voltage would be found across the capacitor at f_{CO} if the input voltage were 10 V?
13. What would be the value of X_C at f_{CO} if C = 10 μF and R = 33 Ω?
14. Which component is the output taken across for a high-pass filter?
15. Less than one-half of the total power appears across the resistor at frequencies above f_{CO}. True or false?

Insertion Loss: A reduction in voltage, current, or power due to the insertion of a filter network.

When a passive filter is placed between a signal source and load there is an **insertion loss.** A portion of the signal is lost across the *RC* network. The op amp can compensate for the insertion loss and even increase the signal strength by its gain. The circuit in Fig. 12-5 is an active low-pass filter.

It is considered a low-pass filter because the output of the *RC* network is taken across the capacitor. The output of the filter is then fed into the noninverting input of the op amp. The high input impedance of the op amp does not effect the performance of the filter. And the low output impedance of the op amp can service just about any size load.

The gain provided by the op amp can be calculated using the gain formula for the noninverting configuration. In this circuit, the gain is

$$A_v = \frac{100 \text{ k}\Omega}{10 \text{ k}\Omega} + 1$$

$$= 11$$

The cutoff frequency is determined by R_2 and C_1. Using the same formula as with passive filters, we find

$$f_{CO} = \frac{1}{2\pi(1.5 \text{ k}\Omega)(0.022 \text{ }\mu\text{F})}$$

$$= 4.82 \text{ kHz}$$

A gain of 11 is supplied by the op amp to every signal that develops across the capacitor. However, an input signal with a frequency above 4.82 kHz drops less than 70.7% of its voltage across the capacitor. This results in less than half the power as well. The end result is that frequencies above f_{CO} produce an unusable signal at the output of the op amp and can be considered as filtered out, or rejected. The frequencies below f_{CO} produce a much larger signal at the output of the op amp and can be considered passed.

Two frequencies are used to demonstrate the change in output voltage. The first frequency is below f_{CO} and the second is above it. The voltage across the

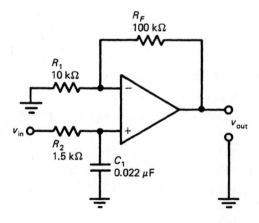

FIGURE 12-5 First-order low-pass active filter.

capacitor is arrived at by the same method as discussed under passive filters. Here are the results:

If v_{in} = 2 V p–p at 1 kHz in the circuit of Fig. 12-5,

$$V_C = 1.96 \text{ V p–p}$$

$$v_{out} = 11 \times 1.96 \text{ V p–p}$$
$$= 21.6 \text{ V p–p}$$

If v_{in} = 2 V p–p at 20 kHz in the circuit of Fig. 12-5,

$$V_C = 469 \text{ mV p–p}$$

$$v_{out} = 11 \times 469 \text{ mV p–p}$$
$$= 5.16 \text{ V p–p}$$

A comparison of these results shows that the output voltage is much lower at frequencies above f_{CO}. These are the frequencies that the low-pass filter rejects.

EXAMPLE 12-2

In the circuit of Fig. 12-5, the values are changed so that R_1 = 27 kΩ, R_2 = 3.3 kΩ, R_F = 75 kΩ, C_1 = 0.1 μF, and v_{in} = 3 V p–p at 1 kHz. Determine (a) f_{CO} and (b) v_{out} for this circuit.

Solution:

(a) This is a low-pass filter and therefore passes all frequencies below the cutoff frequency. The cutoff frequency is found by substituting the values of the filter components into the f_{CO} formula.

$$f_{CO} = \frac{1}{2\pi(3.3 \text{ k}\Omega)(0.1 \text{ }\mu\text{F})}$$
$$= 482 \text{ Hz}$$

(b) The frequency of the input signal is higher than the cutoff frequency just determined. This means that the low-pass filter filters out, or attenuates, this signal. We begin by determining value of the signal that is developed across the capacitor since it is this signal that the op amp amplifies.

$$X_C = \frac{1}{2\pi(1 \text{ kHz})(0.1 \text{ }\mu\text{F})}$$
$$= 1.59 \text{ k}\Omega$$

$$Z = \sqrt{(1.59 \text{ k}\Omega)^2 + (3.3 \text{ k}\Omega)^2}$$
$$= 3.66 \text{ k}\Omega$$

$$I_{R2+C1} = \frac{3 \text{ V p–p}}{3.66 \text{ k}\Omega}$$
$$= 820 \text{ }\mu\text{A}$$

$$V_{C1} = 820 \text{ }\mu\text{A} \times 1.59 \text{ k}\Omega$$
$$= 1.3 \text{ V p–p}$$

Next, we determine the gain of the amplifier. The voltage across the capacitor is amplified by the gain of the amplifier.

$$A_v = \frac{75 \text{ k}\Omega}{27 \text{ k}\Omega} + 1$$
$$= 3.78$$

$$v_{\text{out}} = 1.3 \text{ V p--p} \times 3.78$$
$$= 4.91 \text{ V p--p} \qquad \blacksquare$$

First-Order Filter   The low-pass filter of Fig. 12-5 is called a first-order filter. The order of a filter can be determined by the number of reactive filter networks. Capacitors are reactive components and there is only one RC filter network in the circuit of Fig. 12-5. That is how the first-order filter is physically determined.

First order also describes how the filter responds to frequency. Figure 12-6 shows that gain of the filter decreases by 20 dB for every subsequent tenfold increase in frequency above f_{CO}. The decrease in gain is called roll-off. The maximum allowable decrease in signal strength while still providing a usable signal is −3 dB. This occurs at the cutoff frequency.

In the circuit Fig. 12-5, f_{CO} was found to be 4.823 kHz. A tenfold increase in this frequency is

$$4.823 \text{ kHz} \times 10 = 48.23 \text{ kHz}$$

Applying these values to the graph Fig. 12-6, we find

$$\frac{f}{f_{CO}} = \frac{48.23 \text{ kHz}}{4.823 \text{ kHz}} = 10$$

The value 10 is located on the horizontal axis of this graph and is plotted on the downward sloping line. This point is aligned with −20 dB on the vertical axis. If the point that is 100 times larger than f_{CO} is plotted on the downward sloping line, it will be aligned with the −40 dB mark of the vertical axis. Keep in mind that all frequencies less than f_{CO} are passed by the low-pass filter. These frequencies are located to the left of the point marked 1 on the horizontal axis since this point represents f_{CO}. All the frequencies above f_{CO} decrease at the rate of −20 dB per subsequent tenfold increase in frequency.

If the input voltage were 2 V p–p, the voltage across the capacitor at 48.23 kHz would be

$$V_C = I_T \times X_C$$
$$= 1.33 \text{ mA p--p} \times 150 \text{ }\Omega$$
$$= 0.2 \text{ V p--p}$$

The capacitor is considered the output of the low-pass filter. A 2-V p–p signal was applied to the input of the filter and a 0.2-V p–p signal resulted at the output of the filter. The result is a negative gain. Expressed in **decibels,** it is

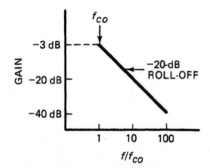

FIGURE 12-6   Typical frequency response of a first-order low-pass filter.

Decibel: A logarithmic means of expressing the gain of a signal.

$$dB = 20 \times \log\left(\frac{v_{out}}{v_{in}}\right)$$

$$= 20 \times \log\left(\frac{0.2 \text{ V p–p}}{2 \text{ V p–p}}\right)$$

$$= -20$$

Since it is actually a negative gain, it is expressed as -20 dB. The first-order filters exhibit a roll-off, or decrease in gain, of -20 dB for every subsequent tenfold increase in frequency above f_{CO}.

Practice Problem: What would be the voltage across the capacitor at 482.3 kHz in the circuit of Fig. 12-5 if the input voltage were 2 V p–p? *Answer:* $V_C =$ 20 mV p–p.

Second-Order Filter   A second-order filter is shown in Fig. 12-7. Notice that there are two reactive filter networks that comprise the low-pass filter. R_1 and C_1 form one network; R_2 and C_2 form the other. Also note that the values of the two RC networks are the same. This makes it easier to design a cutoff frequency. The cutoff frequency is found by

$$f_{CO} = \frac{1}{2\pi R_1 C_1}$$

$$= \frac{1}{2\pi(2 \text{ k}\Omega)(0.01 \text{ }\mu\text{F})}$$

$$= 7.96 \text{ kHz}$$

The gain is determined by the feedback resistor, R_4, and R_3. A flat frequency response can be achieved by designing for a gain of 1.586. This value of gain allows the op amp to control the **damping** factor of the circuit. A filter with a flat

Damping: Energy loss due to absorption by the circuit.

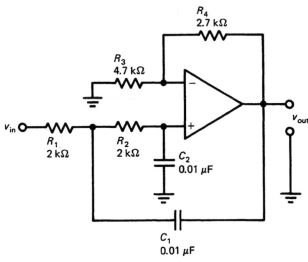

FIGURE 12-7   Second-order low-pass active filter.

frequency response is sometimes called a Butterworth filter. The gain of the noninverting amplifier can be set by choosing a value for R_3 and then calculating the appropriate value for R_4.

$$R_4 = 0.586 \times R_3$$
$$= 0.586 \times 4.7 \text{ k}\Omega$$
$$= 2754 \ \Omega$$

A value of 2.7 kΩ is substituted in the circuit of Fig. 12-7. The actual gain in this case is

$$A_v = \frac{R_4}{R_3} + 1$$
$$= \frac{2.7 \text{ k}\Omega}{4.7 \text{ k}\Omega} + 1$$
$$= 1.574$$

This is close enough to the goal of 1.586 considering the tolerances within the circuit. Negative feedback is provided by R_4 and positive feedback by C_1. The positive feedback maintains the necessary strength of the input signal.

The typical frequency response of a second-order low-pass filter is shown in Fig. 12-8. Negative gain increases at a much steeper rate than it did for the first-order filter. A roll-off of 40 dB for every subsequent tenfold increase in frequency above f_{CO} is provided by the second-order filter. Remember that decibels can be converted to ordinary voltage gain by

$$A_v = \text{inv log} \left(\frac{dB}{20} \right)$$
$$= \text{inv log} \left(\frac{-40 \text{ dB}}{20} \right)$$
$$= 0.01$$

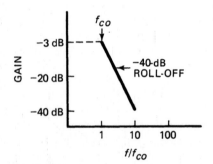

FIGURE 12-8 Typical frequency response of a second-order low-pass filter.

Practice Problem: Convert + 25 dB to an ordinary value of voltage gain. *Answer:* $A_v = 17.8$.

REVIEW QUESTIONS

16. The loss of signal strength across a filter is referred to as _____ loss.
17. The op amp can match the output impedance of a filter to that of a low impedance load. True or false?
18. What would be the gain of the circuit of Fig. 12-5 if R_F and R_1 were both 1-kΩ resistors?
19. Determine f_{CO} of the circuit of Fig. 12-5 if $C_1 = 0.01 \ \mu$F.
20. As frequency increases in the circuit of Fig. 12-5, the output voltage of the op amp _____ .
21. The first-order filter provides a roll-off of _____ for every subsequent _____ increase in frequency.
22. A voltage gain of 0.0501 is equal to _____ dB.

23. How many reactive filter networks does a second-order filter contain?

24. Another name for a filter with a flat frequency response is a _____ filter.

25. What size feedback resistor in the circuit of Fig. 12-7 would provide a gain of 1.586 if R_3 were 12 kΩ?

26. The second-order filter provides a roll-off of _____ for every subsequent _____ increase in frequency.

27. −35 dB is equivalent to a voltage gain of _____ .

12-3 HIGH-PASS ACTIVE FILTERS

High-pass filters pass frequencies above the cutoff frequency while blocking, or attenuating, those below it. The frequencies that a filter passes are called the passband and those that it blocks are called the stop band.

First-Order Filter The high-pass filter of Fig. 12-9 can be recognized as a first-order filter because only one filter network is used. R_1 and C_1 form the high-pass filter network. The output of the filter is taken across R_1 and is fed into the noninverting input of the op amp.

The cutoff frequency for the high-pass filter is calculated the same way as it was for the low-pass filter.

$$f_{co} = \frac{1}{2\pi RC}$$

$$= \frac{1}{2\pi (750 \ \Omega)(0.1 \ \mu F)}$$

$$= 2.12 \ \text{kHz}$$

The gain provided by the noninverting amplifier is

$$A_v = \frac{R_F}{R_2} + 1$$

$$= \frac{4.7 \ \text{k}\Omega}{1 \ \text{k}\Omega} + 1$$

$$= 5.7$$

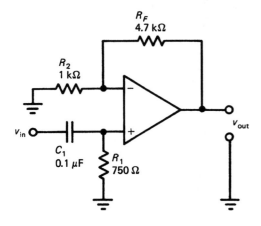

FIGURE 12-9 First-order high-pass active filter.

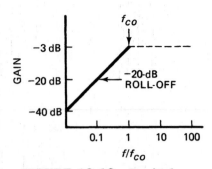

FIGURE 12-10 Typical frequency response of a first-order high-pass filter.

The fact that the high-pass filter of Fig. 12-9 is a first-order filter means that it provides the customary -20 dB roll-off per subsequent tenfold decrease in frequency. At a frequency of 212 Hz, the voltage across the output of the filter, R_1, should show a -20 dB gain. The frequency response graph of Fig. 12-10 illustrates this fact. First, the ratio of the operating frequency to the cutoff frequency must be established.

$$\frac{f}{f_{co}} = \frac{212\ \text{Hz}}{2.12\ \text{kHz}} = 0.1$$

Find 0.1 on the horizontal axis of the graph and mark the point at which it meets the downward slope. Next, follow this point across to the vertical axis and you should arrive at -20 dB. You can prove the validity of this graph by assuming an input voltage of 5 V p–p at 212 Hz and calculating the voltage across the resistor.

$$V_{R1} = I_T \times R_1$$

$$= 663\ \mu\text{A p–p} \times 750\ \Omega$$

$$= 0.5\ \text{V p–p}$$

The input to the filter is 5 V p–p and the output is 0.5 V p–p; therefore, the negative gain is

$$dB = 20 \log \left(\frac{0.5\ \text{V p–p}}{5\ \text{V p–p}} \right)$$

$$= -20$$

Practice Problem: What would be the negative gain in decibels for the circuit of Fig. 12-9 at a frequency of 21.2 Hz? *Answer:* -40 dB.

Second-Order Filter The second-order high-pass filter of Fig. 12-11 is very much like the second-order low-pass filter. The only difference is that the

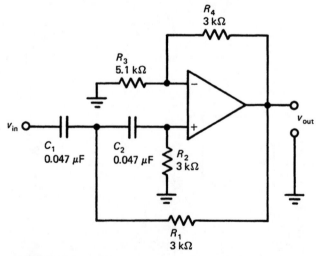

FIGURE 12-11 Second-order high-pass active filter.

resistors and capacitors of the filter networks are interchanged. The output of the filter is taken across the resistor. This signal is then applied to the noninverting input of the op amp and amplified.

Components for each of the filter networks have similar values. This adds to the simplicity of the design. The cutoff frequency is found by

$$f_{CO} = \frac{1}{2\pi R_1 C_1}$$

$$= \frac{1}{2\pi(3\ k\Omega)(0.047\ \mu F)}$$

$$= 1.13\ kHz$$

A flat frequency response is achieved with this circuit. It is, therefore, considered a Butterworth filter. The standard gain of 1.586 is obtained by the combination of the feedback resistor, R_4, and R_2.

$$A_v = \frac{R_4}{R_3} + 1$$

$$= \frac{3\ k\Omega}{5.1\ k\Omega} + 1$$

$$= 1.588$$

A −40-dB roll-off is provided by second-order filters. Figure 12-12 shows the frequency response of the second-order high-pass filter. If the input frequency is one-tenth the value of f_{CO}, then the input voltage will experience a −40-dB attenuation. This means that if 5 V p–p at 113 Hz were applied to the input of the filter, only 50 mV p–p would appear across the output of the filter, R_2. We arrive at this value by multiplying the input signal by the gain, and the value of the gain can be found by converting the decibels.

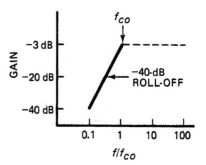

FIGURE 12-12 Typical frequency response of second-order high-pass filter.

$$A_v = \text{inv} \log\left(\frac{dB}{20}\right)$$

$$= \text{inv} \log\left(\frac{-40}{20}\right)$$

$$= 0.01$$

so

$$V_{R2} = A_v \times v_{in}$$

$$= 0.01 \times 5\ V\ p–p$$

$$= 50\ mV\ p–p$$

Each filter network provides a tenfold increase in gain or a −20 dB roll-off. Filter networks can be arranged in different combinations to provide different roll-off characteristics. For instance, a first-order and a second-order filter can be combined to provide a −60-dB roll-off.

EXAMPLE 12-3

In the circuit of Fig. 12-11, the values are changed so that $R_1 = 4.7$ kΩ, $R_2 = 4.7$ kΩ, $R_3 = 3.3$ kΩ, $R_4 = 2$ kΩ, $C_1 = 0.022$ μF, and $C_2 = 0.022$ μF. Determine (a) f_{CO} and (b) the gain of the amplifier expressed in decibels.

Solution:

(a) This is an example of a second-order high-pass filter and, therefore, passes all frequencies above the cutoff frequency. The values of the filter components are substituted into the f_{CO} formula to determine the cutoff frequency.

$$f_{CO} = \frac{1}{2\pi(4.7 \text{ k}\Omega)(0.022 \ \mu\text{F})}$$

$$= 1.54 \text{ kHz}$$

(b) The gain of the amplifier is determined in part by the values of the feedback and inverting input resistors. We find the voltage gain in the usual way and then convert it to decibels.

$$A_v = \frac{2 \text{ k}\Omega}{3.3 \text{ k}\Omega} + 1$$

$$= 1.606$$

$$\text{dB} = 20 \times \log(1.606)$$

$$= 4.11 \qquad \blacksquare$$

REVIEW QUESTIONS

28. Frequencies that are passed by a filter are considered part of the _____ , whereas those that are blocked make up the _____ .

29. The passband of a high-pass filter is comprised of all frequencies (above or below) f_{CO}.

30. What would be the cutoff frequency for the circuit of Fig. 12-9 if C_1 were changed to 0.047 μF?

31. What would be the gain of the circuit of Fig. 12-9 if R_2 were changed to 750 Ω?

32. Determine the output voltage for the circuit of Fig. 12-9 if the input voltage were 3 V p–p at 5 kHz.

33. What would be the cutoff frequency in the circuit of Fig. 12-11 if the resistors of the filter networks were changed to 1 kΩ?

34. How could a gain of 1.586 be maintained in the circuit of Fig. 12-11 if R_3 were substituted with a 2-kΩ resistor?

35. First-order filters provide a roll-off of _____ dB and second-order filters provide a roll-off of _____ dB.

12-4 BAND-PASS FILTERS

The passband of a low-pass filter consists of those frequencies below the cutoff frequency. The passband of a high-pass filter consists of those frequencies above the cutoff frequency. A band-pass filter can be created by combining two of these filters whose passbands overlap.

One use of the band-pass filter is in tuner circuits. Radio stations broadcast over an AM or FM band of frequencies. In order to listen to a particular station (frequency), it is necessary to isolate it from all the others on either side of it. The bandpass filter allows only the band containing that station through while attenuating all the other stations (frequencies).

In Fig. 12-13, a low-pass and high-pass filter are cascaded together. A signal entering the low-pass filter will be passed if its frequency is below 3 kHz. This is the approximate cutoff frequency of the low-pass filter.

$$f_{CO} = \frac{1}{2\pi(2.4 \text{ k}\Omega)(0.022 \text{ }\mu\text{F})}$$

$$= 3.01 \text{ kHz}$$

If the signal is below 3 kHz, it is passed on to the second stage of the filter. The high-pass section of the filter passes the signal if its frequency is above the cutoff frequency of about 1 kHz.

$$f_{CO} = \frac{1}{2\pi(1.6 \text{ k}\Omega)(0.1 \text{ }\mu\text{F})}$$

$$= 995 \text{ Hz}$$

The combination of these two filters results in all signals above 1 kHz and below 3 kHz being passed. All the frequencies between these two points make up a bandwidth equaling 2 kHz. **Bandwidth** is the difference between the upper cutoff frequency and the lower cutoff frequency.

Bandwidth: A range of frequencies between the cutoff frequencies.

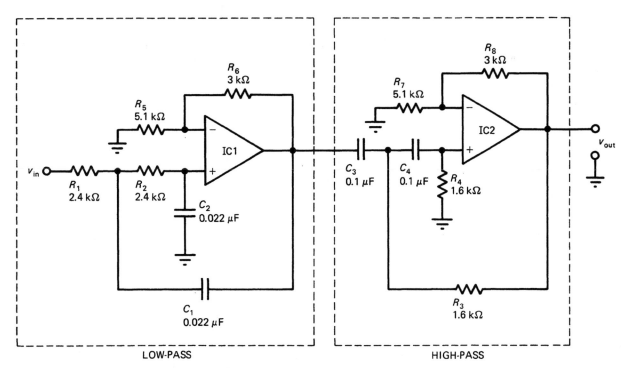

FIGURE 12-13 Band-pass filter.

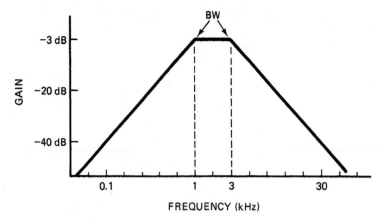

FIGURE 12-14 Frequency-response curve for a band-pass filter.

$$BW = f_{COU} - f_{COL}$$

$$= 3.01 \text{ kHz} - 995 \text{ Hz}$$

$$= 2.02 \text{ kHz}$$

Figure 12-14 is a graphic illustration of the operation of the band-pass filter of Fig. 12-13. The frequencies within the bandwidth have an attenuation of −3 dB or less. As frequencies either increase or decrease beyond the cutoff frequencies, the second-order filters attenuate the signals at a rate of −40 dB per decade.

A flat frequency response is maintained by the band-pass filter because of the Butterworth response of the low-pass and high-pass sections. The gain of each of these stages is found in the usual way.

$$A_v = \frac{3 \text{ k}\Omega}{5.1 \text{ k}\Omega} + 1$$

$$= 1.588$$

The overall gain of the band-pass filter is the product of the individual stage gains.

$$A_v = A_{v1} \times A_{v2}$$

$$= 1.588 \times 1.588$$

$$= 2.52$$

EXAMPLE 12-4

The capacitors in the circuit of Fig. 12-13 are substituted with the following values: $C_1 = 0.01 \ \mu F$, $C_2 = 0.01 \ \mu F$, $C_3 = 0.047 \ \mu F$, and $C_4 = 0.047 \ \mu F$. All the resistor values remain the same. Determine the passband of this filter.

Solution: The passband consists of all frequencies between the cutoff frequencies of the high- and low-pass filters. The cutoff frequency of the low-pass is found first

$$f_{CO} = \frac{1}{2\pi(2.4 \text{ k}\Omega)(0.01 \ \mu F)}$$

$$= 6.63 \text{ kHz}$$

And the cutoff frequency for the high-pass filter is

$$f_{CO} = \frac{1}{2\pi(1.6 \text{ k}\Omega)(0.047 \text{ }\mu\text{F})}$$
$$= 2.12 \text{ kHz}$$

The passband for this filter consists of all frequencies bewteen 2.12 and 6.63 kHz. All frequencies outside this band are attenuated. ∎

Designing the Passband   Designing a band-pass filter for a particular bandwidth is accomplished by interchanging the resistance and f_{CO} in the cutoff-frequency formula. Choose a standard-value capacitor and substitute the desired cutoff frequencies into the formula. The result is the required resistance for each of the filter stages. A value for the resistance is sought because there are many more values available and it is much easier to create a nonstandard value of resistance than capacitance.

Let us say that we want a bandwidth that extends from 20 to 30 kHz and that we have some 0.01-μF capacitors available. The low-pass filter would need an f_{CO} of 30 kHz, so

$$R = \frac{1}{2\pi(30 \text{ kHz})(0.01 \text{ }\mu\text{F})}$$
$$= 531 \text{ }\Omega$$

and the high-pass filter would need an f_{CO} of 20 kHz, so

$$R = \frac{1}{2\pi(20 \text{ kHz})(0.01 \text{ }\mu\text{F})}$$
$$= 796 \text{ }\Omega$$

These resistance values can be formed by connecting some resistors in series or in parallel. The closest standard values may be substituted provided a tolerance in the bandwidth is allowable. Replacing the filter resistors of the circuit of Fig. 12-13 with these creates a bandwidth of 10 kHz. All frequencies above 20 kHz and below 30 kHz are passed.

REVIEW QUESTIONS

36. A band-pass filter can be formed from a _____ filter and a _____ filter.

37. Would an input frequency of 5 kHz be passed or rejected by the filter of Fig. 12-13?

38. What would be the passband of the low-pass filter in Fig. 12-13 if R_1 and R_2 were changed to 1 kΩ?

39. What would be the passband of the high-pass filter in Fig. 12-13 if R_3 and R_4 were changed to 2 kΩ?

40. Determine the bandwidth that would result from the changes of the previous two questions.

41. What would be the overall gain of the circuit of Fig. 12-13 if R_5 and R_7 were changed to 1 kΩ?

42. Calculate the value of resistance necessary to change the f_{CO} of the low-pass section of the circuit of Fig. 12-13 to 4 kHz.

43. Calculate the value of resistance necessary to change the f_{CO} of the high-pass section of the circuit of Fig. 12-13 to 3 kHz.

44. What bandwidth would result from the previous two questions?

12-5 BAND-STOP FILTERS

The general appearance of the band-stop filter is different than that of the band-pass filter. It also differs in its purpose, which is to reject or stop a band of frequencies, as its name implies. Another name for the band-stop filter is the notch filter. Band-stop filters are used most often to eliminate interference.

Figure 12-15 is an example of the band-stop filter. It stops a band of frequencies located around a particular band-stop frequency. Using capacitors of the same size simplifies the design and calculations within this circuit. The center of this band of frequencies can be found by

$$f_S = \frac{1}{2\pi C \sqrt{R_1 R_2}}$$

$$= \frac{1}{2\pi(0.001 \ \mu F) \sqrt{(2.7 \ k\Omega)(330 \ k\Omega)}}$$

$$= 5.33 \ kHz$$

The actual band of frequencies includes a range of frequencies both above and below the particular bandstop frequency. Bandwidth for this circuit can be solved by

$$BW = \frac{1}{\pi R_2 C}$$

$$= \frac{1}{\pi(330 \ k\Omega)(0.001 \ \mu F)}$$

$$= 965 \ Hz$$

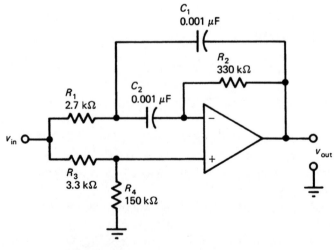

FIGURE 12-15 Band-stop filter.

By adding and subtracting half of this bandwidth to and from f_S, we can determine the band of frequencies that is stopped.

$$\text{Stop Band} = (f_S - 0.5BW) \text{ to } (f_S + 0.5BW)$$
$$= (5.33 \text{ kHz} - 483 \text{ Hz}) \text{ to } (5.33 \text{ kHz} + 483 \text{ Hz})$$
$$= 4.85 \text{ kHz to } 5.81 \text{ kHz}$$

The frequency-response curve for a band-stop filter is shown in Fig. 12-16. Notice that it is in the notch between 4.85 and 5.81 kHz that there is a drastic reduction in gain. The greatest reduction in gain occurs at the center frequency of the stop band (f_S), which was found to be 5.33 kHz. All frequencies above 5.81 kHz and below 4.85 kHz are allowed to pass through this filter.

EXAMPLE 12-5

The circuit of Fig. 12-15 is changed so that $R_1 = 3.3$ kΩ, $R_2 = 470$ kΩ, $R_3 = 180$ kΩ, $R_4 = 180$ kΩ, $C_1 = 0.0022$ μF,, and $C_2 = 0.0022$ μF. What band of frequencies will this filter stop?

Solution: Begin by finding the center frequency of the stop band.

$$f_S = \frac{1}{2\pi(0.0022 \ \mu\text{F}) \ \sqrt{(3.3 \text{ k}\Omega)(470 \text{ k}\Omega)}}$$
$$= 1.84 \text{ kHz}$$

Next, we determine the bandwidth of the stop band.

$$BW = \frac{1}{\pi(470 \text{ k}\Omega)(0.0022 \ \mu\text{F})}$$
$$= 308 \text{ Hz}$$

Finally, the actual range of frequencies is found by adding and subtracting half of the bandwidth from the center band-stop frequency.

$$\text{Stop Band} = (1.84 \text{ kHz} - 154 \text{ Hz}) \text{ to } (1.84 \text{ kHz} + 154 \text{ Hz})$$
$$= 1.69 \text{ kHz to } 1.99 \text{ kHz} \qquad \blacksquare$$

REVIEW QUESTIONS

45.   The purpose of the band-stop filter is to _____ a band of frequencies.

46.   What is another name for the band-pass filter?

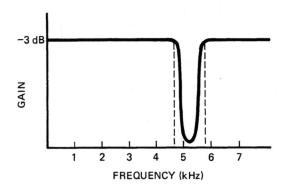

FIGURE 12-16   Frequency-response curve for a band-stop filter.

47. Would an input frequency of 5 kHz be passed or rejected by the filter of Fig. 12-15?

48. What would be the bandwidth of the band-stop filter in Fig. 12-15 if R_2 were changed to 220 kΩ?

49. What would be the bandwidth of the band-stop filter in Fig. 12-15 if R_1 were changed to 3.3 kΩ?

50. Determine the f_S that would result from the changing R_2 to 220 kΩ in the circuit of Fig. 12-15.

51. Which components comprise the filter section in the circuit of Fig. 12-15?

52. Calculate the lower cutoff frequency of the circuit of Fig. 12-15 if R_2 were changed to 220 kΩ.

53. Calculate the upper cutoff frequency of the circuit of Fig. 12-15 if R_2 were changed to 220 kΩ.

54. What bandwidth would result from the previous two questions?

12-6 TROUBLESHOOTING

Troubleshooting the op amp itself is a matter of comparing the inputs and output. If the proper signals exist at the input and there is no signal at the output, then the op amp is most likely the problem. Op amps cannot be repaired but must be replaced. It is a good idea to use IC sockets with op amps. This allows the chips to be inserted and removed without the need for soldering and desoldering. The sockets also eliminate exposing the ICs to the high temperatures of the soldering iron.

The primary components of the filter circuits are the resistor and capacitor networks. These are the frequency-determining components. A filter that is passing or rejecting frequencies outside of the desired range may have a fault in the RC network.

A frequency generator can be used to provide a test signal. Calculate the cutoff frequency of the filter and adjust the generator near this value. Observe the output of the filter with an oscilloscope as you vary the input frequency above and below the cutoff frequency. The output signal should become progressively larger and smaller, respectively, depending on the type of filter. If the filter seems to be out of range, it may be due to one of the following conditions.

Increased Capacitance or Resistance   A lower than usual cutoff frequency results from an increase in either capacitance or resistance. The difference in component values may be due to anything from an error on the assembly line to a much used or abused component within a previously working system. This change can be proven by substituting higher than normal values into the cutoff-frequency formula.

$$f_{CO} = \frac{1}{2\pi RC}$$

The voltage at the output of the filter at the calculated f_{CO} should be 70.7% of the input voltage. It is higher for high-pass filters and lower for low-pass filters when either the resistance or capacitance has increased.

Decreased Capacitance or Resistance   A decrease in resistance or capacitance results in an increase of the cutoff frequency. Capacitors can become leaky with the passage of time and so their capacitance is reduced.

Once again, the signal voltage should measure 70.7% of the input voltage at f_{CO}. The frequency generator should be set to the calculated f_{CO}. A higher than normal measurement is read at the output of a low-pass filter and a lower than normal measurement is found for high-pass filters.

Problems with Feedback   A problem with the negative feedback can affect the flat frequency response of filters. If the negative feedback resistor were to open, the amplifier would operate in an open-loop condition. The output signal would be saturated. If the negative feedback resistor were shorted by a solder bridge, then the circuit would operate as a voltage follower with unity gain.

A problem in the positive-feedback path of the second-order filters affects the overall signal strength as well as the cutoff frequency.

REVIEW QUESTIONS

55. The output signal of a high-pass filter at f_{CO} is larger than normal. Has the resistance increased or decreased?

56. The output signal of a low-pass filter at f_{CO} is larger than normal. Has the capacitance increased or decreased?

57. The output signal of a high-pass filter at f_{CO} is smaller than normal. Has the resistance increased or decreased?

58. The output signal of a low-pass filter at f_{CO} is smaller than normal. Has the resistance increased or decreased?

59. The capacitance of a leaky capacitor is (higher or lower) than normal.

60. Will the gain of the amplifier increase or decrease if the negative-feedback resistor were to open?

SUMMARY

1. Passive filters are constructed of components that do not require an external source of power.

2. The cutoff frequency (f_{CO}) is the point at which the voltage and power of the input signal are evenly divided between the resistor and capacitor.

$$f_{CO} = \frac{1}{2\pi RC}$$

3. Low-pass filters pass frequencies below f_{CO} and attenuate those above it.

4. High-pass filters pass frequencies above f_{CO} and attenuate those below it.

5. Active filters require an external source of power.

6. The order of a filter can be determined by the number of reactive filter components.

7. First-order filters decrease gain by 20 dB for every tenfold change in frequency outside of the passband (beyond f_{CO}).

8. Second-order filters decrease gain by 40 dB for every tenfold change in frequency outside of the passband.

9. Band-pass filters pass a range of frequencies called the bandwidth (BW) and attenuate others outside the passband.

10. The bandwidth consists of all frequencies between the upper and lower cutoff frequencies of the filter.

$$BW = f_{COU} - f_{COL}$$

11. Band-stop filters reject, or stop, a band of frequencies between the upper and lower cutoff frequencies and pass all other frequencies.

12. A frequency generator can be used to provide a test signal when troubleshooting filters.

13. The circuit problem should be localized to either the filter or amplifier in an active filter circuit.

GLOSSARY

Attenuate: To reduce the strength of a signal.

Bandwidth: A range of frequencies between the cutoff frequencies.

Cutoff Frequency: The frequency at which one-half of the total power appears at the output. It defines the limit of usable frequencies.

Damping: Energy loss due to absorption by the circuit.

Decibel: A logarithmic means of expressing the gain of a signal.

Filter: A circuit that passes certain frequencies that make up its passband and rejects others that make up its stop band.

Impedance: The overall opposition offered to alternating current.

Insertion Loss: A reduction in voltage, current, or power due to the insertion of a filter network.

Reactance: The opposition to alternating current offered by inductors or capacitors.

Volt-Ampere Reactive (VAR): The power attributed to reactive components such as capacitors or inductors.

PROBLEMS

SECTION 12-1

The following problems refer to the circuit of Fig. 12-17.

12-1. What type filter is this circuit?

12-2. Calculate the cutoff frequency.

12-3. Find the capacitive reactance.

12-4. Determine the impedance.

12-5. What is the value of voltage at the output?

12-6. Is the input signal passed or blocked? Explain why.

12-7. What would be the capacitive reactance at f_{CO}?

12-8. Define a passive filter.

12-9. Draw a low-pass filter and explain its operation.

12-10. What characteristic values always occur at the cutoff frequency with respect to the resistor and capacitor?

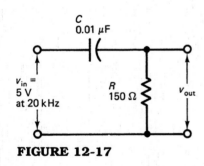

FIGURE 12-17

SECTION 12-2

The following problems refer to the circuit of Fig. 12-18.

12-11. What type of filter is this circuit?

12-12. Describe the meaning of the filter order.

12-13. Find the voltage gain for this circuit.

12-14. Express the voltage gain in decibels.

12-15. Determine the cutoff frequency.

12-16. Calculate the output voltage for this circuit.

12-17. Is the current input frequency passed or blocked?

12-18. Draw a typical response curve for this circuit and label the frequencies along the horizontal axis.

12-19. Explain what a Butterworth filter is.

12-20. Design a second-order Butterworth filter and record the reasons for the selection of each component.

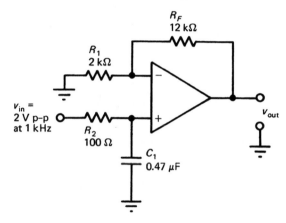

FIGURE 12-18

SECTION 12-3

The following problems refer to the circuit of Fig. 12-19.

12-21. What type of circuit is this circuit?

12-22. Find the cutoff frequency.

12-23. Calculate the value of R_4 that provides the Butterworth response.

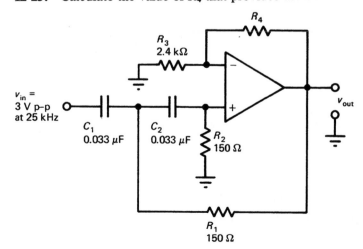

FIGURE 12-19

12-24. Is the input signal passed or blocked? Why?

12-25. Explain the roll-off of this circuit.

SECTION 12-4

The following problems refer to the circuit of Fig. 12-20.

12-26. What type of circuit is this circuit?

12-27. Identify the subsections of this circuit.

12-28. Explain the purpose of this circuit.

12-29. Find the cutoff frequency of the first stage.

12-30. Find the cutoff frequency of the second stage.

12-31. Determine the bandwidth.

12-32. What range of frequencies are included within the bandwidth?

12-33. Calculate the gain of IC1 and IC2.

12-34. Find the overall gain of the circuit.

12-35. Design a circuit whose passband includes frequencies from 5 to 25 kHz.

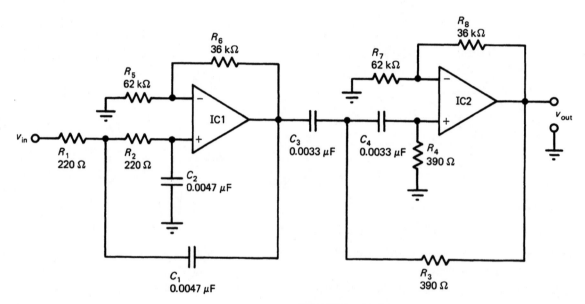

FIGURE 12-20

SECTION 12-5

The following problems refer to the circuit of Fig. 12-21.

12-36. What type of circuit is this circuit?

12-37. Identify the filter components of this circuit.

12-38. Explain the purpose of this circuit.

12-39. Calculate the f_S for this circuit.

12-40. Determine the bandwidth.

12-41. Find the lower cutoff frequency of the stop band.

12-42. Find the upper cutoff frequency of the stop band.

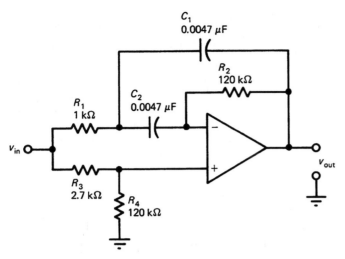

FIGURE 12-21

12-43. What range of frequencies are included within the bandwidth?

12-44. What effect will increasing R_2 have on the bandwidth?

12-45. Design a circuit whose stop band includes frequencies from 3 to 5 kHz.

SECTION 12-6

12-46. Explain the result of a decrease in the capacitance of C_1 in the circuit of Fig. 12-18.

12-47. Explain the result of an increase in the resistance of R_2 in the circuit of Fig. 12-18.

12-48. How would the circuit of Fig. 12-19 be affected if R_4 were to open?

12-49. List the possible cause for no output signal in the circuit of Fig. 12-19.

12-50. What effect would an open in C_3 have on the circuit of Fig. 12-20?

ANSWERS TO REVIEW QUESTIONS

1. passive	**2.** Capacitive reactance	**3.** decreases
4. $X_C = 7.23\ \Omega$	**5.** $Z = 345\ \Omega$	**6.** decreases
7. Cutoff frequency	**8.** High-pass filter	**9.** half
10. capacitor	**11.** VARs	**12.** 7.07 V
13. 33 Ω	**14.** Resistor	**15.** False
16. insertion	**17.** True	**18.** $A_v = 2$
19. $f_{CO} = 10.6$ kHz	**20.** decreases	**21.** −20 dB, tenfold
22. −26	**23.** Two	**24.** Butterworth
25. 7032 Ω	**26.** −40 dB, tenfold	**27.** 0.0178
28. passband, stop band	**29.** above	**30.** 4515 Hz
31. 7.27	**32.** 15.7 V p–p	**33.** 3386 Hz
34. Change R_4 to 1172 Ω	**35.** −20, −40	**36.** low-pass, high-pass
37. Rejected	**38.** Frequencies < 7234 Hz	**39.** Frequencies > 796 Hz
40. BW = 6438 Hz	**41.** $A_v = 16$	**42.** R_1 and $R_2 = 1809\ \Omega$

43. R_3 and $R_4 = 531 \; \Omega$ 44. BW = 1 kHz   45. stop
46. Notch filter   47. Rejected   48. BW = 1.45 kHz
49. BW = 965 Hz   50. $f_S = 6.53$ kHz   51. R_1, R_2, C_1, and C_2
52. 5.81 kHz   53. 7.25 kHz   54. 1.44 kHz
55. Increased   56. Decreased   57. Decreased
58. Increased   59. lower   60. Increase

Oscillators

When you have completed this chapter, you should be able to:

- Analyze the requirements for oscillation.
- Study how frequencies for oscillators are designed.
- Investigate the concept of phase shifts.
- Define resonance.
- Recognize the differences between sinusoidal and nonsinusoidal oscillators.

INTRODUCTION

Oscillator: A self-starting circuit that produces a periodic waveform.

An **oscillator** is often used to produce a clock signal. Clock signals are used to provide the timing for a circuit. The waveform of the oscillator must be periodic and occur at precise intervals if the timing of a circuit is to be dependable.

Oscillators are often defined as circuits that turn dc into ac. Although this is true of many oscillator circuits, there are exceptions. The relaxation oscillator discussed under UJTs (Chapter 10), for instance, provides a pulsating dc waveform. A broader definition of an oscillator is a circuit that generates a periodic waveform.

Sinusoidal: Like a sine wave.
Nonsinusoidal: Not like a sine wave.

Oscillators can be divided into those that provide a **sinusoidal** output and those that provide a **nonsinusoidal** output. The relaxation oscillator is an example of the nonsinusoidal variety. Its output is produced by abrupt changes within the circuit and either a pulse or sawtooth waveform results. The majority of oscillator circuits discussed in this chapter are of the sinusoidal variety.

13-1 OSCILLATOR CHARACTERISTICS

Oscillators can best be understood if our focus is maintained on their three major characteristics. These are amplification, positive feedback, and frequency determination. Each of these characteristics is discussed separately, although all are included in the circuits that follow.

Amplification Oscillators are considered to be self-starting. This means that even with the absence of an applied input signal, the oscillator begins to oscillate and produces a periodic waveform at the output. If there is no input signal, then what is there to amplify? The answer is wide-band noise.

Noise exists at some level in most places at most times, including at the input of an amplifier. It is called wide-band noise because the frequencies of the noise cover a wide band of frequencies. The first step is to increase the strength of this noise to a usable level. Isolating the right frequency is discussed under frequency determination.

Amplification is provided by a BJT, FET, or op amp. Once the noise triggers the oscillator into operation, the circuit must increase the signal to a usable level and replace energy losses due to the components within the oscillator circuit itself. The amplification must be controlled so that saturation does not occur. Once the required signal strength is reached, a gain of $+1$ is desired.

A block diagram of an oscillator circuit is shown in Fig. 13-1. The triangle is a universal symbol for an amplifier and does not necessarily mean an op amp. The gain control can be achieved through the use of negative feedback for op amps,

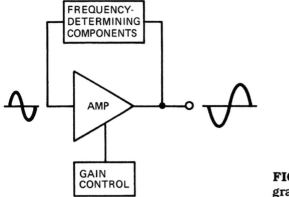

FIGURE 13-1 Block diagram of an oscillator circuit.

through the biasing components of BJTs and FETs, or through the use of variable voltage dividers (potentiometers).

Positive Feedback   Once the initial signal has been amplified, it is necessary to reinforce and sustain that signal. Feedback that reinforces, or aids, the input signal is called positive or **regenerative feedback.** The path of positive feedback in Fig. 13-1 is through the frequency-determining components themselves. A portion of the signal is lost across these components and is replaced by the gain of the amplifier.

In Fig. 13-1, the input and output signals are out of phase by 180°. If the output signal were returned in its present phase, it would be considered negative feedback because it would oppose, or subtract, from the input signal. The output signal must be shifted another 180° before being returned to the input in order for it to be considered positive feedback. As a matter of fact, feedback that is in phase with the input and an overall gain of 1 make up what is called the Barkhausen criteria for oscillation.

The frequency-determining components not only provide the physical path for positive feedback, but normally provide the necessary phase shift. Phase shift is accomplished by the use of lead or lag networks. Leading phase-shift networks cause the output to lead the input by a certain number of degrees and lagging phase-shift networks cause the output to lag behind the input by a certain number of degrees. Capacitors provide a voltage-lagging-current phase shift and inductors provide a voltage-leading-current phase shift. The voltage across a resistor in an *RL* or *RC* circuit depends on the current of the capacitor or inductor. In other words, the resistor provides a leading-voltage phase shift when used with a capacitor and a lagging-voltage phase shift when used with an inductor.

Frequency Determination   The actual frequency of oscillation is determined by the combination of resistors, capacitors, inductors, and quartz crystals. Different combinations are found in each of the following circuits. The frequency-determining nature of passive components was covered in Chapter 12. It was found that a resistor and capacitor could be combined to provide a specified cutoff frequency. In a similar manner, components are combined to select a single oscillation frequency.

Regenerative Feedback: Another name for positive feedback. A signal that reinforces the original signal.

REVIEW QUESTIONS

1. An oscillator provides a _____ waveform.
2. The relaxation oscillator is an example of a (sinusoidal or nonsinusoidal) oscillator.

3. Some oscillator circuits convert _____ into _____.

4. The oscillator is triggered into operation by wide-band _____ .

5. The _____ within the oscillator replaces energy losses.

6. How is the gain of op amps controlled?

7. What is another name for positive feedback?

8. A gain of +1 and positive feedback make up the _____ criteria for oscillation.

9. For feedback to be positive it must be (in or out of) phase with the input.

10. The frequency of _____ is determined by the combination of resistors, capacitors, inductors, and crystals.

13-2 *RC* PHASE-SHIFT OSCILLATOR

The phase-shift oscillator utilizes three or more *RC* phase-shift networks to provide the necessary 180° phase shift as well as the oscillation frequency. Either leading or lagging phase-shift networks can be used. A brief discussion of phase shifting itself and its application to an oscillator circuit follow.

Phase-Shift Network   A leading phase-shift network is shown in Fig. 13-2. It is considered leading because the output is taken off the resistor. The voltage across the resistor, V_R, is shown to lead or precede the input voltage, v_{in}, by 60°.

The angle of the leading phase shift can be determined by calculating a ratio of capacitive reactance to resistance and then finding the inverse tangent of that value. Capacitive reactance is found by

$$X_C = \frac{1}{2\pi fC}$$

$$= \frac{1}{2\pi(1 \text{ kHz})(4.7 \ \mu\text{F})}$$

$$= 33.9 \ \Omega$$

The phase angle, θ (theta), is equal to the phase shift between the input voltage and the voltage across the resistor. It can be found by using either the voltages or the reactance and resistance.

$$\theta = \text{inv tan}\left(\frac{X_C}{R}\right)$$

$$= \text{inv tan}\left(\frac{33.9 \ \Omega}{20 \ \Omega}\right)$$

$$= 59.5°$$

The phase angle between the resistor and capacitor is always 90°. Therefore, the phase shift between v_{in} and V_C is equal to the difference between theta and 90°. It is 30.5° in this example.

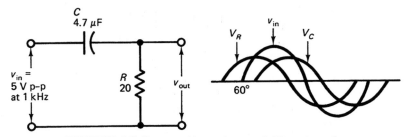

FIGURE 13-2 Leading phase-shift network.

A phase shift of 60° is the maximum recommended for a single *RC* network. Increasing the phase shift beyond this point reduces the strength of the output signal too much. When phase shifts greater than 60° are required, several *RC* networks are placed in series, or cascaded. A 180° phase shift can be achieved by cascading three 60° phase-shift networks or four 45° networks.

RC Phase-Shift Operation   Three *RC* phase-shift networks are used in Fig. 13-3 to provide the necessary 180° phase shift. The *RC* networks also determine the frequency of oscillation. When the resistors and capacitors are of equal value, the oscillation frequency can be found by

$$f_O = \frac{1}{2\pi RC\sqrt{6}}$$

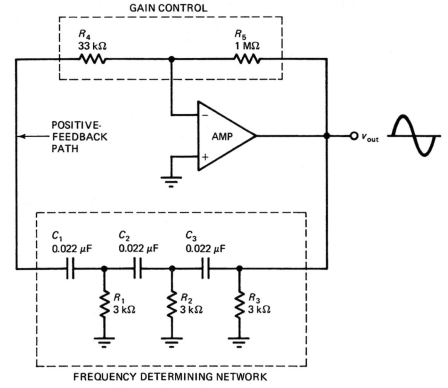

FIGURE 13-3 *RC* phase-shift oscillator.

or

$$f_O = \frac{1}{15.39RC}$$

$$= \frac{1}{(15.39)(3 \text{ k}\Omega)(0.022 \ \mu\text{F})}$$

$$= 985 \text{ Hz}$$

The 15.39 of the second formula is the product of $2\pi \sqrt{6}$. The oscillation frequency for the circuit of Fig. 13-3 is 985 Hz.

Designing a particular frequency of oscillation is achieved by transposing R and f_O. The reason for solving the resistance is because of the wide range of resistor values available as compared to capacitor values. The original frequency desired for this circuit was 1 kHz.

$$R = \frac{1}{15.39fC}$$

$$= \frac{1}{(15.39)(1 \text{ kHz})(0.022 \ \mu\text{F})}$$

$$= 2954 \ \Omega$$

The closest standard value of resistance to 2954 Ω is 3 kΩ. Substituting slightly different resistances for R_1 through R_3 in the circuit of Fig. 13-3 results in an oscillation frequency of 985 Hz rather than 1 kHz. A variable resistor could be used to fine tune the circuit.

The positive-feedback path in the circuit of Fig. 13-3 is indicated. A signal that is now out of phase with the op amp's output travels along this path. Remember that the inverting amplifier provides a 180° phase shift and the RC network provides another 180° phase shift. The total is 360°, which means that the output is in phase with the input.

The open-loop gain of an op amp is extremely high. We spoke of an open-loop gain of 100,000 in the last chapter. This produces a saturated output signal if it is not controlled. The gain must be reduced so that the overall gain of the oscillator is equal to or slightly greater than 1.

Attenuation: The reduction in the amplitude of a signal.

The RC networks of Fig. 13-3 **attenuate** the output of the op amp by a factor of 29. This is true for any three-section RC network that provides a 180° phase shift. The amplifier must replace this loss by providing a gain of at least 29. Gain for the op amp is controlled by R_4 and R_5 in the circuit of Fig. 13-3. The feedback resistor is R_5; therefore,

$$A_v = \frac{R_5}{R_4}$$

$$= \frac{1 \text{ M}\Omega}{33 \text{ k}\Omega}$$

$$= 30.3$$

Always round upward when seeking the closest standard value of resistance. It is necessary to have a gain slightly greater than 29.

EXAMPLE 13-1

In the circuit of Fig. 13-3, the values are changed so that R_1, R_2, and R_3 are 4.7-kΩ resistors and C_1, C_2, and C_3 are 0.01-μF capacitors. Determine (a) the oscillating frequency and (b) the value of R_5 if R_4 were a 2.7-kΩ resistor.

Solution:

(a) The oscillation frequency is found by substituting the new values into the formula.

$$f_O = \frac{1}{2\pi(4.7 \text{ k}\Omega)(0.01 \ \mu\text{F})(\sqrt{6})}$$

$$= \frac{1}{(15.39)(4.7 \text{ k}\Omega)(0.01 \ \mu\text{F})}$$

$$= 1.38 \text{ kHz}$$

(b) The RC network attenuates the output signal by a factor of 29; therefore, the gain of the amplifier must replace this loss. The feedback resistor (R_5) must be approximately 30 times larger than the inverting input resistor.

$$R_5 = 30 \times 2.7 \text{ k}\Omega$$

$$= 81 \text{ k}\Omega \qquad \blacksquare$$

REVIEW QUESTIONS

11. What is the minimum number of RC phase-shift networks recommended to create a 180° phase shift?

12. The voltage across the resistor (leads or lags) the input voltage in an RC network.

13. The voltage across the resistor (leads or lags) the voltage across the capacitor in an RC network by _____ .

14. What is the phase shift between V_R and v_{in} when $R = 100 \ \Omega$ and $X_C = 220 \ \Omega$?

15. What is the phase shift between V_C and v_{in} when $R = 100 \ \Omega$ and $X_C = 220 \ \Omega$?

16. Determine the oscillating frequency of the circuit of Fig. 13.3 if the resistances of R_1 through R_3 were changed to 4.7-kΩ resistors.

17. What is the attenuation factor for RC phase-shift networks such as that of Fig. 13-3?

18. What size feedback resistor would be needed to overcome the attenuation factor of the circuit of Fig. 13-3 if $R_4 = 2.7$ kΩ?

13-3   COLPITTS OSCILLATOR

The Colpitts oscillator relies on a parallel resonant tank circuit to determine its frequency of oscillation. A **tank circuit** combines capacitance and inductance to form a storage tank for energy. The voltage of a capacitor lags current by 90° and the voltage of an inductor leads current by 90°. The capacitor and inductor store and return this current to the circuit on an alternating basis.

A phase angle of 180° exists between the capacitor and inductor; the capacitor and inductor directly oppose one another. At a frequency called the **resonant** frequency, the capacitive reactance (X_C) is equal in magnitude to the inductive

Tank Circuit: A circuit formed by a capacitive and an inductive component for the purpose of storing energy.

Resonance: The condition wherein a capacitive and inductive device compliment each other perfectly. $X_C = X_L$ and $\theta = 0°$.

reactance (X_L). Therefore, the resonant tank circuit provides almost no reactance to the signal developed within it. Total reactance is determined by

$$X_T = X_L - X_C$$

If there were absolutely no resistance in the tank circuit, then the oscillation within the tank would continue forever. All components, however, have some resistance and so energy is lost in overcoming that resistance. The result is that the signal eventually loses its strength and is reduced to zero. This is called **damping.** Let us observe the action that takes place within a tank circuit.

Damping: The diminishing of a signal's power due to resistance within the circuit.

Tank-Circuit Action The tank-circuit action illustrated in Fig. 13-4 assumes that a voltage has been applied to the tank circuit and has charged the capacitor. After the voltage is taken away, the following action takes place.

- Figure 13-4(A): The capacitor has been charged and is now in the process of discharging. It provides a current that is used to build the magnetic field of the inductor. Notice the polarity of the inductor as it initially opposes the current flow within the tank. Eventually, the capacitor finishes discharging and the magnetic field is completely formed.

- Figure 13-4(B): When the capacitor ceases to provide current, the inductor collapses its magnetic field in an attempt to maintain current flow. The polarity of the inductor changes as the field collapses. Energy from the magnetic field is dumped back into the circuit and current continues to flow in the same direction. As current flows, it charges the capacitor with the opposite polarity as before. Eventually, the magnetic field completely collapses and the capacitor is charged once again.

- Figure 13-4(C): At this point, the capacitor begins to discharge again, but notice how the direction of current flow has changed. The inductor begins by opposing this current while its magnetic field develops. The capacitor finishes discharging and the magnetic field of the inductor is completely formed.

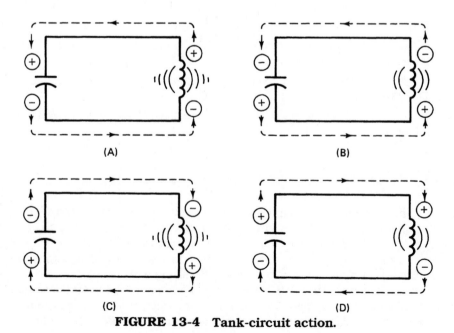

FIGURE 13-4 Tank-circuit action.

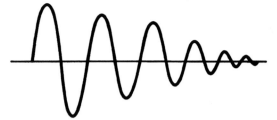

FIGURE 13-5 Effect of damping on a sine wave.

- Figure 13-4(D): In the last phase, the coil collapses its magnetic field, which keeps current flowing in the same direction. The current charges the capacitor back to its original polarity of Fig. 13-4(A) and the whole process begins over again.

Examine all four phases of the tank circuit in Fig. 13-4 and notice that current alternates in direction. The action of the alternating current within the tank is called oscillation. Oscillation would continue forever if power were not lost due to the resistance of the components.

The loss of signal strength due to resistance within the circuit is called damping. Figure 13-5 shows the effect of damping on a sine wave. The signal becomes weaker and weaker until it finally is reduced to zero.

Colpitts Operation   The main sections of the Colpitts oscillator are shown in Fig. 13-6. The tank circuit consists of C_1, C_2, and L. The oscillation frequency is found by determining the total capacitance of the tank and applying it along with the value of inductance to the resonant-frequency formula. Resonant frequency is equal to the frequency of oscillation.

$$C_T = \frac{C_1 \times C_2}{C_1 + C_2}$$

$$= \frac{22\ \mu F \times 3.3\ \mu F}{22\ \mu F + 3.3\ \mu F}$$

$$= 2.87\ \mu F$$

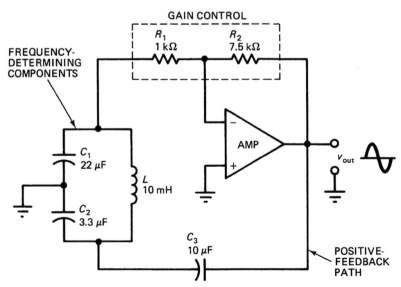

FIGURE 13-6   Colpitts oscillator.

and

$$f_0 = \frac{1}{2\pi\sqrt{LC_T}}$$

$$= \frac{1}{2\pi\sqrt{(10 \text{ mH})(2.87 \ \mu\text{F})}}$$

$$= 939 \text{ Hz}$$

As the tank oscillates, the signal is applied to the inverting input of the amplifier. The inverted signal at the output is fed back through the coupling capacitor, C_3, to the tank, which provides the necessary 180° phase shift. Some of the signal strength is lost as it makes its way back through the tank. The amount of the attenuation caused by the tank must be replaced by the gain of the amplifier if oscillation is to be sustained. Therefore, the gain of the circuit should be

$$A_v > \frac{C_1}{C_2}$$

$$> \frac{22 \ \mu\text{F}}{3.3 \ \mu\text{F}}$$

$$> 6.67$$

In the circuit of fig. 13-6, the gain is controlled by R_1 and R_2. It is found by using the gain formula for inverting amplifiers.

$$A_v = \frac{7.5 \text{ k}\Omega}{1 \text{ k}\Omega}$$

$$= 7.5$$

This value of gain is slightly higher than the required 6.67 and, therefore, sustains oscillation.

EXAMPLE 13-2

In the circuit of Fig. 13-6, the values are changed so that $C_1 = 10 \ \mu\text{F}$, $C_2 = 1 \ \mu\text{F}$, and $L = 4.7$ mH. What is (a) the oscillation frequency and (b) the gain needed to replace circuit losses.

Solution:
(a) Capacitors C_1 and C_2 are in series with each other, so the overall capacitance of the tank is

$$C_T = C_1 \ \| \ C_2$$
$$= 0.909 \ \mu\text{F}$$

The value of this total capacitance is next inserted into the formula for finding the frequency of oscillation.

$$f_0 = \frac{1}{2\pi\sqrt{(4.7 \text{ mH})(0.909 \ \mu\text{F})}}$$
$$= 2.43 \text{ kHz}$$

(b) The approximate attenuation caused by the tank circuit is a ratio of C_1 to C_2. A gain greater than this ratio must be provided if oscillation is to be sustained.

$$A_v > \frac{10\ \mu F}{1\ \mu F}$$
$$> 10 \qquad\qquad \blacksquare$$

Practice Problem: Determine the oscillating frequency and gain necessary for the circuit of Fig. 13-6 if C_1 were changed to a 33-μF capacitor. *Answer:* $f_O = 919$ Hz and $A_v > 10$.

Designing the Oscillation Frequency

The frequency of the oscillator can be adjusted by changing the values of the capacitors. If a frequency of 300 kHz is desired in the circuit of Fig. 13-6, then substitute the tank capacitors with new values. The first step is to find the value of total tank capacitance needed to produce 300 kHz when coupled with a 10-mH inductor.

$$C_T = \frac{1}{4\pi^2 f^2 L}$$

$$= \frac{1}{(39.5)(300\ kHz)^2(10\ mH)}$$

$$= 28.13\ pF$$

A **feedback fraction** (β) of about 0.1 and 0.5 is normal. The feedback fraction in the circuit of Fig. 13-6 is the ratio of C_2 to C_1. Therefore, a good starting point for the design is to choose a value of C_1 that is approximately 10 times larger than C_T.

Feedback Fraction: The fraction of the output signal that is fed back to the input. It is represented by the Greek letter β (beta).

$$C_1 = 10 \times C_T$$

$$= 10 \times 28.13\ pF$$

$$= 281.3\ pF$$

$$= 291\ pF\ (available)$$

The closest available value of capacitance is 291 pF. Finding a value for C_2 requires that we subtract the reciprocal of C_1 from the reciprocal of C_T and then find the reciprocal of the result. In a formula, it looks like this:

$$C_2 = \frac{1}{1/C_T - 1/C_1}$$

$$= \frac{1}{1/28.13\ pF - 1/291\ pF}$$

$$= 31.3\ pF$$

$$= 31.6\ pF\ (available)$$

Notice that the nearest available value of 291 pF was used in the last formula. Once again, a value of 31.6 pF is the closest available value to 31.1 pF.

Substituting these capacitors into the circuit of Fig. 13.6 changes the frequency of oscillation to approximately 300 kHz.

$$C_T = \frac{291\ \text{pF} \times 31.6\ \text{pF}}{291\ \text{pF} + 31.6\ \text{pF}}$$

$$= 28.5\ \text{pF}$$

and

$$f_O = \frac{1}{2\pi \sqrt{(10\ \text{mH})(28.5\ \text{pF})}}$$

$$= 298\ \text{kHz}$$

A trim capacitor could be used in place of C_2. This would allow the capacitance to be adjusted between 2 and 50 pF.

Only a fraction of the output signal reaches the input. The feedback fraction resulting from the use of these capacitors is

$$\beta = \frac{C_2}{C_1}$$

$$= \frac{31.6\ \text{pF}}{291\ \text{pF}}$$

$$= 0.1096$$

Therefore, the gain must be

$$A_v > \frac{C_1}{C_2}$$

$$> 9.21$$

The gain-control resistors should be changed. A larger value for R_1, about 100 kΩ, has less of a loading effect on the signal coming off the tank. The feedback resistor is, therefore, changed to 1 MΩ or more to provide a gain greater than 9.21.

EXAMPLE 13-3

Adjust the values of the components of the circuit of Fig. 13-6 for a 100-kHz oscillation frequency using a 4.7-mH coil in the tank.

Solution: We begin by determining the total value of capacitance needed.

$$C_T = \frac{1}{4\pi^2 (100\ \text{kHz})^2 (4.7\ \text{mH})}$$

$$= 539\ \text{pF}$$

Next, a value for C_1 that is approximately 10 times larger than total capacitance is chosen.

$$C_1 = 10 \times 539\ \text{pF}$$

$$= 0.005\ \mu\text{F}$$

Having chosen a value for C_1 allows us to find a value for C_2.

$$C_2 = \frac{1}{1/C_T - 1/C_1}$$

$$= \frac{1}{1/539 \text{ pF} - 1/0.005 \text{ } \mu\text{F}}$$

$$= 604 \text{ pF}$$

Finally, we determine the necessary gain.

$$A_v > \frac{0.005 \text{ } \mu\text{F}}{604 \text{ pF}}$$

$$> 8.28$$

This means that if a 200-kΩ resistor were chosen for R_1, the feedback resistor (R_2) should be at least 8.28 times larger.

$$R_1 = 200 \text{ k}\Omega$$
$$R_2 > 8.28 \times 200 \text{ k}\Omega$$
$$> 1.66 \text{ M}\Omega \qquad \blacksquare$$

Colpitts Oscillator Using a BJT   A bipolar junction transistor is used as the amplifier for the Colpitts oscillator of Fig. 13-7. The voltage-divider bias is used to establish the dc operating voltages. A radio-frequency coil (RFC) acts as a low-resistance path for dc and as a high ac impedance to the oscillator signal.

Positive feedback is provided by the tank and coupling capacitor C_3. A signal that is in phase with the input is then applied to the base of the transistor.

The frequency of oscillation is determined by the components within the tank circuit. The same formulas from the op-amp circuit can be used for determining f_O in this circuit.

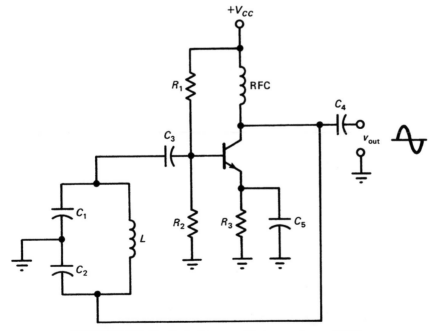

FIGURE 13-7   Colpitts oscillator using a BJT.

Finally, gain is controlled by the ratio of r_c to r_e'. The bypass capacitor, C_5, provides a low-resistance path to ground for the ac signal, which, in turn, increases the gain of the amplifier.

REVIEW QUESTIONS

19. The voltage across the inductor (leads or lags) the current within the circuit.
20. A _____ phase shift exists between the capacitor and the inductor.
21. Capacitive and inductive reactances are equal at the _____ frequency.
22. When energy is lost due to absorption by the circuit, it is called _____ .
23. What would be the total capacitance of the tank circuit in Fig. 13-6 if C_2 were changed to a 2.2-μF capacitor?
24. What would be the resonant frequency of the tank in Fig. 13-6 if C_2 were changed to a 2.2-μF capacitor?
25. What would be the gain of the circuit of Fig. 13-6 if C_2 were changed to a 2.2-μF capacitor?
26. Determine the total tank capacitance needed in the circuit of Fig. 13-6 if an oscillating frequency of 150 kHz were desired.

13-4 HARTLEY OSCILLATOR

The Hartley oscillator of Fig. 13-8 differs from the Colpitts oscillator in that its tank circuit is tapped between the two inductors rather than between two capacitors. The operation of the circuit is the same.

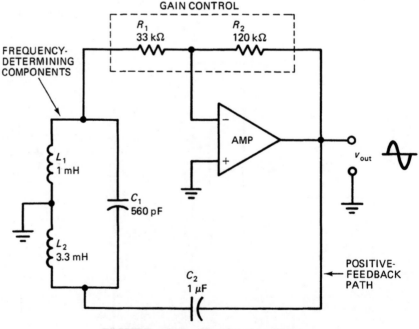

FIGURE 13-8 Hartley oscillator.

The total inductance within the tank is equal to the sum of the inductors.

$$L_T = L_1 + L_2$$
$$= 1 \text{ mH} + 3.3 \text{ mH}$$
$$= 4.3 \text{ mH}$$

This value is now used to determine the frequency of oscillation.

$$f_O = \frac{1}{2\pi\sqrt{L_T C}}$$
$$= \frac{1}{2\pi\sqrt{(4.3 \text{ mH})(560 \text{ pF})}}$$
$$= 103 \text{ kHz}$$

Once the signal is amplified, it is fed back through the coupling capacitor and tank circuit, which causes a phase shift of 180°. On its way back to the inverting input, the signal is attenuated by the tank circuit. The loss of energy must be replaced by the amplifier. Therefore, the gain should be

$$A_v > \frac{L_2}{L_1}$$
$$> \frac{3.3 \text{ mH}}{1 \text{ mH}}$$
$$> 3.3$$

The gain control of the circuit of Fig. 13-8 provides a gain of

$$A_v = \frac{R_2}{R_1}$$
$$= \frac{120 \text{ k}\Omega}{33 \text{ k}\Omega}$$
$$= 3.64$$

EXAMPLE 13-4 _____

In the circuit of Fig. 13-8, the values are changed so that $L_1 = 4.7$ mH and $L_2 = 10$ mH. Determine (a) the frequency of oscillation and (b) the necessary gain.

Solution:

(a) Inductors in series add directly.

$$L_T = 10 \text{ mH} + 4.7 \text{ mH}$$
$$= 14.7 \text{ mH}$$

and

$$f_O = \frac{1}{2\pi\sqrt{(14.7 \text{ mH})(560 \text{ pF})}}$$
$$= 55.5 \text{ kHz}$$

(b) The attenuation that must be overcome is the ratio of L_2 to L_1.

$$A_v > \frac{10 \text{ mH}}{4.7 \text{ mH}}$$

$$> 2.13 \qquad \blacksquare$$

Practice Problem: Determine the oscillator frequency and the gain necessary for the circuit of Fig. 13-8 if L_2 were changed to a 4.7-mH inductor. *Answer:* $f_O =$ 89.1 kHz and $A_v > 4.7$.

Designing the Oscillation Frequency

Designing for a particular frequency of oscillation requires changing the components within the tank. Generally, capacitors are available in a wider range of values than inductors. Formulas for substituting either component are given. Specific circumstances best determine which formula to use.

$$C_T = \frac{1}{4\pi^2 f^2 L}$$

$$L_T = \frac{1}{4\pi^2 f^2 C}$$

The $4\pi^2$ is equal to 39.5, which may be substituted in either formula. Finding capacitance was discussed with the Colpitts oscillator; therefore, finding inductance is the topic of this discussion.

An oscillation frequency of 300 kHz is desired and we have a 0.022-μF capacitor. Therefore, the value of total inductance is needed. The values of frequency and capacitance are substituted into the formula for inductance.

$$L_T = \frac{1}{(39.5)(300 \text{ kHz})^2 (0.022 \ \mu\text{F})}$$

$$= 12.8 \ \mu\text{H}$$

Here is a list of some standard values of inductance (all values are in μH): 1, 2.2, 2.7, 3.3, 3.9, 4.7, 5.6, 6.8, 7.5, 8.2, and 10. A pair of inductors must be chosen whose sum is equivalent to 12.8 μH. We might choose 10 μH for L_2 and 2.7 μH for L_1. In that case, the total inductance is 12.7 μH.

$$L_T = 2.7 \ \mu\text{H} + 10 \ \mu\text{H}$$

$$= 12.7 \ \mu\text{H}$$

so

$$f_o = \frac{1}{2\pi \sqrt{(12.7 \ \mu\text{H})(0.022 \ \mu\text{F})}}$$

$$= 301 \text{ kHz}$$

and

$$A_v > \frac{10 \ \mu\text{H}}{2.7 \ \mu\text{H}}$$

$$= 3.7$$

Substituting these inductors into the circuit of Fig. 13-8 produces a frequency of 301 kHz. The gain control has to be adjusted slightly so that the gain of the amplifier is brought above 3.7.

REVIEW QUESTIONS

27. What would be the total inductance of the tank circuit in Fig. 13-8 if L_2 were changed to a 5.6-mH inductor?

28. What would be the resonant frequency of the tank circuit in Fig. 13-8 if L_2 were changed to a 5.6-mH inductor?

29. What would be the necessary gain of the circuit of Fig. 13-8 if L_2 were changed to a 5.6-mH inductor?

30. Determine the total tank inductance needed in the circuit of Fig. 13-8 if an oscillating frequency of 150 kHz were desired.

13-5 CRYSTAL OSCILLATOR

LC oscillators are susceptible to temperature changes and other forms of outside disturbance. A fluctuation in the frequency of oscillation can occur. Precision can be gained through the use of a quartz crystal. Undoubtedly, you have heard the term quartz crystal associated with wristwatches. The accuracy of the watch is emphasized by use of the term in advertisements. Oscillators can be made to run more accurately through use of a quartz crystal.

Each of the oscillators discussed so far have had some combination of components that determined the frequency. The quartz crystal can be used as a resonant circuit by itself. Figure 13-9 shows the schematic symbol for a quartz crystal as well as an equivalent circuit to help describe its operation.

Quartz is a mineral that has piezoelectric qualities. The **piezoelectric effect** displayed by a crystal is that it begins to vibrate when voltage is applied. The frequency of the vibration is dependent upon the cut and thickness of the crystal (see Section 15-4, "Crystals"). Crystals come in a variety of frequency ratings.

Piezoelectric Effect: The condition when certain minerals produce electric current under pressure or vibrate when subjected to an applied voltage.

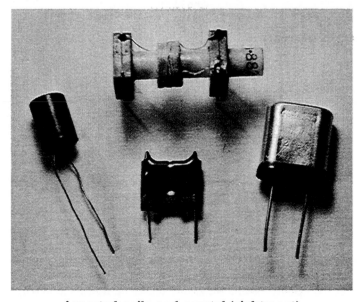

Assorted coils and crystal (rightmost).

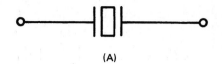

(A)

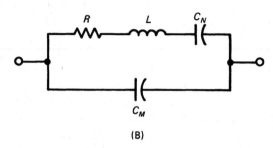

(B)

FIGURE 13-9 Crystal: (A) schematic and (B) equivalent circuit.

Each quartz crystal has a certain amount of internal resistance, inductance, and capacitance, as shown in Fig. 13-9(B). The natural capacitance of the mineral itself is labeled C_N. When the quartz is mounted between two electrodes, it also contains a mounted capacitance, C_M. The quartz acts like a dielectric between the two electrode plates. The natural inductance and resistance are represented by L and R, respectively.

The resonant frequency of a crystal can be determined by taking these values into consideration. Crystals can range in frequency from under 1 kHz up to 200 MHz. We use a low-frequency crystal of about 10 kHz as an example. It has the following specifications: $R = 50$ kΩ, $L = 8450$ H, $C_N = 0.03$ pF, and $C_M = 6$ pF. From this information, the series and parallel resonant frequencies of the crystal can be determined. The series frequency can be found by

$$f_S = \frac{1}{2\pi \sqrt{LC_N}}$$

$$= 9996 \text{ Hz}$$

and the parallel frequency is

$$f_P = \frac{1}{2\pi \sqrt{LC_{N+M}}}$$

$$C_{N+M} = \frac{C_N \times C_M}{C_N + C_M}$$

$$= 0.02985 \text{ pF}$$

$$f_P = \frac{1}{2\pi \sqrt{(8450 \text{ H})(0.02985 \text{ pF})}}$$

$$= 10{,}021 \text{ Hz}$$

This crystal operates at 9996 Hz in a series mode and 10,021 Hz in a parallel mode. An example of a parallel mode is if the crystal were substituted in place of the inductor of the Colpitts oscillator. It would then be connected in parallel to the tank. Figure 13-10 shows an example of the crystal connected in a series mode.

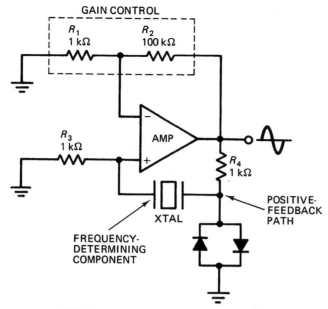

FIGURE 13-10 Crystal oscillator.

The crystal oscillator of Fig. 13-10 contains all the necessary characteristics for oscillation. An op amp is used to provide the amplification. Positive feedback is provided by the crystal to the noninverting input. Notice that the abbreviation XTAL is used to denote a crystal on a schematic diagram. The diodes are used to prevent the output voltage from saturating. One diode conducts on each alternation of the sine wave. Therefore, the voltage across the crystal and R_3 is limited to 0.7 V. Using the resistance of the crystal stated earlier, we can find the input voltage at the noninverting input of the amplifier. It is equal to the voltage across R_3.

$$V_{R3} = \frac{R_3}{R_3 + R_X} V_D$$

$$= \frac{1 \text{ k}\Omega}{1 \text{ k}\Omega + 50 \text{ k}\Omega} 0.7 \text{ V}$$

$$= 13.7 \text{ mV}$$

The gain of the noninverting amplifier is controlled by R_1 and R_2. Gain for this circuit is equal to

$$A_v = \frac{R_2}{R_1} + 1$$

$$= \frac{100 \text{ k}\Omega}{1 \text{ k}\Omega} + 1$$

$$= 101$$

The output voltage is a sine wave with a peak voltage of

$$v_{\text{out}} = A_v \times V_{R3}$$

$$= 101 \times 13.7 \text{ mV}$$

$$= 1.38 \text{ V pk.}$$

The frequency of oscillation using this particular crystal is 9996 Hz, as stated earlier.

EXAMPLE 13-5 _____

Determine the peak output voltage of the circuit of Fig. 13-10 if R_3 were substituted with a 5-kΩ resistor. Note that the resistance of the crystal is 50 kΩ.

Solution: The diodes maintain 0.7 V across the combination of R_3 and the XTAL. A voltage divider is formed by the crystal and the resistor. Voltage at the noninverting input can be found by

$$V_{R3} = \frac{5 \text{ k}\Omega}{5 \text{ k}\Omega + 50 \text{ k}\Omega} \, 0.7 \text{ V}$$
$$= 0.0636 \text{ V pk.}$$

Output voltage is found by multiplying the input voltage across R_3 times the voltage gain of the circuit.

$$v_{\text{out}} = 0.636 \text{ V pk.} \times 101$$
$$= 6.42 \text{ V pk.} \qquad \blacksquare$$

REVIEW QUESTIONS _____

31. What type of mineral was used for the crystals in this chapter?
32. Crystals are less susceptible to heat and other environmental factors than their LC equivalents. True or false?
33. The ability of the crystal to vibrate when a voltage is applied is called the _____ _____ .
34. Determine the series resonant frequency of a crystal with the following specifications: $R = 50$ kΩ, $L = 8000$ H, $C_N = 0.03$ pF, and $C_M = 30$ pF.
35. Determine the parallel resonant frequency of a crystal with the following specifications: $R = 50$ kΩ, $L = 8000$ H, $C_N = 0.03$ pF, and $C_M = 30$ pF.
36. What is the abbreviation for crystal?
37. Calculate the output voltage for the circuit of Fig. 13-10 if R_3 were changed to 3 kΩ and R of the crystal were 50 kΩ.

13-6 WIEN BRIDGE OSCILLATOR _____

The Wien bridge oscillator uses two RC networks to determine its frequency of oscillation. These RC networks form what is called a lead-lag network. Maximum signal strength occurs at the resonant frequency of this network.

Lead-Lag Network The lead-lag network of Fig. 13-11 shows a series RC network and a parallel RC network. Recall that frequency and capacitive reactance are inversely related. In other words, when frequency is increased, the capacitive reactance decreases, and vice versa.

At frequencies above resonant frequency, C_2 tends to short the output signal. When frequencies are lower than the resonant frequency, C_1 acts more like an open. Both of these conditions reduce the signal strength at the output. The maximum signal strength is achieved at a frequency somewhere in between. This frequency is called resonant frequency.

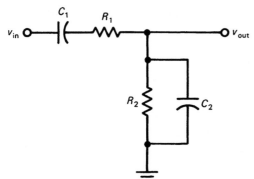

FIGURE 13-11 Lead-lag network.

The circuit is called a lead-lag network because at frequencies above resonance, a leading phase shift exists at the output, and at frequencies below resonance, a lagging phase shift exists. A phase shift of zero degrees is necessary for oscillation to be sustained. The output signal in the upcoming circuit is applied to the noninverting input because a 180° phase shift is no longer necessary. Resonant frequency produces a 0° phase shift.

Wien Bridge Operation

The Wien bridge oscillator of Fig. 13-12 contains a new component. The circled resistor with a T next to it is a thermistor. Its name is a contraction of thermal resistor. Thermal conditions (temperature) cause a change in its resistance. The thermistor in this circuit has a positive **temperature coefficient,** meaning that its resistance increases as its temperature increases.

The resonant frequency in this circuit continues to be equal to the frequency of oscillation. The same formula for finding f_O is used here as with the other

Temperature Coefficient: A positive temperature coefficient means that resistance increases with an increase in temperature, and a negative temperature coefficient means that resistance decreases with an increase in temperature.

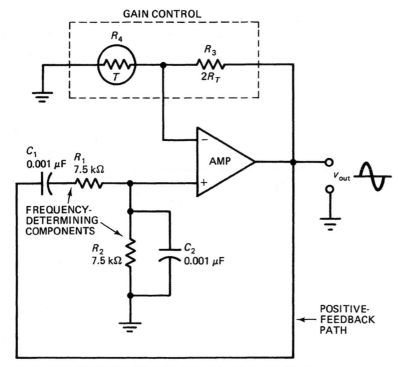

FIGURE 13-12 Wien bridge oscillator.

oscillators, and we can design a particular frequency by exchanging R and f_O as was done earlier. When similar value components are used to form the RC networks of Fig. 13-12., the f_O is determined by

$$f_O = \frac{1}{2\pi RC}$$

$$= \frac{1}{2\pi(7.5 \text{ k}\Omega)(0.001 \text{ }\mu\text{F})}$$

$$= 21.2 \text{ kHz}$$

The noninverting amplifier amplifies this signal, which is reinforced via the positive-feedback path. However, it is attenuated by a factor of 3 as it passes through the lead-lag networks. The gain of the amplifier must replace this loss. The attenuation factor of 3 is a standard value for the Wien bridge.

Gain is controlled by R_3 and the thermistor, R_4. Initially, the thermistor's resistance is rather low, resulting in a high gain for the amplifier. This provides the necessary boost to start the oscillations. The amplified signal continues to increase in strength due to positive feedback and so does the negative feedback. This increases the temperature of the thermistor. As the temperature of the thermistor increases, so does its resistance and voltage.

The resistance of the thermistor increases until it reaches a value we call R_T. Feedback resistor R_3 is designed to be twice the value of R_T. When the thermistor reaches the value of R_T, the gain of the amplifier is 3.

$$A_v = \frac{R_3}{R_4} + 1$$

$$= \frac{2R_T}{R_T} + 1$$

$$= 3$$

The gain of 3 is attenuated by a factor of 3, which results in an overall gain of 1, or unity. A balance at the inputs of the op amp is provided by the Wien bridge at this point.

The amplitude of the output signal can be adjusted by changing the value of R_3. A larger value of R_3 causes a larger thermistor resistance before the circuit stabilizes. Remember that stabilization within the circuit is reached when the resistance of the thermistor is one-half the value of R_3 (i.e., when R_3 is twice the resistance of the thermistor). Since the resistance of the thermistor is larger, its voltage drop is larger. The input signal is equal to the stabilized voltage across the thermistor. A larger input signal results in a larger output signal.

If R_3 were a 2-kΩ resistor in the circuit of Fig. 13-12, then the circuit would stabilize when the thermistor reached one-half this value.

$$R_4 = 0.5 \times R_3$$

$$= 0.5 \times 2 \text{ k}\Omega$$

$$= 1 \text{ k}\Omega$$

Let us assume that a 1-kΩ increase in resistance by the thermistor is accompanied by a 1-V rms increase in voltage across it. The thermistor is actually not a linear device, so these values are used for the sake of simplifying the following explanation.

Since a 1-V rms/1 kΩ increase for the thermistor is assumed, the voltage across the thermistor when the circuit stabilizes is equal to

$$V_{R4} = \frac{R_4}{1\ k\Omega} \times 1\ V\ rms$$

$$= \frac{1\ k}{1\ k\Omega} \times 1\ V\ rms$$

$$= 1\ V\ rms$$

The current through the thermistor is equal to the current through the feedback resistor. Use the previous information to find the feedback current and voltage across the feedback resistor, R_3.

$$I_{R3} = I_{R4} = \frac{V_{R4}}{R_4}$$

$$= \frac{1\ V\ rms}{1\ k\Omega}$$

$$= 1\ mA\ rms$$

and

$$V_{R3} = I_{R3} \times R_3$$

$$= 1\ mA\ rms \times 2\ k\Omega$$

$$= 2\ V\ rms$$

Earlier, while studying noninverting amplifiers, we learned that the output voltage is equal to the sum of the voltages across R_3 and R_4. Therefore,

$$v_{out} = V_{R3} + V_{R4}$$

$$= 2\ V\ rms + 1\ V\ rms$$

$$= 3\ V\ rms$$

Another way of determining the output signal is to realize that the value of voltage across the thermistor is equal to the voltage at the noninverting input. Now, multiply this value by the gain of the circuit.

$$v_{in} = V_{R4} = 1\ V\ rms$$

$$v_{out} = v_{in} \times A_v$$

$$= 3 \times 1\ V\ rms$$

$$= 3\ V\ rms$$

Another device that can be used in place of the thermistor is the tungsten lamp. It, too, changes its resistance with temperature. This theory holds true of any device displaying a positive temperature coefficient.

EXAMPLE 13-6

The values of the circuit of Fig. 13-12 are changed so that $R_1 = 4.7$ kΩ, $R_2 = 4.7$ kΩ, $C_1 = 0.01$ μF, $C_2 = 0.01$ μF, $R_3 = 3$ kΩ, and the thermistor provides a 1 V rms/kΩ change in its voltage. Determine the value of (a) f_O and (b) v_{out}.

Solution:

(a) The frequency of an oscillation in this circuit is found by

$$f_O = \frac{1}{2\pi(4.7 \text{ k}\Omega)(0.01 \text{ }\mu\text{F})}$$

$$= 3.39 \text{ kHz}$$

(b) The circuit is balanced when the thermistor reaches one-half of the value of R_3.

$$R_4 = 0.5 \times 3 \text{ k}\Omega$$

$$= 1.5 \text{ k}\Omega$$

According to the rating of the thermistor given before, its voltage is 1.5 V rms at this point. This value is arrived at by realizing that the thermistor drops approximately 1 V rms for every 1 kΩ of resistance.

$$V_{R4} = \frac{1.5 \text{ k}\Omega}{1 \text{ k}\Omega} \times 1 \text{ V rms}$$

$$= 1.5 \text{ V rms}$$

The voltage at the noninverting input of the op amp is equal to the voltage across R_4. Since it is standard for the Wien bridge to produce a gain of 3, the output is

$$v_{out} = 1.5 \text{ V rms} \times 3$$

$$= 4.5 \text{ V rms} \qquad \blacksquare$$

Practice Problem: What would be the output voltage for the circuit of Fig. 13-12 if $R_3 = 5$ kΩ and the thermistor provided a 1 V rms/kΩ change in its voltage? *Answer:* $v_{out} = 7.5$ V rms.

REVIEW QUESTIONS

38. What determines the frequency of oscillation for the Wien bridge?

39. When does the maximum signal strength occur at the output of a lead-lag network?

40. Frequency and capacitive reactance are (directly or inversely) related.

41. Frequencies above resonance cause a (leading or lagging) phase shift in a lead-lag network.

42. What is the phase shift in a lead-lag network at resonance?

43. Determine the frequency of oscillation in the circuit of Fig. 13-12 if C_1 and C_2 were changed to 0.047-μF capacitors.

44. What is the attenuation factor of the lead-lag network in a Wien bridge oscillator?

45. What gain is the amplifier of Fig. 13-12 designed to provide once it stabilizes?

46. What must be the resistance of R_4 in the circuit of Fig. 13-12 for the circuit to stabilize if $R_3 = 3$ kΩ?

47. Determine the output voltage for the circuit of Fig. 13-12 if $R_3 = 3$ kΩ and the thermistor provides a 1-V rms/kΩ change in voltage.

48. What device can be used to replace a thermistor?

49. A thermistor has a (positive or negative) temperature coefficient.

13-7   RELAXATION OSCILLATOR

The relaxation oscillator is a nonsinusoidal oscillator. This means that the output is not a sine wave. Two methods of constructing a relaxation oscillator are discussed. One uses a UJT and the other uses the op amp. Notice the difference in the output waveform of each.

Using a UJT   A unijunction transistor is used to build the relaxation oscillator of Fig. 13-13. The UJT's emitter to base 1 starts out as a very high resistance. Conduction does not occur until a certain level of voltage, called the peak voltage, occurs at its emitter (see Chapter 10, "Thyristors and Unijunction Devices"). This voltage is controlled by the charging and discharging of the capacitor.

The RC time constant provided by R_1 and C_1 controls the frequency of oscillation. As C_1 charges, it eventually reaches the peak voltage necessary to "turn on" the UJT. Once the UJT is turned on, it begins to conduct heavily between its emitter and base 1. The sudden surge of current is responsible for the spiked output waveform. The emitter to base 1 also provides a discharge path for the capacitor. As current eventually falls below a value called the valley current, the UJT "turns off." This process is repeated over and over again.

The result of the UJT turning on and off is a series of current surges. As the changing current passes through R_3, it produces the nonsinusoidal waveform that is shown in Fig. 13-13.

The frequency of oscillation can be determined if the value of the intrinsic standoff ratio is known. It is normally listed on the spec sheet of a UJT and is symbolized by the Greek letter η (eta). The eta of Fig. 13-13 is listed as 0.65. We begin by finding the period, or time of the oscillations.

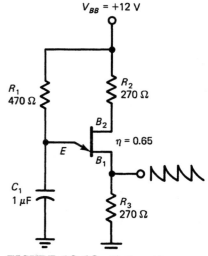

FIGURE 13-13   Relaxation oscillator using a UJT.

$$T = RC \ln \left(\frac{1}{1 - n} \right)$$

$$= (470 \ \Omega)(1 \ \mu F) \ln \left(\frac{1}{1 - 0.65} \right)$$

$$= 493 \ \mu s$$

Frequency is the reciprocal of time.

$$f_O = \frac{1}{T}$$

$$= \frac{1}{493 \ \mu s}$$

$$= 2.03 \text{ kHz}$$

The frequency of the spiked waveform in Fig. 13-13 is 2.03 kHz.

EXAMPLE 13-7

The circuit of Fig. 13-13 is changed so that $R_1 = 2.7$ kΩ, $C_1 = 0.22$ μF, and $\eta = 0.75$ for the UJT. Determine the frequency of oscillation.

Solution:
$$T = (2.7 \text{ kΩ})(0.22 \text{ μF}) \ln \left(\frac{1}{1 - 0.75} \right)$$

$$= 824 \text{ μs}$$

and

$$f_O = \frac{1}{824 \text{ μs}}$$

$$= 1.21 \text{ kHz} \qquad ■$$

Practice Problem: What would be the frequency of oscillation in the circuit of Fig. 13-13 if R_1 were changed to the 1-kΩ resistance? *Answer:* $f_O = 953$ Hz.

Using an Op Amp   Another version of the relaxation oscillator uses the op amp and is depicted in Fig. 13-14. It is a variation of the Schmitt trigger discussed in Chapter 11. The output is saturated positively or negatively by the positive feedback provided by the voltage divider, R_2 and R_3.

The frequency of oscillation is determined by R_1 and C_1. If the output starts off in positive saturation, then the capacitor begins to charge toward the positive saturated value through R_1. Eventually, it reaches a value slightly more positive than the voltage across R_3. At that point, the output is forced into negative saturation. Once again, the capacitor begins to charge toward the negative value. When it charges slightly past the value of negative voltage that is now across R_3, it forces the op amp back into positive saturation.

The positive and negative saturations of the op amp result in a square wave at the output. This circuit is sometimes called a square-wave generator.

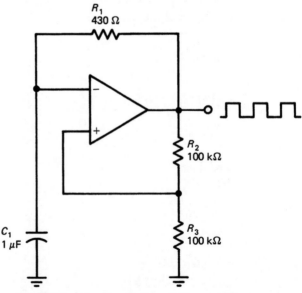

FIGURE 13-14   Relaxation oscillator using op amp.

The frequency of oscillation can be determined by first considering the feedback fraction (β) of the circuit of Fig. 13-14. Only a fraction of the output signal reaches the noninverting input. In this case, it is one-half since the voltage divider, R_2 and R_3, is made up of equal value resistors.

$$\beta = \frac{R_3}{R_2 + R_3}$$

$$= \frac{100 \text{ k}\Omega}{100 \text{ k}\Omega + 100 \text{ k}\Omega}$$

$$= 0.5$$

Substituting this information and the *RC* time constant into the following formula allows us to find the period or time of the waveform.

$$T = 2RC \ln \left(\frac{1 + \beta}{1 - \beta} \right)$$

$$= (2)(430 \ \Omega)(1 \ \mu\text{F}) \ln \left(\frac{1 + 0.5}{1 - 0.5} \right)$$

$$= 945 \ \mu\text{s}$$

Frequency is the reciprocal of time.

$$f_O = \frac{1}{T}$$

$$= \frac{1}{945 \ \mu\text{s}}$$

$$= 1.06 \text{ kHz}$$

EXAMPLE 13-8

The values of the circuit of Fig. 13-14 are changed so that $R_1 = 680 \ \Omega$, $R_2 = 1$ MΩ, $R_3 = 220 \text{ k}\Omega$, and $C_2 = 0.22 \ \mu\text{F}$. Determine the frequency of oscillation.

Solution: We begin by finding the feedback fraction.

$$\beta = \frac{220 \text{ k}\Omega}{1 \text{ M}\Omega + 220 \text{ k}\Omega}$$

$$= 0.18$$

Next, the time of the output signal is determined.

$$T = (2)(680 \ \Omega)(0.22 \ \mu\text{F}) \ln \left(\frac{1 + 0.18}{1 - 0.18} \right)$$

$$= 109 \ \mu\text{s}$$

and

$$f_O = \frac{1}{109 \ \mu\text{s}}$$

$$= 9.17 \text{ kHz}$$ ∎

Practice Problem: What would be the frequency of oscillation in the circuit of Fig. 13-14 if R_1 were substituted with a 1-kΩ resistor? *Answer:* $f_O = 455$ Hz.

50. Oscillators that do not produce sine waves are called _____.

51. The frequency of oscillation for both the UJT and op amp relaxation oscillator is determined by the _____.

52. The emitter of the circuit of Fig. 13-13 conducts when the capacitor charges to the _____ voltage.

53. Emitter current in the circuit of Fig. 13-13 ceases when the current falls below the specified _____ _____ of the UJT.

54. The output of the circuit of Fig. 13-14 saturates to the opposite polarity when V_{C1} is greater than _____.

13-8 TROUBLESHOOTING _____

The oscillator is divided into four subsections. Keep in mind the overall function of the oscillator, which is to produce a periodic waveform at its output. When troubleshooting a specific circuit, try to find the specifications for its frequency, output voltage level, and whether or not the waveform should be sinusoidal. Isolate the oscillator from all other circuitry if possible.

No Output Voltage   The purpose of the amplifier is to replace any energy loss and produce a waveform of sufficient size. No output voltage is one symptom of amplifier trouble. Check the supply voltage to the amplifier. All active components require a source voltage to operate.

Since no other signals can exist without the amplifier, there is not much else to check in the way of signals. A close visual inspection may reveal an open in the circuit trace or a damaged component. If the amplifier is socketed, it would be an easy matter to replace it with a new component. Piggybacking an IC might also work in certain circumstances. When all else fails, it may be necessary to solder in a new amplifier.

Saturated Output Voltage   If the output voltage is constantly saturated in a single direction, then the problem may lie in the positive-feedback loop or frequency-determining components. The amplifier itself may be defective. Replacing the op amp or transistor may be necessary.

On the other hand, suppose the output is oscillating but saturated in both directions. In this case, the positive feedback and frequency-determining sections are most likely all right. A prime suspect for this fault is the gain-control section. An open in the negative-feedback path is a possibility. This could create an open-loop-condition.

Knowing how the circuit should operate is very important. The relaxation oscillator using the op amp produces a saturated square-wave output. So in the case of nonsinusoidal oscillators, saturation in both directions is not a problem.

Incorrect Oscillating Frequency   The starting point for this fault is the frequency-determining components. Perhaps a leaky capacitor or a partially shorted coil is the culprit. An oscilloscope or frequency counter should be used to verify the incorrect frequency. If either the inductor or capacitor is variable, it should be adjusted. *LC* oscillators tend to drift with time and may need occasional adjustments. The crystal oscillator is more precise and only minor adjustments are possible.

The measuring equipment itself may be loading down the oscillator. Make sure that the test equipment has a very high input impedance. Also, the amplifier itself may be the problem at very high frequencies. Carefully check the specifications when substituting components with different part numbers. Their frequency responses may not be the same.

REVIEW QUESTIONS

The following symptoms may be caused by a fault in one of the following sections. Choose the letter of the most probable cause. Each letter may be used only once.

A. Amplifier

B. Frequency-determining components

C. Gain control

55. The output signal is saturated in both directions.

56. There is no output.

57. The output frequency is lower than normal.

SUMMARY

1. Oscillators provide periodic waveforms.

2. Oscillator circuits require amplification, positive feedback, and frequency-determining components.

3. A specific frequency within the wide-band noise at the input of an amplifier is amplified and reinforced through positive feedback.

4. A phase-shift oscillator utilizes three or more RC phase-shift networks to provide a 180° phase shift as well as an oscillation frequency.

5. A three-section RC network attenuates the output signal of the amplifier by a factor of 29.

6. Important formulas for the RC phase-shift oscillator are

$$X_C = \frac{1}{2\pi f C} \qquad f_o = \sqrt{6}\frac{1}{2\pi RC\sqrt{6}}$$

$$\theta = \text{inv tan}\left(\frac{X_C}{R}\right) \qquad A_v = \frac{R_5}{R_4}$$

7. The Colpitts oscillator relies on a parallel resonant tank circuit (tapped to ground between the capacitors) to determine its frequency of oscillation.

8. A tank circuit stores energy.

9. Important formulas for the Colpitts oscillator are

$$C_T = \frac{C_1 \times C_2}{C_1 + C_2}$$

$$f_o = \frac{1}{2_\pi \sqrt{LC_T}}$$

$$A_v > \frac{C_1}{C_2}$$

10. The Hartley oscillator relies on a parallel resonant tank circuit (tapped to ground between the inductors) to determine its frequency of oscillation.

11. Important formulas for the Hartley oscillator are

$$L_T = L_1 + L_2$$

$$f_O = \frac{1}{2\pi\sqrt{L_T C}}$$

$$A_v > \frac{L_2}{L_1}$$

12. Quartz crystal oscillators provide precise oscillation frequencies.

13. Crystals have natural frequencies of vibration.

14. Important formulas for the crystal oscillator are

$$f_S = \frac{1}{2\pi\sqrt{LC_N}} \qquad A_v = \frac{R_2}{R_1} + 1$$

$$f_P = \frac{1}{2\pi\sqrt{LC_{N+M}}} \qquad v_{out} = A_v \times V_{R3}$$

$$V_{R3} = \frac{R_3}{R_3 + R_X} V_D$$

15. The Wien bridge oscillator uses a lead-lag network to determine the resonant frequency.

16. The resonant frequency for a Wien bridge oscillator produces a 0° phase shift.

17. The lead-lag network attenuates the output signal by a factor of 3 in a Wien bridge oscillator.

18. A thermistor changes its resistance with changes in temperature.

19. Important formulas for the Wien bridge oscillator are

$$f_O = \frac{1}{2\pi RC} \qquad V_{R4} = R_4 \times \frac{V_{rms}}{k\Omega}$$

$$A_v = \frac{R_3}{R_4} + 1 \qquad v_{in} = V_{R4}$$

$$\qquad\qquad\qquad v_{out} = V_{in} \times A_v$$

$$R_4 = 0.5 \times R_3$$

20. The relaxation oscillator is a nonsinusoidal oscillator.

21. The oscillation frequency of a relaxation oscillator using a UJT is found by

$$T = RC \ln\left(\frac{1}{1 - \eta}\right)$$

$$f_O = \frac{1}{T}$$

22. The oscillation frequency of a relaxation oscillator using an op amp is found by

$$\beta = \frac{R_3}{R_2 + R_3}$$

$$T = 2RC \ln\left(\frac{1 + \beta}{1 - \beta}\right)$$

$$f_O = \frac{1}{T}$$

23. When troubleshooting an oscillator, reduce the circuit to its four subsections (amplifier, frequency-determining components, gain control, and positive-feedback path) and examine the section that directly relates to the symptom.

GLOSSARY

Attenuation: The reduction in the amplitude of a signal.

Damping: The diminishing of a signal's power due to resistance within the circuit.

Feedback Fraction: The fraction of the output signal that is fed back to the input. It is represented by the Greek letter β (beta).

Nonsinusoidal: Not like a sine wave.

Oscillator: A self-starting circuit that produces a periodic waveform.

Piezoelectric Effect: The condition when certain minerals produce electric current under pressure or vibrate when subjected to an applied voltage.

Regenerative Feedback: Another name for positive feedback. A signal that reinforces the original signal.

Resonance: The condition wherein a capacitive and inductive device compliment each other perfectly. $X_C = X_L$ and $\theta = 0°$.

Sinusoidal: Like a sine wave.

Tank Circuit: A circuit formed by a capacitive and an inductive component for the purpose of storing energy.

Temperature Coefficient: A positive temperature coefficient means that resistance increases with an increase in temperature, and a negative temperature coefficient means that resistance decreases with an increase in temperature.

PROBLEMS

SECTION 13-1

13-1. Describe the difference between sinusoidal and nonsinusoidal waveforms.

13-2. Define an oscillator.

13-3. Explain the importance of amplification within the oscillator circuit.

13-4. Why is controlling gain so important in oscillator circuits?

13-5. What criteria constitute the Barkhausen criteria for oscillation?

13-6. Explain the difference between positive and negative feedback.

13-7. How would you describe a phase shift?

13-8. How is the frequency of oscillation determined?

SECTION 13-2

The following problems refer to the circuit of Fig. 13-15.

13-9. What type of circuit is this circuit?

13-10. Describe the phase shift provided by the *RC* networks.

13-11. What is the frequency of oscillation?

13-12. Calculate the gain provided by the amplifier.

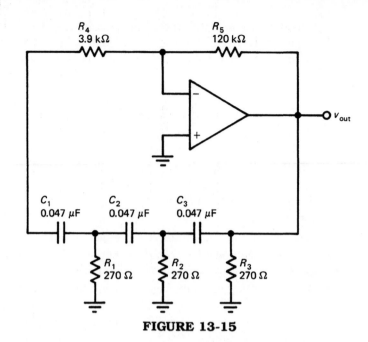

FIGURE 13-15

13-13. Label the gain control, amplifier, frequency-determining components, and positive-feedback sections of this circuit.

13-14. Determine the value of R_4 necessary to overcome the attenuation factor of this circuit if $R_5 = 100$ kΩ.

13-15. Design an *RC* phase-shift oscillator with an oscillation frequency of 20 kHz using 0.001-μF capacitors.

SECTION 13-3

The following problems refer to the circuit of Fig. 13-16.

13-16. What type of circuit is this circuit?

13-17. Explain the phase relationships within this circuit.

13-18. Describe the action of the tank circuit.

13-19. What is damping?

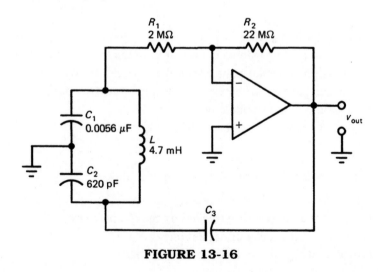

FIGURE 13-16

13-20. Determine the total capacitance of the tank circuit.

13-21. Find the oscillating frequency.

13-22. Calculate the gain.

13-24. Label the gain control, amplifier, positive feedback, and frequency-determining components of this circuit.

13-25. Determine the total tank capacitance needed in this circuit if an oscillating frequency of 500 kHz were desired.

13-26. Design an oscillator similar to the circuit of Fig. 13-16 but with a resonant frequency of 200 kHz. Use the proper amount of gain.

SECTION 13-4

The following problems refer to the circuit of Fig. 13-17.

13-27. What type of circuit is this circuit?

13-28. Describe the meaning of attenuation.

13-29. Determine the total inductance of the tank circuit.

13-30. Find the oscillating frequency.

13-31. Calculate the gain.

13-32. Label the gain control, amplifier, positive feedback, and frequency-determining components of this circuit.

13-33. Determine the total tank inductance needed in this circuit if an oscillating frequency of 500 kHz were desired.

13-34. Design an oscillator similar to the circuit of Fig. 13-17 but with a resonant frequency of 200 kHz. Use the proper amount of gain.

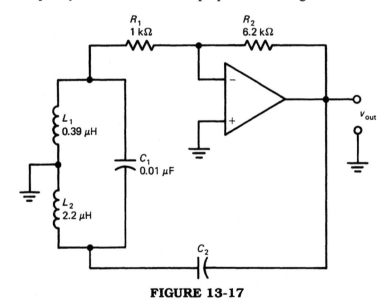

FIGURE 13-17

SECTION 13-5

The following problems refer to the circuit of Fig. 13-18. The specifications for the crystal in this circuit are $R = 20$ kΩ, $L = 100$ H, $C_N = 0.05$ pF, and $C_M = 15$ pF.

13-35. What is the advantage of using crystals?

13-36. Determine the series and parallel frequencies of oscillation for the crystal in this circuit.

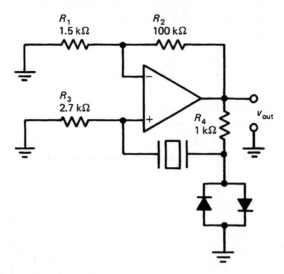

FIGURE 13-18

13-37. Calculate the gain.

13-38. What is the output voltage?

13-39. Label the function of each section of this circuit.

SECTION 13-6

The following problems refer to the circuit of Fig. 13-19. Note that the thermistor provides a 2-V rms/kΩ change in the voltage.

13-40. How does a lead-lag network operate?

13-41. How does signal strength relate to resonant frequency?

13-42. Explain what is meant by a positive temperature coefficient.

13-43. What is the frequency of oscillation?

13-44. Determine the maximum gain of the noninverting amplifier.

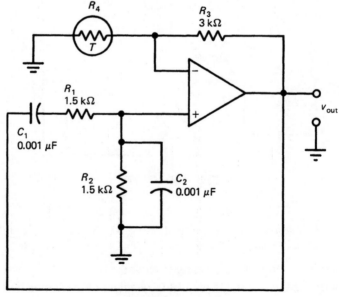

FIGURE 13-19

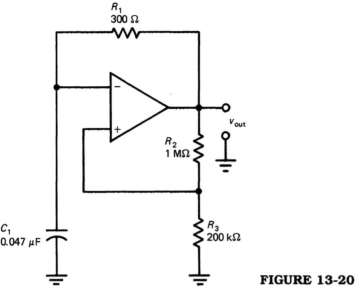

R_1
300 Ω

R_2
1 MΩ

v_{out}

C_1
0.047 μF

R_3
200 kΩ

FIGURE 13-20

13-45. Calculate the maximum resistance of R_4.

13-46. Find the value of the output voltage.

13-47. Label the function of each section of this circuit.

13-48. Provide an example of how the output voltage can be increased.

13-49. Design a circuit with an oscillation frequency of 30 kHz and an output voltage of between 5 and 8 V rms.

SECTIONS 13-7 AND 13-8

The following problems refer to the circuit of Fig. 13-20.

13-50. What type of oscillator is this circuit?

13-51. Is the output sinusoidal? Explain why or why not.

13-52. Determine the feedback fraction.

13-53. Find the frequency of oscillation.

13-54. Draw the output waveform.

13-55. List the faults that would be associated with each of the following sections of an oscillator: amplifier, positive feedback, gain control, and frequency-determining components.

ANSWERS TO REVIEW QUESTIONS

1.	periodic	**2.**	nonsinusoidal	**3.**	dc, ac
4.	noise	**5.**	amplifier	**6.**	By negative feedback
7.	Regenerative feedback	**8.**	Barkhausen	**9.**	in
10.	oscillation	**11.**	Three	**12.**	leads
13.	leads, 90°	**14.**	65.6°	**15.**	24.4°
16.	628 Hz	**17.**	29	**18.**	Minimum of 78.3 kΩ
19.	leads	**20.**	180°	**21.**	resonant
22.	damping	**23.**	2 μF	**24.**	1125 Hz
25.	$A_v > 10$	**26.**	112.5 pF	**27.**	6.6 mH

28. 82.8 kHz	**29.** $A_v > 5.6$	**30.** $L_T = 2.01$ mH
31. Quartz	**32.** True	**33.** piezoelectric effect
34. 10,273 Hz	**35.** 10,279 Hz	**36.** XTAL
37. 4 V	**38.** *RC* networks	**39.** At resonance
40. inversely	**41.** leading	**42.** 0°
43. 452 Hz	**44.** 3	**45.** 3
46. 1.5 kΩ	**47.** 4.5 V rms	**48.** Tungsten lamp
49. positive	**50.** nonsinusoidal	**51.** *RC* time constant
52. peak	**53.** valley current	**54.** V_{R3}
55. C	**56.** A	**57.** B

555
Timer

When you have completed this chapter, you should be able to:

- Analyze the monostable operation of the 555 timer.
- Calculate and observe how the RC time constant relates to one-shot and oscillator timing.
- Analyze the astable operation of the 555 timer.
- Recognize the symptoms of a faulty 555 timer.

INTRODUCTION

The 555 timer is a versatile device. It can be operated in the **monostable** mode, which means it maintains one stable state (ground or 0 V) at its output until triggered. Upon receiving a trigger pulse, its output enters a semistable state (a positive voltage) for a fixed period of time determined by an RC time constant. In the **astable** mode, the 555 timer produces a periodic-pulse waveform whose frequency is determined by an RC time constant. Each of these modes has many applications within the realm of electronics.

  The 555 timer is an integrated circuit and is produced in either a DIP or metal can package. There are over two dozen transistors and a dozen resistors that form the internal circuitry. We use a block diagram to aid in understanding the internal workings of the timer. We also review topics such as the transistor switch and comparators. A specification sheet is included here for reference (see page 379).

14-1 INSIDE VIEW

Since the 555 timer uses the dual in-line package, it looks very much like the op amp. The 555 timer has eight pins that are numbered from 1 to 8 in a counterclockwise direction. Pin 1 is located at the end of the IC that has a notch or small indented circle. Figure 14-1 shows a block diagram of the timer in a typical IC layout (see page 380).

555 timer.

Resistors R_1 to R_3 A voltage divider is formed by R_1 through R_3. All three resistors are the same value; therefore, they divide $+V_{CC}$ into thirds. $+V_{CC}$ can be any value from 4.5 to 18 V. Notice that R_1, which drops $\frac{1}{3} +V_{CC}$, is connected to the noninverting input of IC1, and R_2, which drops $\frac{2}{3} +V_{CC}$, is connected to the inverting input of IC2. These two resistors maintain reference voltages for the comparators.

Comparator IC1 Comparator IC1 compares the reference voltage at the noninverting input ($\frac{1}{3}$ of $+V_{CC}$) with the voltage at the trigger input. If the voltage at the noninverting input is more positive than the trigger voltage, then the output of IC1 will saturate positively to $+V_{CC}$. If the voltage at the noninverting input is less positive than the trigger voltage, then the output of IC1 will drop to zero volts.

Comparator IC2 Comparator IC2 compares the threshold voltage at the noninverting input against the reference voltage at the inverting input ($\frac{2}{3}$ of $+V_{CC}$). If the threshold voltage is more positive than the reference voltage, then the output of IC2 will saturate positively. On the other hand, if the threshold voltage is

D1669, SEPTEMBER 1973—REVISED AUGUST 1985

- Timing from Microseconds to Hours
- Astable or Monostable Operation
- Adjustable Duty Cycle
- TLL-Compatible Output Can Sink or Source up to 200 mA
- Functionally Interchangeable with the Signetics SE555, SE555C, SA555, NE555; Have Same Pinout

SE555C FROM TI IS NOT RECOMMENDED FOR NEW DESIGNS

NE555, SE555, SE555C . . . JG DUAL-IN-LINE PACKAGE
SA555, NE555 . . . D, JG, OR P DUAL-IN-LINE PACKAGE
(TOP VIEW)

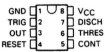

SE555, SE555C . . . FK CHIP CARRIER PACKAGE
(TOP VIEW)

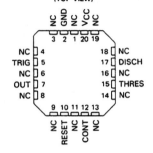

NC—No internal connection

description

These devices are monolithic timing circuits capable of producing accurate time delays or oscillation. In the time-delay or monostable mode of operation, the timed interval is controlled by a single external resistor and capacitor network. In the astable mode of operation, the frequency and duty cycle may be independently controlled with two external resistors and a single external capacitor.

The threshold and trigger levels are normally two-thirds and one-third, respectively, of V_{CC}. These levels can be altered by use of the control voltage terminal. When the trigger input falls below the trigger level, the flip-flop is set and the output goes high. If the trigger input is above the trigger level and the threshold input is above the threshold level, the flip-flop is reset and the output is low. The reset input can override all other inputs and can be used to initiate a new timing cycle. When the reset input goes low, the flip-flop is reset and the output goes low. Whenever the output is low, a low-impedance path is provided between the discharge terminal and ground.

The output circuit is capable of sinking or sourcing current up to 200 milliamperes. Operation is specified for supplies of 5 to 15 volts. With a 5-volt supply, output levels are compatible with TTL inputs.

The SE555 and SE555C are characterized for operation over the full military range of −55°C to 125°C. The SA555 is characterized for operation from −40°C to 85°C, and the NE555 is characterized for operation from 0°C to 70°C.

functional block diagram

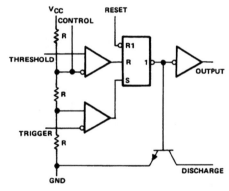

Reset can override Trigger, which can override Threshold.

less positive than the reference voltage, then the output of IC2 will drop to zero volts.

Flip-Flop IC3 The flip-flop has two inputs, (S)et and (C)lear, that control the outputs, Q and Q' (Q prime or NOT Q). If the set input receives a positive voltage, then Q will be set to $+V_{CC}$. A positive voltage on the clear input clears Q to 0 V. Set and clear should not receive positive voltages simultaneously or the

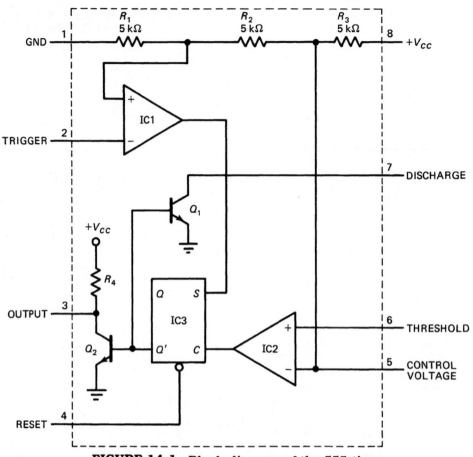

FIGURE 14-1 Block diagram of the 555 timer.

state of Q and Q' will be ambiguous. When neither set nor clear receives a positive voltage, the outputs maintain their previous states.

Since Q is not used in the block diagram, our focus is on Q'. The Q' output is always the opposite state of Q. Therefore, the positive voltage on set, which sets Q to $+V_{CC}$, also clears Q'. And, likewise, a positive voltage on clear clears Q to 0 V, which means that Q' is set to $+V_{CC}$. Notice that Q' controls both transistors in the block diagram.

The **reset** input of the flip-flop is used to immediately change the Q output to 0 V and Q' to $+V_{CC}$ regardless of the S and C inputs. A small circle or bubble on the reset input indicates that the reset is activated by 0 V. Reset and clear have the same meaning in terms of indicating a low-voltage level (0 V). As a matter of fact, the S and C inputs could also be labeled as the S and R inputs, respectively. Clear is used to avoid confusion between it and the reset input. Reset is seldom used and is normally tied to $+V_{CC}$.

Transistor Q_1 Transistor Q_1 is used as a switch to discharge a capacitor connected between the discharge pin and ground. It is closed by a positive voltage on its base, which it receives from the Q' output of the flip-flop. When the Q' output of the flip-flop is 0 V, the transistor acts like an open switch and the capacitor connected to pin 7 of the timer is given a chance to charge.

Transistor Q_2 Transistor Q_2 is also controlled by the Q' output of the flip-flop. It acts as an inverter between Q' and the output of the timer. When Q' is 0 V, the transistor acts like an open switch and V_{CC} appears at the output. A positive voltage at Q' causes the transistor switch to close and 0 V are read at the output of the timer.

EXAMPLE 14-1 _____

If 10 V were applied between the V_{CC} and the GND inputs of the circuit of Fig. 14-1, then what voltage would appear at (a) the noninverting input of IC1 and (b) the inverting input of IC2?

Solution:
(a) Recall that the three resistors form a voltage divider and each drops one-third of the supply voltage. The voltage at the noninverting input of IC1 is equal to the voltage across R_1, or one-third of V_{CC}. One-third of 10 V is 3.33 V.

(b) The voltage at the inverting input of IC2 is equal to the sum of the voltage drops across R_1 and R_2, or two-thirds of the supply voltage. Two-thirds of V_{CC} is 6.66 V. ∎

REVIEW QUESTIONS _____

1. What is the name given to the operating mode that has only one stable state?
2. What is the name given to the operating mode that provides a periodic waveform at its output?
3. What does DIP represent?
4. The 555 timer is an integrated circuit. True or false?
5. If $+V_{CC} = 5$ V, then what voltage will be found at the noninverting input of IC1 in the circuit of Fig. 14-1?
6. If $+V_{CC} = 5$ V, then what voltage will be found at the inverting input of IC2 in the circuit of Fig. 14-1?
7. The circuit formed by IC1 is called a _____ .
8. If $+V_{CC} = 5$ V and the trigger input were 5 V, would the output of IC1 be 5 or 0 V?
9. If $+V_{CC} = 5$ V and the threshold input were 5 V, would the output of IC2 be 5 or 0 V?
10. If $+V_{CC} = 5$ V, $S = 5$ V and $C = 0$ V, would the Q' output of IC3 be 5 or 0 V?
11. If $+V_{CC} = 5$ V, trigger $= 5$ V, and threshold $= 5$ V, would the Q' output of IC3 be 5 or 0 V?
12. What voltage level is needed to activate the reset input?
13. How are the outputs of the flip-flop affected when reset is activated?
14. Does the capacitor connected between pin 7 and ground of the timer discharge when the flip-flop is set or cleared?
15. Does the output of the timer equal $+V_{CC}$ when the flip-flop is set or cleared?

14-2   MONOSTABLE OPERATION _____

The monostable operation of the 555 timer involves one stable state and one **quasistable** state. The stable state consists of a constant low output. It remains low until the trigger voltage drops below $+\frac{1}{3}V_{CC}$. At that point, the output changes state to approximately $+V_{CC}$. This is the quasistable state, so named because it

Quasistable: An input or output state that appears to be stable at certain times.

appears to be stable only at certain times. The output remains high for a time determined by the *RC* time constant and then returns to its low stable state.

A 555 timer connected in the monostable mode is sometimes referred to as a **one-shot.** Figure 14-2 shows how the timer is connected to provide monostable operation. We use this diagram to explain the operation of the timer in the monostable mode and then consider the way it would be drawn on an actual schematic.

The starting condition of the circuit is as follows:

(1) GND = 0 V	(8) $+V_{CC}$ = 6 V
(2) TRIGGER = 6 V	(7) DISCHARGE = 0 V
(3) OUTPUT = 0 V	(6) THRESHOLD = 0 V
(4) RESET = 6 V	(5) VOLTAGE CONTROL = 4 V

Step 1 Circuit action begins when the trigger input is pulsed low momentarily. When the trigger voltage falls below the reference voltage across R_1, which is 2 V in this case, the output of IC1 saturates positively. This positive voltage sets the flip-flop. Since the voltage at the trigger is pulsed low only momentarily, it eventually returns to 6 V. When the trigger returns to 6 V, it causes the output of IC1 to drop back to 0 V. This removes the positive voltage off the set input of the flip-flop; however, the flip-flop still remains set.

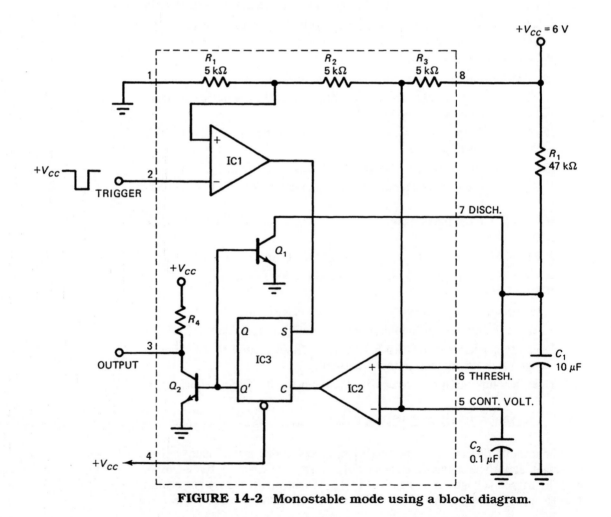

FIGURE 14-2 Monostable mode using a block diagram.

Step 2 The Q output of the flip-flop is now set; therefore, the Q' output is cleared or 0 V. The Q' output is tied to the base of both transistors and causes them to cut off. The collector–emitter region of each transistor acts like an open switch. Opening Q_2 causes the output of the timer to increase to +6 V. (The output of the timer is now in its quasistable state.) Opening Q_1 allows the external capacitor, C_1, to charge through R_1.

Step 3 As the capacitor charges, its voltage is compared against that of the inverting input of IC2. The voltage at the inverting input is 4 V and the output of IC2 is currently 0 V. Eventually, the voltage across C_1 surpasses 4 V and causes the output of IC2 to saturate positively. Since 4 V is 66.6% of V_{CC} and it takes one time constant for the capacitor to charge to 63%, the time for C_1 to surpass 4 V is slightly longer than one time constant. It takes approximately 1.1 time constants to exceed this value. The output of the timer remains high for this period of time. Therefore, the pulse width of the output is

$$P_W = 1.1 \times RC$$
$$= 1.1 \times (47 \text{ k}\Omega)(10 \text{ }\mu\text{F})$$
$$= 517 \text{ ms}$$

It takes a little over a one-half second for the threshold voltage of C_1 to rise to 4 V. As soon as it increases slightly past that point, it causes the output of IC2 to saturate positively.

Step 4 The positive output of IC2 clears the Q output of the flip-flop. Q' returns to a positive voltage level and turns on both transistors. Turning on Q_2 means the collector–emitter switch closes and the output of the timer returns to 0 V. (Keep in mind that the output of the timer remained high for 517 ms in this example.) Q_1 is also turned on and when its collector–emitter switch closes, it discharges the capacitor (C_1). The circuit is now back to the starting condition.

The waveforms of Fig. 14-3 illustrates the external voltage conditions of the timer during monostable-mode operation. The response of the output to the trigger input is immediate. Notice how the pulse width of the output is directly related

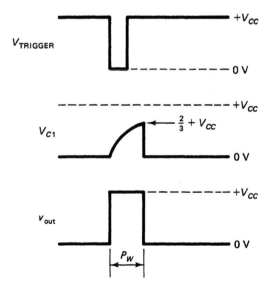

FIGURE 14-3 Monostable waveforms.

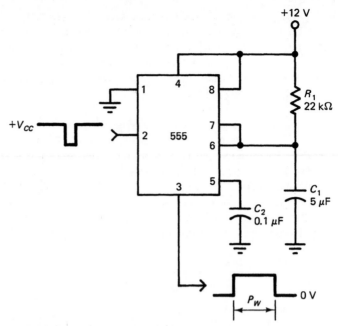

FIGURE 14-4 Schematic representation of the 555 timer in the monostable mode.

to the time it takes for the capacitor to charge to slightly more than $\frac{2}{3}$ of $+V_{CC}$. Figure 14-4 shows an actual schematic diagram of the 555 timer connected in the monostable mode. The pin numbers are not arranged in numerical order. This is very common in the schematic representation of the 555 timer. The objective in drawing the schematic is to avoid crossing lines when possible.

EXAMPLE 14-2 _____

The circuit of Fig. 14-4 is modified so that R_1 = 3.3 kΩ and C_1 = 470 μF. What is the pulse width if this circuit were triggered?

Solution: The output pulses high for a period of time determined by the RC time constant of the circuit.

$$P_W = 1.1 \times (3.3 \text{ k}\Omega \times 470 \text{ }\mu\text{F})$$
$$= 1.71 \text{ s} \quad \blacksquare$$

Practice Problem: What is the pulse width of the output in the circuit of Fig. 14-4? *Answer:* P_W = 121 ms.

Designing the Pulse Width   The time of the pulse width at the output can be changed by substituting different values into the RC time constant. Since resistors come in a larger range of values, resistance is the variable in solving for the RC time constant. It is wise to use a larger capacitance for longer pulse widths to avoid extremely large values for the resistance. Likewise, use smaller values of capacitance for shorter pulse widths to avoid extremely low resistances.

If a pulse width of 2 s is desired and a 470-μF capacitor is available, then

$$R = \frac{2 \text{ s}}{(1.1)(470 \text{ }\mu\text{F})}$$
$$= 3.87 \text{ k}\Omega$$

R_1 and C_1 in the circuit of Fig. 14-4 is now substituted with a 3.9-kΩ resistor and a 470-μF capacitor to produce a pulse width of approximately 2 s.

Practice Problem: What size resistance would be necessary to produce a 20-ms pulse width in the circuit of Fig. 14-4 if C_1 were a 2.2-μF capacitor? *Answer:* $R = 8.26$ kΩ.

REVIEW QUESTIONS

16. What is the voltage level of the stable state?
17. What term is used to express the state that only appears to be stable?
18. What value of voltage must the trigger voltage drop below in the circuit of Fig. 14-4 in order to change the output state?
19. Another name for the monostable 555 timer is the _____ .
20. Determine the value of threshold voltage necessary to change the output of the circuit of Fig. 14-4 back to its stable state.
21. What is the purpose of pin 7 in the circuit of Fig. 14-4?
22. Pin 4 in the circuit of Fig. 14-4 is the _____ input and is currently (active or inactive) in this circuit.
23. What is the voltage at pin 5 in the circuit of Fig. 14-4?
24. Calculate the pulse width of the output in the circuit of Fig. 14-4 if C_1 were substituted with a 33-μF capacitor.
25. Determine the value of resistance that would be necessary to provide a 1-s pulse width in the circuit of Fig. 14-4.
26. What is the approximate value of the output voltage during the time the pulse width is high in the circuit of Fig. 14-4?

14-3 ASTABLE OPERATION

Astable operation means that there are no stable states. In other words, the output is periodically changing states. The astable 555 timer operates as a pulse generator whose frequency is determined by an *RC* time constant. The circuit of Fig. 14-5 looks different from that of the monostable mode of operation. For one thing, the trigger input is controlled by the charging and discharging of C_2. Also, the threshold and discharge pins are not tied together. They are separated by a resistor that provides a longer discharge time for the capacitor.

Step 1 When the power is initially turned on, the voltage across C_2 is 0 V. This voltage is applied to both the trigger and threshold inputs. Comparator IC2 compares the 6.66 V on its inverting input to the 0 V on its noninverting input and produces 0 V at its output. The 0 V on the trigger input causes the output of IC1 to saturate positively. A positive voltage on the set input of the flip-flop sets the Q output.

Step 2 With Q set, Q' is clear, or 0 V. The 0 V of Q' turns off both transistors. Transistor Q_2 causes 10 V to appear at the output. Transistor Q_1 is also cut off and, therefore, allows the external capacitor, C_2, to charge through R_1 and R_2.

Step 3 The output is still high at this point and remains high until the voltage at the threshold input exceeds $+\frac{2}{3}V_{CC}$, which is 6.66 V in this case. It takes less

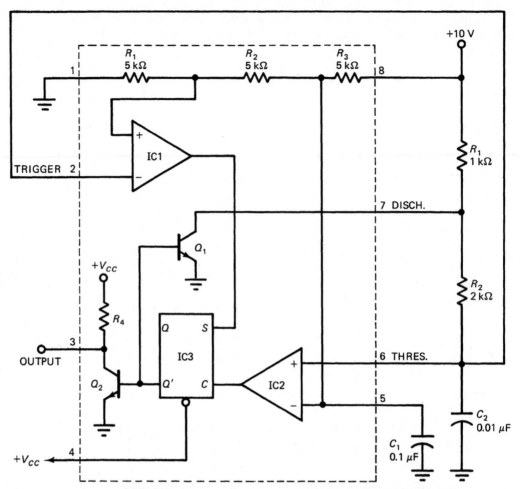

FIGURE 14-5 Astable mode using a block diagram.

than one time constant for the capacitor to charge to this value for every pulse after the first pulse. (This is because the starting point of the capacitor is 3.33 V instead of 0 V, as will be explained shortly.) As a matter of fact, it takes 0.693 time constant for the capacitor to charge to this value. The period of time that the pulse is high can be found by

$$P_{WH} = 0.693(R_1 + R_2)C_2$$

$$= 0.693(1 \text{ k}\Omega + 2 \text{ k}\Omega)0.01 \ \mu\text{F}$$

$$= 20.8 \ \mu\text{s}$$

Notice that both R_1 and R_2 figure into the RC time constant for the charge time. The output waveform remains at 10 V for 20.8 μs before dropping to 0 V.

Step 4 As the capacitor charges during the period of the 20.8 μs, the trigger voltage eventually rises above 3.33 V. The output of IC1 changes to 0 V.

After 20.8 μs, the capacitor has charged to slightly more than 6.66 V and this threshold voltage causes the output of comparator IC2 to saturate positively. The positive voltage coming off of IC2 clears the Q output of the flip-flop. The Q' output increases to a positive value and turns on both transistors.

Step 5   As transistor Q_2 is turned on, it causes the output of the timer to drop to 0 V. Turning on Q_1 causes the capacitor to discharge, but notice that both R_2 and C_2 are part of the discharge path. This means that the capacitor will not discharge instantly, but at a rate determined by the RC time constant of these two components.

Here is a very important point: the capacitor, C_2, does not discharge all the way to 0 V. It discharges to approximately $+\frac{1}{3}V_{CC}$. A slight decrease below this value triggers IC1, which eventually turns off transistors Q_1 and Q_2, allowing the capacitor to begin charging again. The end result is that the capacitor charges from $+\frac{1}{3}V_{CC}$ to $+\frac{2}{3}V_{CC}$ and then back to $+\frac{1}{3}V_{CC}$. This process is repeated over and over again to produce a periodic waveform at the output of the timer.

The output of the timer is 0 V during the discharge time of the capacitor. The discharge path includes both R_2 and C_2, and it takes less than one time constant for the discharge to occur. Therefore, the time that the pulsating output waveform is low can be found by

$$P_{WL} = 0.693 R_2 C_2$$
$$= 0.693(2 \text{ k}\Omega)(0.01 \ \mu\text{F})$$
$$= 13.9 \ \mu\text{s}$$

The waveform at the output of the astable timer of Fig. 14-5 remains high for 20.8 μs and low for 13.9 μs. It continues to pulse high and low for these lengths of time as long as power is applied.

Frequency and Duty Cycle of the Output   The charging of the capacitor and the frequency of the output are directly related. Figure 14-6 compares the voltage across the capacitor to the output voltage. Since a periodic waveform is formed with a high pulse width of 20.8 μs and a low pulse width of 13.9 μs, the overall time of the waveform is equal to the sum of these two values.

$$T = P_{WH} + P_{WL}$$
$$= 20.8 \ \mu\text{s} + 13.9 \ \mu\text{s}$$
$$= 34.7 \ \mu\text{s}$$

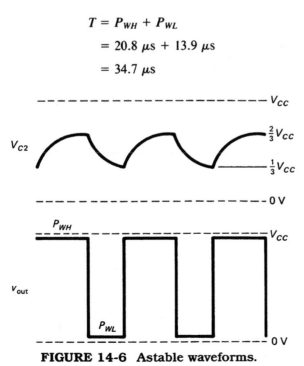

FIGURE 14-6   Astable waveforms.

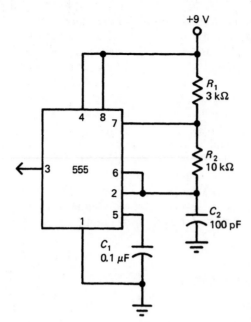

FIGURE 14-7 Schematic diagram of a 555 timer in the astable mode.

Frequency is the reciprocal of time.

$$f = \frac{1}{T}$$

$$= \frac{1}{34.7\ \mu s}$$

$$= 28.8\ kHz$$

Duty Cycle: The period of a cycle when work is done.

The **duty cycle** is the percentage of time that work is being done. It is assumed that work is done when the output of the timer is high. Using the preceding values, we find that the duty cycle of the circuit of Fig. 14-5 is

$$D = \frac{P_{WH}}{T} \times 100$$

$$= \frac{20.8\ \mu s}{34.7\ \mu s} \times 100$$

$$= 59.9\%$$

The schematic representation of the astable 555 timer is shown in Fig. 14-7. Once again, the pin numbers are rearranged for convenience in drawing the circuit.

EXAMPLE 14-3 _____

The circuit of Fig. 14-7 is modified so that $R_1 = 3\ k\Omega$, $R_2 = 5.1\ k\Omega$, and $C_2 = 0.1\ \mu F$. Determine the frequency of the output signal.

Solution: The time that the output pulses high is determined by the charge time of C_2. Capacitor C_2 charges through both R_1 and R_2; therefore,

$$P_{WH} = 0.693 \times (3\ k\Omega + 5.1\ k\Omega) \times 0.1\ \mu F$$

$$= 561\ \mu s$$

The output remains low for the time it takes capacitor C_2 to discharge. This capacitor discharges through R_2 only; therefore,

$$P_{WL} = 0.693 \times 5.1 \text{ k}\Omega \times 0.1 \text{ }\mu\text{F}$$
$$= 353 \text{ }\mu\text{s}$$

The overall time of the output signal is

$$T = 561 \text{ }\mu\text{s} + 353 \text{ }\mu\text{s}$$
$$= 914 \text{ }\mu\text{s}$$

and the frequency is

$$f = \frac{1}{914 \text{ }\mu\text{s}}$$
$$= 1.09 \text{ kHz} \qquad \blacksquare$$

Practice Problem: What are the frequency and duty cycle of the output of the circuit of Fig. 14-7? *Answer:* $f = 627$ kHz and $D = 56.5\%$.

REVIEW QUESTIONS

27. What term is used to describe a condition in which there is no stable state?
28. What would be P_{WH} in the circuit of Fig. 14-7 if R_2 were replaced with a 5-kΩ resistor?
29. What would be P_{WL} in the circuit of Fig. 14-7 if R_2 were replaced with a 5-kΩ resistor?
30. Determine the frequency of the output of the circuit of Fig. 14-7 if R_2 were replaced with a 5-kΩ resistor.
31. Calculate the duty cycle of the output of the circuit of Fig. 14-7 if R_2 were replaced with a 5-kΩ resistor.
32. What is the voltage at pin 5 in the circuit of Fig. 14-7?

14-4   DEBOUNCE CIRCUIT

Switch bounce occurs when a switch is closed and the switching element literally bounces on the contacts within the switch. The bouncing action creates a series of open and closed switch conditions. Although the bounce may not be apparent to the user because of its high speed, it is recognizable to electronic equipment working in microsecond time frames.

A counter, for example, can increment its count by 1 for every transition of 0 to 5 V. If the output of the switching arrangement of Fig. 14-8 were connected to a

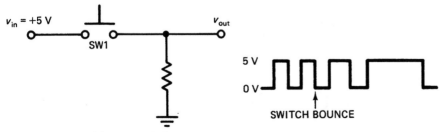

FIGURE 14-8   **Switch with contact bounce.**

counter, the count would have incremented by 4 when the switch was pressed only once. The contact bounce may not be felt by the user, but the illustration shows that the output waveform has four positive-going transitions. Contact bounce must be overcome if accurate control of the circuit is to be achieved.

Although contact bounce is inherent in the switch itself, we can devise a circuit that ignores its effects. One method is to use the 555 timer in the monostable mode, as shown in Fig. 14-9. As soon as SW1 makes contact for the first time, ground is applied to the trigger input and the output goes high. The output remains high for a certain period of time regardless of what is happening at the trigger input. The switch may be bouncing, but the timer ignores it.

There are two considerations regarding the user that must be addressed in designing a debounce circuit like Fig. 14-9. The first is to make the output delay long enough to ignore the bounce. The second consideration is to make the output delay short enough to allow the user to press it repeatedly without having to wait for extended periods of time.

A delay of about one-half second was designed for the circuit of Fig. 14-9. This allows the user to press and release the switch twice per second. The actual value of the delay can be found by using the pulse-width formula.

$$P_W = 1.1 R_1 C_1$$
$$= 1.1(430 \text{ k}\Omega)(1 \text{ }\mu\text{F})$$
$$= 0.473 \text{ ms}$$

When the user presses SW1, the output goes high for 473 ms, during which time the user releases SW1. A single positive-going transition (and negative-going transition) appears at the output each time SW1 is pressed. The contact bounce is ignored while the output is high.

Practice Problem: What should be the value of R_1 in the circuit of Fig. 14-9 in order to provide a 300-ms pulse width at the output? *Answer:* $R_1 = 272.7$ kΩ.

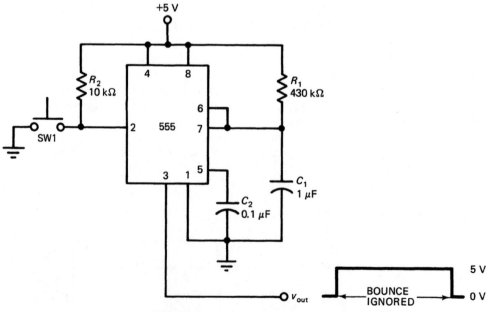

FIGURE 14-9 Debounced switch.

33. The unwanted opening and closing of a switch when it is pressed is called _____ _____ .

34. A circuit that overcomes the problem described in Review Question 33 is called a _____ circuit.

35. In which mode is the 555 timer used in the circuit of Fig. 14-9?

36. What would be the pulse width of the output of the circuit of Fig. 14-9 if R_1 were replaced with a 100-kΩ resistor?

37. What should be the value of R_1 if a pulse width of 20-ms were desired?

14-5 VOLTAGE-CONTROLLED OSCILLATOR

The astable mode of the 555 timer was discussed earlier. Frequency of oscillation was varied by changing the value of the RC time constant. Voltage-controlled oscillators (VCOs) vary the frequency of oscillation by a change in voltage.

During the discussion of the astable mode, it was found that the capacitor charges from $\frac{1}{3}V_{CC}$ to $\frac{2}{3}V_{CC}$ due to the reference voltages of the internal voltage divider of the 555. Up to this point, the voltage-control input of the timer has not been used and, therefore, its voltage defaulted to the internal value of $\frac{2}{3}V_{CC}$. The control voltage input will now be used to vary the frequency of the timer.

The voltage-controlled oscillator is illustrated in Fig. 14-10. It looks similar to the astable mode of operation previously discussed with the exception of the potentiometer, which is now connected to the control voltage input (pin 5) of the timer. The potentiometer is used to vary the voltage at pin 5 in this example.

A key to understanding this circuit is recalling the internal layout of the timer. Figure 14-11 shows the effect that the control voltage has on the internal operation of the timer. Without the control voltage, IC2 would normally use $\frac{2}{3}V_{CC}$ as a reference voltage for comparison with the threshold voltage. The application of V_{CON} changes the reference voltage to the value of V_{CON} itself. This new value of control voltage is now split across the two 5-kΩ resistors and a reference voltage of $0.5V_{CON}$ is applied to the noninverting input of IC1. The value at the trigger input is compared against this new value rather than the previous $\frac{1}{3}V_{CC}$.

Examining these facts reveals that the charging distance of the capacitor from the trigger voltage to the threshold voltage is always equal to $0.5V_{CON}$. If

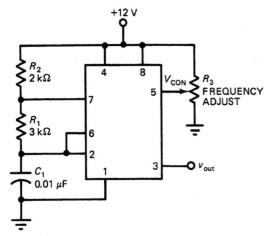

FIGURE 14-10 Voltage-controlled oscillator.

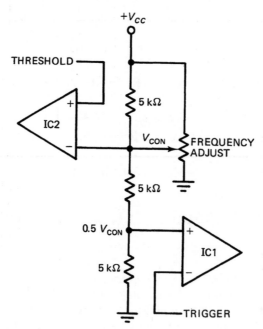

FIGURE 14-11 Application of control voltage.

V_{CON} is equal to 6 V, then $0.5V_{CON}$ is 3 V. The capacitor charges and discharges between 3 and 6 V. Figure 14-12 illustrates this condition.

If V_{CON} were increased, then the difference between the two voltages would be greater. Consequently, it takes the capacitor longer to charge and discharge. The output, therefore, changes less frequently. Summarizing, increasing V_{CON} decreases the frequency, and decreasing V_{CON} increases the frequency.

The following approximations are used to illustrate how the frequency changes with V_{CON}. A value of 6 V for V_{CON} is used in the first analysis of the charge time. The output pulses high during the charge time of the capacitor.

$$P_{WH} = \ln \left(\frac{V_{CC} - 0.5V_{CON}}{V_{CC} - V_{CON}} \right) \times (R_1 + R_2)C_1$$

$$= \ln \left(\frac{12\ V - 3\ V}{12\ V - 6\ V} \right) \times (3\ k\Omega + 2\ k\Omega)(0.01\ \mu F)$$

$$= 20.3\ \mu s$$

Eventually, the threshold value of 6 V is reached by the capacitor, which causes the discharge action to begin. The capacitor discharges from 6 V down to 3

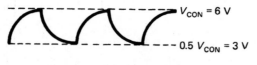

FIGURE 14-12 Capacitor charge–discharge diagram.

V, at which point the timer is triggered into recharging the capacitor. The output pulses low during the discharge time.

$$P_{WL} = \ln\left(\frac{V_{CON}}{0.5V_{CON}}\right) \times R_1 C_1$$

$$= \ln\left(\frac{6\text{ V}}{3\text{ V}}\right) \times (3\text{ k}\Omega)(0.01\ \mu\text{F})$$

$$= 20.8\ \mu\text{s}$$

The overall time of the output waveform is equal to the time that it pulses high plus the time that it pulses low.

$$T = P_{WH} + P_{WL}$$

$$= 20.3\ \mu\text{s} + 20.8\ \mu\text{s}$$

$$= 41.1\ \mu\text{s}$$

Frequency is the reciprocal of time.

$$f = \frac{1}{T}$$

$$= \frac{1}{41.1\ \mu\text{s}}$$

$$= 24.3\text{ kHz}$$

If the control voltage were increased to 9 V, then the voltage of the capacitor would swing between 9 V and 4.5 V. The extra time needed to charge and discharge the extra voltage decreases the frequency of the output.

$$P_{WH} = \ln\left(\frac{12 - 4.5\text{ V}}{12\text{ V} - 9\text{ V}}\right) \times (5\text{ k}\Omega)(0.01\ \mu\text{F})$$

$$= 45.8\ \mu\text{s}$$

and

$$P_{WL} = \ln\left(\frac{9\text{ V}}{4.5\text{ V}}\right) \times (3\text{ k}\Omega)(0.01\ \mu\text{F})$$

$$= 20.8\ \mu\text{s}$$

Therefore,

$$T = 45.8\ \mu\text{s} + 20.8\ \mu\text{s}$$

$$= 66.6\ \mu\text{s}$$

so

$$f = \frac{1}{66.6\ \mu\text{s}}$$

$$= 15\text{ kHz}$$

An increase in control voltage resulted in a decrease of frequency from 24.3 kHz to 15 kHz.

EXAMPLE 14-4

Determine the output frequency in the circuit of Fig. 14.10 if V_{CON} were adjusted to 8 V.

Solution: We calculate the high and low pulse widths in order to determine the time of the output signal. The reciprocal of this time is equal to the frequency.

$$P_{WL} = \ln\left(\frac{8\ V}{4\ V}\right) \times (3\ k\Omega)(0.01\ \mu F)$$

$$= 20.8\ \mu s$$

$$P_{WH} = \ln\left(\frac{12\ V - 4\ V}{12\ V - 8\ V}\right) \times (3\ k\Omega + 2\ k\Omega)(0.01\ \mu F)$$

$$= 34.7\ \mu s$$

$$T = 20.8\ \mu s + 34.7\ \mu s$$

$$= 55.5\ \mu s$$

$$f = \frac{1}{55.5\ \mu s}$$

$$= 18\ kHz \qquad \blacksquare$$

Practice Problem: Determine the output frequency if V_{CON} were adjusted to 7 V in the circuit of Fig. 14-10. *Answer:* f = 21.1 kHz.

REVIEW QUESTIONS

38. How is the frequency of a VCO controlled?
39. What is the name of the pin used for frequency control in the circuit of Fig. 14-10?
40. What mode of operation is used with the timer in the circuit of Fig. 14-10?
41. What value of voltage must the trigger voltage drop below in order to pulse the output high?
42. What value of voltage must the threshold voltage reach before the capacitor begins to discharge and the output drops low?
43. Decreasing V_{CON} (increases or decreases) the output frequency.

14-6 555 ORGAN

A more entertaining application of the 555 timer is now considered. Music or sound is made up of a variety of frequencies. The 555 timer can be used to create different frequencies by changing the RC time constant in the astable mode. The organ in Fig. 14-13 does just that.

The frequency at the output depends on the charging and discharging of C_1. Charging occurs through R_1 and one of the eight resistors whose switch is pressed closed, and the output waveform goes high during this time. Eventually, the

394

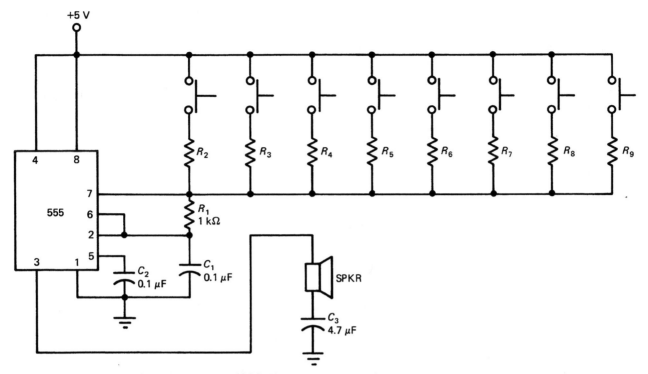

FIGURE 14-13   555 Timer organ.

capacitor reaches the threshold voltage ($\frac{2}{3}V_{CC}$) and the output waveform drops low for the duration of the discharge time. The capacitor discharges through R_1 only.

By providing different charging resistors (R_2 to R_9), assorted output frequencies can be produced. The following table shows the frequency for each note along with the necessary value of resistance.

Note	Frequency (Hz)	Resistance (kΩ)
C	523	25.6 (R_2)
D	587	22.6 (R_3)
E	659	19.9 (R_4)
F	698	18.7 (R_5)
G	784	16.4 (R_6)
A	880	14.4 (R_7)
B	988	12.6 (R_8)
C	1047	11.8 (R_9)

The time of the waveform is composed of P_{WH} and P_{WL}. P_{WL} is the same in every case. This is because R_1 is the only resistance involved in discharging the capacitor.

$$P_{WL} = 0.693R_1C_1$$

$$= 0.693(1 \text{ k}\Omega)(0.1 \ \mu\text{F})$$

$$= 69.3 \ \mu\text{s}$$

P_{WH} of the waveform depends upon which key is pressed. Consider the middle C note (523 Hz). It is generated with the aid of R_2. If another key were

---

pressed, the value of the associated resistor would be substituted in place of R_2 in the following equation.

$$P_{WH} = 0.693(R_1 + R_2)C_1$$
$$= 0.693(1 \text{ k}\Omega + 25.6 \text{ k}\Omega)(0.1 \text{ }\mu\text{F})$$
$$= 1.84 \text{ ms}$$

The time of the waveform is equal to the sum of the high and low pulse-width values. Frequency is equal to the reciprocal of this sum.

$$T = P_{WH} + P_{WL}$$
$$= 1.84 \text{ ms} + 69.3 \text{ }\mu\text{s}$$
$$= 1.91 \text{ ms}$$

and

$$f = \frac{1}{T}$$
$$= \frac{1}{1.91 \text{ ms}}$$
$$= 524 \text{ Hz}$$

This is close to the target frequency of 523 Hz. The listed resistances are not standard values, so an equivalent resistance can be created by either combining resistors or by using rheostats. Rheostats have the advantage of allowing you to fine tune the particular note.

Designing a Tone   The first step is to decide on a frequency. In this example, we design for the low A note, which has a frequency of 220 Hz. Knowing the frequency allows us to determine the time.

$$T = \frac{1}{f}$$
$$= \frac{1}{220 \text{ Hz}}$$
$$= 4.55 \text{ ms}$$

It was mentioned earlier that the discharge time for every note in the circuit of Fig. 14-13 is the same. The discharge time is equivalent to P_{WL}, which is 69.3 μs. This means that P_{WH} must make up the difference between the time of the note and P_{WL}.

$$P_{WH} = T - P_{WL}$$
$$= 4.55 \text{ ms} - 69.3 \text{ }\mu\text{s}$$
$$= 4.48 \text{ ms}$$

At this point, we need to determine what size resistor produces a 4.48-ms high pulse width in the circuit of Fig. 14-13. Through rearranging the formula for

396

P_{WH}, we arrive at

$$R = \frac{P_{WH}}{0.693 \times C_1} - R_1$$

$$= \frac{4.48 \text{ ms}}{0.693 \times 0.1 \text{ uF}} - 1 \text{ k}\Omega$$

$$= 63.6 \text{ k}\Omega$$

Now we double check to make sure that this value of resistance gives us the desired 4.48-ms high pulse width.

$$P_{WH} = 0.693 \times (R_1 + R)C_1$$

$$= 0.693 \times (1 \text{ k}\Omega + 63.6 \text{ k}\Omega)(0.1 \text{ }\mu\text{F})$$

$$= 4.48 \text{ ms}$$

Find the overall time of the waveform:

$$T = P_{WH} + P_{WL}$$

$$= 4.48 \text{ ms} + 69.3 \text{ }\mu\text{s}$$

$$= 4.55 \text{ ms}$$

and the frequency:

$$f = \frac{1}{T}$$

$$= \frac{1}{4.55 \text{ ms}}$$

$$= 220 \text{ Hz}$$

EXAMPLE 14-5

A resistor and switch is placed in parallel to R_9 in the circuit of Fig. 14-13. It is desired that when this switch is pressed, a tone with a frequency of 550 Hz be produced. What size resistor should be used?

Solution: We start with the premise that R_1 is not changed and, therefore, P_{WL} remains at 69.3 μs, as determined before. The time of the desired frequency is

$$T = \frac{1}{550 \text{ Hz}}$$

$$= 1.82 \text{ ms}$$

P_{WH} must make up the remainder of the time of this frequency.

$$P_{WH} = 1.82 \text{ ms} - 69.3 \text{ }\mu\text{s}$$

$$= 1.75 \text{ ms}$$

so

$$R = \frac{1.75 \text{ ms}}{0.693 \times 0.1 \text{ }\mu\text{F}} - 1 \text{ k}\Omega$$

$$= 24.3 \text{ k}\Omega \qquad \blacksquare$$

Practice Problem: Determine the value of the accompanying charging resistance in the circuit of Fig. 14-13 necessary to produce a low C note, which has a frequency of 130.8 Hz. *Answer:* $R = 108$ kΩ.

REVIEW QUESTIONS

44. Differences in sound are determined by their _____ .
45. Which has a higher pitch: 500 Hz or 1 kHz?
46. The volume of a sound is determined by the voltage and not the frequency. True or false?
47. What is the P_{WH} produced by R_6 of the listing as applied to the circuit of Fig. 14-13?
48. Calculate the overall time of the middle E note from the listing.

14-7 TROUBLESHOOTING

The detection system of Fig. 14-14 is used in the discussion of troubleshooting 555 timers. It uses both the monostable and astable modes of operation. The purpose of the circuit is to produce a tone when SW1 closes.

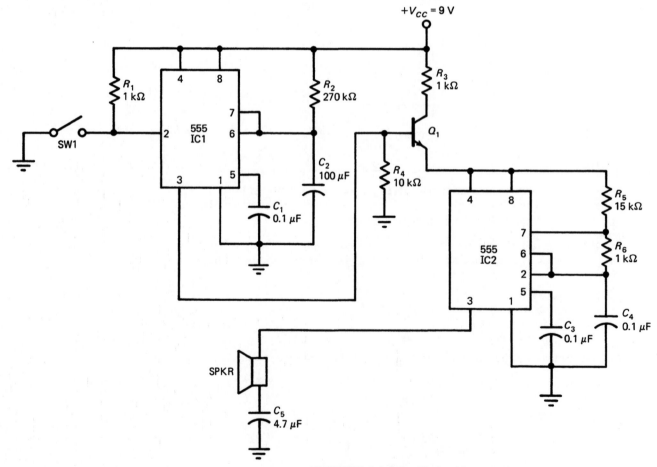

FIGURE 14-14 Detection system.

CHAP. 14 / 555 TIMER

IC1 uses the monostable mode of operation. The output of this IC (pin 3) is normally low. Once SW1 is closed, a low is applied to the trigger input, which causes the output to change to approximately 9 V. It maintains a high output for about 30 seconds.

The high at the output of IC1 is applied to the base of Q_1. This transistor is normally turned off and prevents V_{CC} from being applied to IC2. When it receives a positive voltage on its base, it turns on, thus providing a path to V_{CC} for IC2.

As power is applied to IC2, it begins to oscillate at a frequency of about 850 Hz and a tone is emitted from the speaker. So, the chain of events begins with closing SW1 and ends with a tone from the speaker, which lasts for approximately 30 seconds (regardless of opening the switch).

Two of the more obvious symptoms of this circuit are either no tone or one that is continuous.

No Tone   First, make sure that the switch is actually closed. Next, a measurement of V_{CC} should be made. If there is power, then the circuit can be divided in half by measuring the voltage at the base of Q_1. A positive reading indicates that the problem lies between the transistor and the speaker output. If no voltage is read at the base when the switch is closed, then the problem lies before the transistor.

Zero volts on the trigger of IC1 should produce a high at the output. If it does not, then IC1 should be replaced. An oscilloscope should be used to check for activity. A short pulse will not register on a DVM. If only a short pulse exists, then the problem lies with the RC time constant. Remember that the pulse should remain high for approximately 30 s.

A 30-s positive pulse at the base of Q_1 indicates IC1 is functional. The voltage at pins 4 and 8 of IC2 should be measured during this time. If they are not receiving a sufficient level of voltage from V_{CC}, then the transistor is bad.

At last we come to IC2. The output should show a pulsating waveform (850 Hz). If no waveform is visible and the passive components appear all right, then replace the timer. On the other hand, if a waveform does exist, then the speaker itself may be bad. Another possibility is an open capacitor, C_5.

Continuous Tone   A tone that does not turn off can be quite annoying. Start by testing SW1 and if it seems to be operating properly, then disconnect the speaker to silence the noise while troubleshooting. The root of the problem lies in IC2 constantly being supplied with power. This eliminates the power supply and IC2 as possible causes.

Check the base of Q_1. If it shows zero volts, then the transistor is bad. It should not be conducting under this condition. If it shows a continuous positive voltage, then the problem lies with IC1.

REVIEW QUESTIONS

Indicate whether the following conditions are true (correct) or false (faulty) for the circuit of Fig. 14-14.

49. Closing the switch causes a tone to be heard.

50. Opening the switch stops the tone immediately.

51. The transistor is turned on by zero volts on its base.

52. A stream of pulses should be measured at the output of IC1 when SW1 is closed.

53. The voltage at pin 2 of IC1 is normally a positive voltage.

54. A square wave should be measured at the output of IC2 when SW1 is closed.

55. The transistor acts as a switch that connects V_{CC} to IC2 when turned on.

SUMMARY

1. The 555 timer can be operated in the monostable mode (like the one-shot) or in the astable mode (like the oscillator).

2. The monostable mode provides a stable state (0 V) at the output until triggered, whereupon the output enters its quasi-stable state (+ V) for a period of time determined by an RC time constant.

3. The astable mode provides a periodic waveform at the output whose frequency is determined by the time constant of an RC network.

4. The trigger voltage for monostable operation is any voltage less than $\frac{1}{3} + V_{CC}$.

5. The pulse width of the output in the monostable mode is determined by

$$P_W = 1.1 \times RC$$

6. The astable mode of operation produces an output square wave that is high for a period of time (P_{WH}) and low for a period of time (P_{WL}).

$$P_{WH} = 0.693(R_1 + R_2)C_2$$
$$P_{WL} = 0.693R_2C_2$$

7. The frequency of output for the astable mode of operation is found by

$$T = P_{WH} + P_{WL}$$
$$f = \frac{1}{T}$$

8. A debounce circuit eliminates the contact bounce that occurs when closing a mechanical switch.

9. Voltage-controlled oscillators (VCOs) vary the frequency of oscillation by a change in voltage (V_{CON}) at the control voltage input (pin 5) of the timer.

10. Important formulas for the VCO are

$$P_{WH} = \ln\left(\frac{V_{CC} - 0.5\,V_{CON}}{V_{CC} - V_{CON}}\right) \times (R_1 + R_2)C_1$$
$$P_{WL} = \ln\left(\frac{V_{CON}}{0.5\,V_{CON}}\right) \times R_1C_1$$
$$T = P_{WH} + P_{WL}$$
$$f = \frac{1}{T}$$

11. The 555 organ is designed to change the RC time constant by switching in different resistors, which in turn changes the frequency at the output.

12. Designing a tone for the 555 organ requires that P_{WL} is less than the value(s) of P_{WH} that you pick. Next, choose a frequency and follow the steps for finding the value of the charge-time resistor R.

$$T = \frac{1}{f}$$
$$P_{WH} = T - P_{WL}$$
$$R = \frac{P_{WH}}{0.693 \times C_1} - R_1$$

GLOSSARY

Astable: A mode of operation having no stable states.

Duty Cycle: The period of a cycle when work is done.

Monostable: A mode of operation having one stable state.

One-Shot: Another name for a device displaying one stable state.

Quasistable: An input or output state that appears to be stable at certain times.

Reset: 0 V or logic zero, it produces the same condition as clear.

PROBLEMS

SECTION 14-1

The following problems refer to the circuit of Fig. 14-15.

14-1. Label the purpose of each component.

14-2. Calculate the value of the reference voltage found at the noninverting input of IC1.

14-3. Determine the output of IC1 based on the trigger voltage.

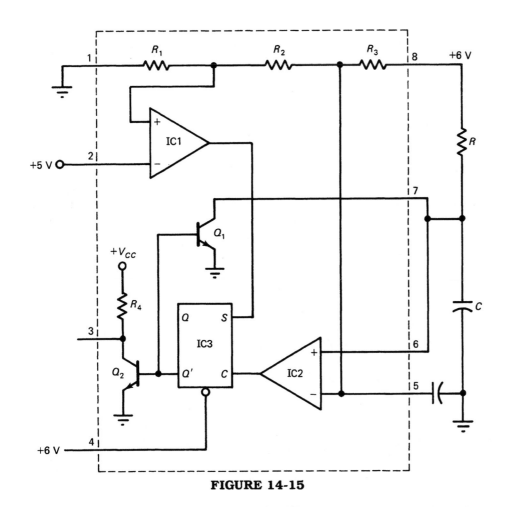

FIGURE 14-15

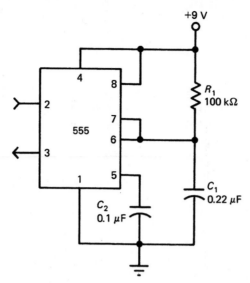

FIGURE 14-16

14-4. Calculate the value of the reference voltage found at the inverting input of IC2.

14-5. Will the flip-flop be set or cleared if Pin 6 = 5 V?

14-6. Is Q_1 in a condition to discharge the capacitor based on the information provided in Problem 5?

14-7. What is the condition of the output (pin 3)?

14-8. What would be needed to change the state of the output?

14-9. Is the reset of the flip-flop activated? Explain.

SECTION 14-2

The following problems refer to the circuit of Fig. 14-16.

14-10. What is the difference between the stable state and the quasistable state?

14-11. Label each of the pins in this circuit.

14-12. What mode of operation is used in this circuit?

14-13. Determine the value of voltage necessary to trigger the output into its quasistable state.

14-14. Calculate the pulse width of the output.

14-15. Determine the value of threshold voltage necessary to return the output to its stable state once it is triggered.

14-16. What is the purpose of the reset pin?

14-17. What is the purpose of pin 7?

14-18. Determine the voltage at pin 5.

14-19. Draw waveforms of the voltage at the trigger, C_1, and the output. Explain their relationship.

14-20. Design a circuit similar to that of Fig. 14-16, but with a pulse width of 5 s using a 100-μF capacitor in place of C_1.

SECTION 14-3

The following problems refer to the circuit of Fig. 14-17.

14-21. What type of circuit is this circuit?

14-22. Explain the oscillation action of this circuit.

402

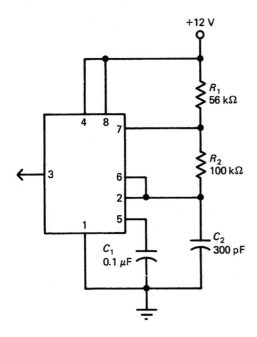

+12 V

R_1
56 kΩ

R_2
100 kΩ

C_2
300 pF

C_1
0.1 μF

FIGURE 14-17

14-23. Determine the length of time that the pulse remains high.

14-24. Determine the length of time that the pulse remains low.

14-25. Calculate the frequency of the output?

14-26. What is the duty cycle of the output?

14-27. Draw an approximate waveform of the output.

14-28. What is the purpose of pin 6 in this circuit?

14-29. Can the duty cycle be made less than 50% by substituting different values of resistance?

14-30. Design a circuit similar to that of 14-17 with an output frequency of 100 kHz.

SECTION 14-4

The following problems refer to the circuit of Fig. 14-18.

14-31. Explain the purpose of the debounce circuit.

14-32. Determine the pulse width of the output.

14-33. What is meant by a "positive-going transition."

14-34. How can the pulse width be increased?

14-35. Design a circuit with a 400-ms pulse width using a 10-μF capacitor in place of C_1.

SECTION 14-5

The following problems refer to the circuit of Fig. 14-19.

14-36. What type of circuit is this circuit?

14-37. Explain how frequency is controlled.

14-38. Determine P_{WH} if $V_{CON} = 5$ V.

14-39. Determine P_{WL} if $V_{CON} = 5$ V.

14-40. What is the frequency when $V_{CON} = 5$ V?

14-41. Calculate the frequency if V_{CON} were adjusted to 3 V.

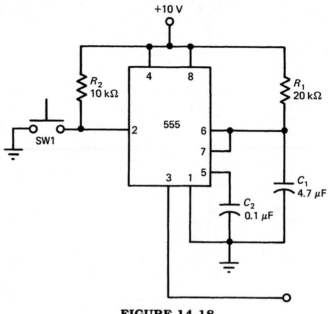

FIGURE 14-18

14-42. Calculate the frequency if V_{CON} were adjusted to 7 V.

14-43. Draw the output waveform for each of the above results. Use Fig. 14-12 as a guide.

14-44. What are the charge and discharge paths for the capacitor?

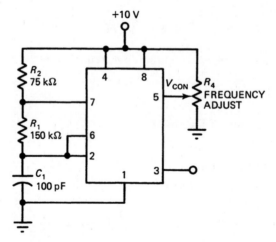

FIGURE 14-19

SECTION 14-6

The following problems refer to the circuit of Fig. 14-20.

14-45. What mode of operation is used in this circuit?

14-46. Determine the length of time of the low pulse width.

14-47. Calculate the frequency that is produced when SW1 is pressed. ($R_2 = 44.1$ kΩ.) This is a low C note.

14-48. Calculate the frequency that is produced when SW8 is pressed. ($R_9 = 19.1$ kΩ.) This is the upper C note of this scale.

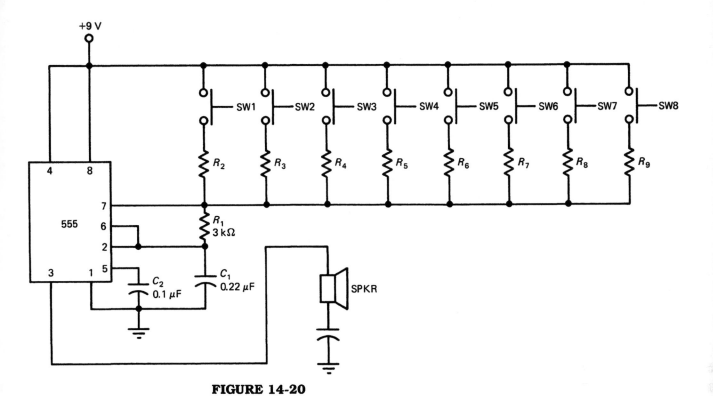

FIGURE 14-20

14-49. Find the resistances necessary to complete the notes of the scale: R_3 for 147 Hz (D), R_4 for 165 Hz (E), R_5 for 175 Hz (F), R_6 for 196 Hz (G), R_7 for 220 Hz (A), and R_8 for 247 Hz (B).

14-50. What would happen if two switches were pressed at the same time? Try to predict the results of pressing SW2 and SW3 simultaneously.

SECTION 14-7

14-51. Explain what you would expect to measure at the output of a 555 timer connected in the monostable mode. How does the trigger affect it?

14-52. Explain what you would expect to measure at the output of a 555 timer connected in the astable mode. How can a circuit operating in this mode by physically differentiated from the monostable mode circuit?

14-53. What is the purpose of the trigger input of the 555 timer and how is it operated?

14-54. What is the purpose of the threshold input of the 555 timer and how is it normally used?

14-55. What is the purpose of the discharge pin of the 555 timer?

ANSWERS TO REVIEW QUESTIONS

1.	Monostable	**2.**	Astable	**3.**	Dual in-line package
4.	True	**5.**	1.67 V	**6.**	3.3 V
7.	comparator	**8.**	0 V	**9.**	5 V
10.	0 V	**11.**	5 V	**12.**	0 V
13.	$Q' = V_{CC}$ and $Q = 0$ V	**14.**	Cleared ($Q' = V_{CC}$)	**15.**	Set ($Q' = 0$ V)

16.	0 V	**17.**	Quasistable	**18.**	4 V
19.	one-shot	**20.**	$V_{C1} > 8$ V	**21.**	To discharge C_1
22.	Reset, inactive	**23.**	8 V	**24.**	$P_W = 799$ ms
25.	182 kΩ	**26.**	12 V	**27.**	Astable
28.	554 ns	**29.**	347 ns	**30.**	1.11 MHz
31.	61.5%	**32.**	6 V	**33.**	contact bounce
34.	debounce	**35.**	Monostable	**36.**	110 ms
37.	18.2 kΩ	**38.**	By voltage	**39.**	Voltage control
40.	Astable	**41.**	0.5 V_{CON}	**42.**	V_{CON}
43.	increases	**44.**	frequency	**45.**	1 kHz
46.	True	**47.**	1.21 ms	**48.**	1.52 ms
49.	True	**50.**	False	**51.**	False
52.	False	**53.**	True	**54.**	True
55.	True				

Miscellaneous Devices

When you have completed this chapter, you should be able to:

- Investigate the principle and application of solar cells.
- Analyze the difference between electronics and photonics with reference to fiber optics.
- Investigate the operation of an LCD.
- Determine the frequency capabilities of crystals.
- Analyze the magnetic principles on which relays, motors, and speakers depend.

INTRODUCTION

This chapter includes a broad cross section of devices. They are included here because of their association with semiconductor devices. You have probably come in contact with most, if not all, of these devices in your everyday life.

The first category of devices depends on light. Solar cells convert light into electrical energy. Fiber optics uses light to transmit and receive information. The liquid-crystal display requires certain angles of light to display information.

The second category of devices involve sound. The crystal is used in oscillators because of its selective frequency response. Speakers convert the electrical signals into sound waves.

Finally, the category of movement is represented by the motor and relay. Early computers used relays to provide the switching action that is now provided by transistors. Relays are still important in control applications. Motors come in a variety of packages. Some dc motors run at a continuous rate, whereas stepper motors rotate a step at a time. Each provides its own unique type of movement.

15-1 SOLAR CELLS

Photovoltaic Process: The process of converting light to voltage.
Photon: A packet of light.

Solar cells are **photovoltaic** cells that produce electricity from sunlight. Light can be viewed as coming in packets of energy called **photons**. The photons are absorbed by the solar cell and a voltage develops. This voltage can be used to power anything from a pocket calculator to a space station.

The solar cell is actually a PN junction device. A silicon semiconductor is doped to create the P-type and N-type regions. The point at which these regions meet is called the PN junction. Ionic bonding (see Chapter 1) occurs and a situation similar to that depicted in Fig. 15-1 results.

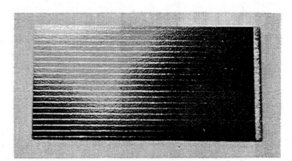

Solar cell.

Light enters through the P-type surface of the cell and is absorbed by the atoms within. The energy level of the valence electrons increases to the point where they break free of their orbits, creating electron–hole pairs. They are now available for current flow.

A difference in potential develops between the P-type and N-type regions that is proportional to the light's intensity. The amounts of voltage, current, and power are directly related to the surface area of the cell itself. Larger surface areas can absorb more light. A surface area of 2.5 cm × 5 cm (approximately 1 in. × 2 in.) can produce about 0.55 V with no load connected.

The voltage level of the cell is also determined in part by the amount of load current being drawn. Figure 15-2 shows the relationship between voltage and current. Voltage is maximum when no current is drawn and current is maximum under shorted load conditions. Unless the circuit draws more than 150 mA, we assume that the cell provides 0.5 V.

The small voltage supplied by the cell explains why large solar panels are needed to provide commercial power. Although it may seem economical to make use of sunlight, it is still rather expensive to manufacture solar cells. There are also the elements to contend with such as clouds and heat, which diminish the effectiveness of the solar cell. Storage facilities are necessary to conserve energy for the night and adverse weather conditions. The maximum efficiency of solar cells in converting sunlight to electricity is only about 15%.

Solar cells act like ordinary batteries when connected in series or parallel. Figure 15-3 shows the schematic symbol for a solar cell. Two groups of 10 series cells are connected in parallel with each other. Cells connected in series cause their voltage to add, but current remains the same. If it is assumed that each cell provides 0.5 V at 150 mA, then the overall value of one solar bank is 5 V at 150 mA. Since the two bank of cells are connected in parallel, the currents add, but the voltages remain the same. Therefore, the total solar power source provides 5 V at 300 mA.

The solar power source provides the necessary current and voltage to the load resistance and also charges the nickel–cadmium (Ni–Cd) battery. The Ni–Cd battery provides power in the absence of light. A blocking diode is used to isolate the solar cells from the Ni–Cd battery. It is biased so that current flows from the cells to the battery and load when there is light. Yet, in the absence of light, the diode prevents the Ni–Cd battery from discharging through the cells and causing possible damage.

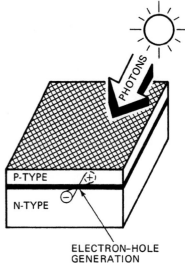

FIGURE 15-1 Solar cell turning light into electricity.

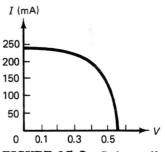

FIGURE 15-2 Solar cell *I–V* curve.

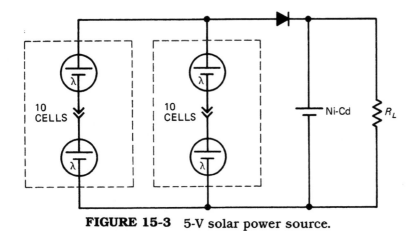

FIGURE 15-3 5-V solar power source.

A load resistance of 1 kΩ draws only 4.3 mA of current. The diode drops 0.7 V of the solar bank's voltage.

$$I_{R_L} = \frac{V_{in}}{R_L}$$

$$= \frac{4.3 \text{ V}}{1 \text{ k}\Omega}$$

$$= 4.3 \text{ mA}$$

This is much less than the 150 mA that the solar cell is able to provide. The solar cell of this discussion can drive loads larger than 28 Ω.

EXAMPLE 15-1 _____

What is the overall voltage and current when four solar cells that provide 0.5 V at 100 mA each are connected in parallel.

Solution: When solar cells are connected in parallel, voltage remains the same and current adds. The voltage in this example is 0.5 V and the current is 400 mA. ∎

Practice Problem: Determine the overall voltage and current if three solar cells that provide 0.4 V at 200 mA each were connected in series aiding.   *Answer:* 1.2 V at 200 mA.

REVIEW QUESTIONS _____

1. Another name for the solar cell is the _____ cell.
2. Solar cells convert _____ into electrical energy.
3. What is a photon?
4. The voltage capability of a solar cell increases as the surface area increases. True or false?
5. Will the voltage of the solar cell increase or decrease with an increase in load current? (See Fig. 15-2.)
6. The voltages of solar cells _____ when connected in series aiding.
7. What value of voltage and current will two 0.5-V solar cells at 200 mA provide when connected in parallel?
8. What type of battery provides rechargeable storage?

15-2 FIBER OPTICS _____

Telephone companies are increasing their use of fiber-optic lines. The benefits of using fiber optics are great. Fiber optics allows for very high transmission speeds, uses less space, and is less affected by noise. The high speed is a result of using light as the medium for transmission. Light is not affected by the interference of fluorescent lights, fans, static electricity, etc.; therefore, communication is cleaner.

The optical fiber itself is simply a strand of glass or plastic surrounded by cladding that aids in the reflection of the light beam along the cable. A protective coating constitutes the outermost layer of the cable. The ends of the cable are cut squarely to reduce signal loss. Typical signal power losses for optical fiber is about 1 or 2 dB/km (1 km = 0.62137 mi).

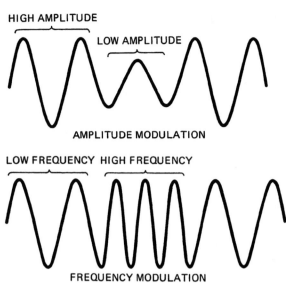

FIGURE 15-4 Amplitude modulation and frequency modulation.

Information can be transmitted on a carrier frequency of between 10^{14} and 10^{15} Hz. This range of frequencies spans from the infrared to ultraviolet light, with visible light right in between. High frequencies result in a larger bandwidth and a larger bandwidth increases the information-carrying capability.

Information is transmitted over the optical fiber by modulating the light. Modulation consists of changing the operation of a medium (such as light) in order to convey information. Two of the more popular forms of modulation are **amplitude modulation** (AM) and **frequency modulation** (FM). Figure 15-4 illustrates each of these.

Amplitude modulation involves changing the voltage level to represent different information. It is used in low- and medium-frequency radio transmission. Frequency modulation changes the frequency to represent different information. It is used in very high-frequency (VHF) radio and television transmissions. Another type of modulation that is popular in communications networks is phase modulation. It is similar to frequency modulation except that the phase angle of the signal is adjusted to represent different information. In both frequency and phase modulation, the amplitude of the signal remains unchanged.

The circuit of Fig. 15-5 performs a form of amplitude modulation. A switch is used to turn the light-emitting diode (D_1) on or off. An infrared-emitting diode is used here. It is normally packaged within a connector that is ready to use with fiber optics. The optical fiber is simply plugged into the connector.

Each time SW1 is pressed, the transistor conducts and the LED emits a pulse of light into fiber-optic cable. When SW1 is released, the pulldown resistor (R_1) holds the base at ground potential and the transistor turns off. No light is emitted when SW1 is released.

On the other end of the cable is a **phototransistor**. It, too, is seated within a connector to which the fiber-optic cable is attached. The base lead of the transistor is replaced by a lens. As light strikes the lens, the phototransistor conducts and the ordinary LED (D_2) lights. The transmitter sends pulses of light when SW1 is pressed and the LED in the receiver lights on receiving each pulse of data.

Other transmitting and receiving devices include the injection laser diode (ILD) and PIN diode. The ILD uses laser light and is, therefore, considerably

Amplitude Modulation (AM): Modulation that varies voltage levels to convey information.
Frequency Modulation (FM): Modulation that varies frequencies to convey information.

Phototransistor: A transistor that conducts when exposed to light.

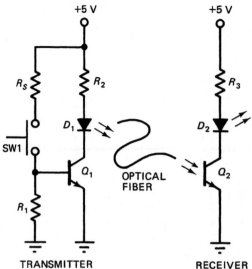

FIGURE 15-5 Simple fiber-optic communications link.

TRANSMITTER   RECEIVER

more powerful. But it is also more expensive. It emits a laser beam when furnished with the proper level of current. The PIN (positive-intrinsic-negative) diode is used in communications and works on the same principle as the phototransistor. It conducts when it receives light.

The optical fiber is flexible and can be bent without affecting the transmission of light within. This feature makes it convenient to install for communications networks. Fiber optics is still rather expensive, but its conservation of space and light weight makes it very attractive.

REVIEW QUESTIONS

9. What is the medium of transmission in fiber optics?
10. What are optical fibers composed of?
11. Changing the operation of a medium in order to convey information is called _____.
12. What does AM represent?
13. Voltage levels are changed to convey different information in FM. True or false?
14. What type of diode is used for D_1 in the circuit of Fig. 15-5?
15. Pressing SW1 closed in the circuit of Fig. 15-5 applies approximately (0 V, 5 V) to the base of Q_1 and D_1 (lights or remains unlit).
16. What is Q_2 called?
17. Is a PIN diode used for transmitting or receiving?
18. Does D_2 in the receiver of Fig. 15-5 light when SW1 at the transmitter is pressed or released?

15-3 LIQUID-CRYSTAL DISPLAY

The LCD is an alphanumeric digital display. It can be found anywhere from wristwatches to laptop computers. The letters and numbers are formed by black bars or dots on a light gray background. The formation of the characters is similar

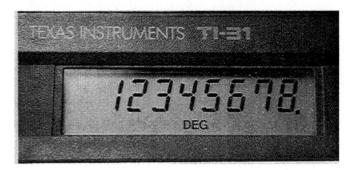

LCD as found on calculators.

to the seven-segment LED display discussed earlier in the book except that no light is emitted.

The display is made from nematic crystal or nematic fluid that is sealed between two plates of glass. When no electric field is applied, the crystals remain transparent. Upon applying the field, the crystals align themselves perpendicular to the front viewing glass and become opaque.

A seven-segment LCD is shown in Fig. 15-6. A lead is supplied for each of the segments. There is also a lead labeled *BP*, which is the backplane common to all the segments. The electric field must be applied between the backplane and the segment that is to be activated. The segment and backplane can be likened to a capacitor. Very little current is drawn when low frequencies are applied. However, if the frequency becomes too low (below 30 Hz), the flicker in the display becomes noticeable.

If an ac signal is needed to enable the LCD to work, then how is a dc battery-powered watch able to use the display? The answer is that an oscillator is used to create the appearance of ac. There are special digital chips that are designed to work with the displays. Normally, CMOS (complementary metal-oxide semiconductor) chips are employed. They use very little power and virtually eliminate the small dc voltages that occur with TTL logic chips. There are also chips with oscillators built into them.

The circuit of Fig. 15-7 demonstrates the use of an oscillator to enable the display. Obviously, there must also be circuitry to select various combinations of segments. This is normally handled by a single digital decoder IC.

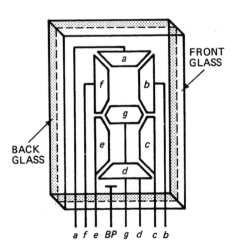

FIGURE 15-6 A seven-segment liquid-crystal display.

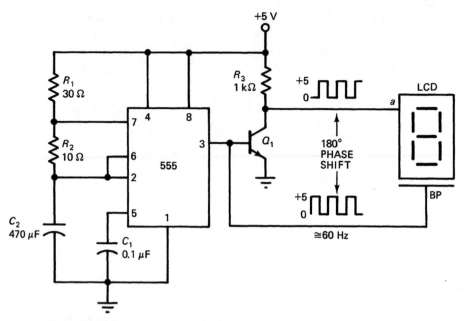

FIGURE 15-7 An oscillator and inverter enable segments to form.

An oscillator is formed by using the astable mode operation of the 555 timer. The frequency of the oscillator is roughly 60 Hz in this example. The output of the oscillator (pin 3) is tied to the base of Q_1 (which acts as an inverter) and the backplane of the LCD. The inverter (Q_1) provides a 180° phase shift from the base to the collector and this signal is applied to the "a" segment of the display.

By observing the situation from the viewpoint of the backplane, it is found that when BP is +5 V, the segment voltage (0 V) seems more negative. When BP is 0 V, the segment voltage (+5 V) now seems more positive. So from the backplane's perspective, the applied voltage seems to fluctuate between +5 V and −5 V. The manufactured ac signal provides the electric field necessary to cause the "a" segment to become opaque.

EXAMPLE 15-2 _____

Which segments must receive an ac signal in order to form the number 5 in Figs. 15.6 and 15.7?

Solution: The number 5 is formed by providing the ac signal to the a, c, d, g, and f segments simultaneously. ∎

REVIEW QUESTIONS _____

19. What does LCD represent?
20. LCDs emit black light when a dc voltage is applied. True or false?
21. What type of material is used to form the actual characters?
22. Which lead is common to all segments of a LCD?
23. Is ac or dc used to activate the crystal?
24. What is the voltage at the collector of Q_1 in the circuit of Fig. 15-7 when the base receives a low pulse (0 V)?
25. CMOS ICs use less power than their TTL equivalents. True or false?

CHAP. 15 / MISCELLANEOUS DEVICES

Crystals are literally minerals that are mined from the earth. They include tourmaline, Rochelle salt, and quartz, with quartz being the most economical. Crystals in their raw state are hexagonal in shape, as shown in Fig. 15-8. The crystals are then cut into wafers. The thickness and angle of cut are important in determining the frequency of the crystal. The thinner the cut, the higher the frequency.

The ability of a crystal to vibrate when pressure is applied is called the piezoelectric effect. A crystal also vibrates at the frequency of an applied signal, but vibrates most efficiently at its own resonant frequency. The resonant characteristic of the crystal is represented by the *RLC* circuits of Fig. 15-9.

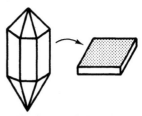

FIGURE 15-8 Wafer cut from crystal.

Resonant Frequency   When a crystal is used in an application such as an oscillator and is connected in series with the feedback signal, it resonates at its series resonant frequency. This frequency can be determined if the internal resistance, inductance, and capacitance of the crystal are known. If the internal values of the crystal are $R = 200 \ \Omega$, $L = 1$ H, $C_N = 0.0063$ pF, and $C_M = 10$ pF, then the series resonant frequency is

$$f_S = \frac{1}{2\pi\sqrt{LC_N}}$$

$$= \frac{1}{2\pi\sqrt{(1 \text{ H})(0.0063 \text{ pF})}}$$

$$= 2.0051 \text{ MHz}$$

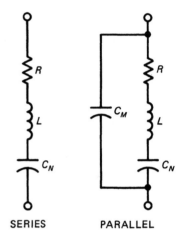

SERIES   PARALLEL

FIGURE 15-9 Equivalent circuits of a crystal.

The mounting capacitance (C_M) that is caused by the attached leads is not considered in the series resonant formula. This is because the series resonant resistance of the crystal is very low (just about equal to R) and C_M is shunted out of the picture.

The parallel resonant frequency is usually very close in value to the series resonant frequency. It is found by

$$f_P = \frac{1}{2\pi\sqrt{LC_{N+M}}}$$

and

$$C_{N+M} = \frac{C_N \times C_M}{C_N + C_M}$$

$$= \frac{0.0063 \text{ pF} \times 10 \text{ pF}}{0.0063 \text{ pF} + 10 \text{ pF}}$$

$$= 0.006296 \text{ pF}$$

Therefore,

$$f_P = \frac{1}{2\pi\sqrt{(1 \text{ H})(0.006296 \text{ pF})}}$$

$$= 2.0058 \text{ MHz}$$

The series and parallel resonant frequencies can both be rounded off to 2.01 MHz.

Practice Problem: Determine the series and parallel resonant frequencies for a crystal whose internal characteristics are $R = 300\ \Omega$, $L = 3$ H, $C_N = 0.0084$ pF, and $C_M = 5$ pF. *Answer:* $f_S = 1.00258$ MHz and $f_P = 1.00342$ MHz.

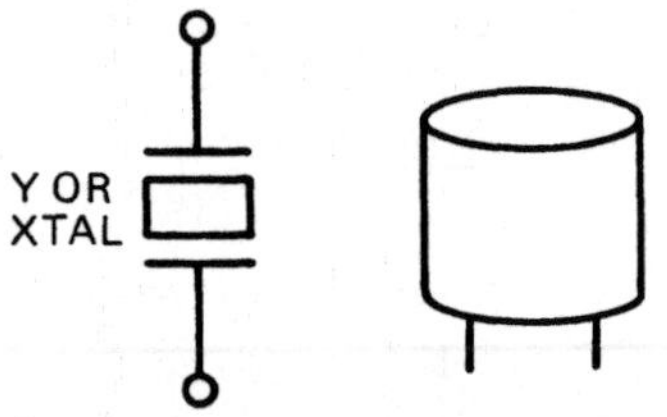

FIGURE 15-10 Schematic symbol and typical package for a crystal.

The schematic symbol for a crystal is depicted in Fig. 15-10. XTAL or the letter Y is often used to denote the crystal on schematic diagrams. One type of packaging is shown next to the crystal. The crystal in this case is sealed inside the metal canister.

Q Factor The Q factor, or figure of merit, is very high for crystals. This means that they waste very little power. The Q of a typical LC oscillator at its resonant frequency may have a value of a few hundred, whereas the Q of a crystal is usually upwards of ten thousand. Q for the crystal previously discussed can be approximated as

$$Q = \frac{X_L}{R}$$

where

$$\begin{aligned} X_L &= 2\pi f L \\ &= 2\pi(2.01\ \text{MHz})(1\ \text{H}) \\ &= 12.6\ \text{M}\Omega \end{aligned}$$

and

$$\begin{aligned} Q &= \frac{12.6\ \text{M}\Omega}{200\Omega} \\ &= 63{,}000 \end{aligned}$$

The higher the value of Q, the better. A crystal provides a very efficient and stable means of supplying accurate frequencies. Frequency drift for a crystal is typically 1/10,000 of 1%. It is not greatly affected by environmental conditions.

Overtones The crystal used in the circuit of Fig. 15-11 has a fundamental frequency of 2 MHz. The circuit itself is designed to take advantage of the overtones of the crystal. Overtones are multiples of the resonant frequency. The third overtone of this crystal is 6 MHz and the fourth overtone is 8 MHz. Using overtones allows crystals to be used in high-frequency applications.

The frequency of oscillation in the circuit of Fig. 15-11 is determined by the tank circuit.

$$\begin{aligned} f_O &= \frac{1}{2\pi\sqrt{LC}} \\ &= \frac{1}{2\pi\sqrt{(4.7\ \mu\text{H})(150\ \text{pF})}} \\ &= 5.99\ \text{MHz} \end{aligned}$$

This is very close to the second overtone of the crystal. The circuit oscillates at approximately 6 MHz. The crystal is connected in series with the positive-feedback path, so a multiple of its fundamental series frequency is maintained.

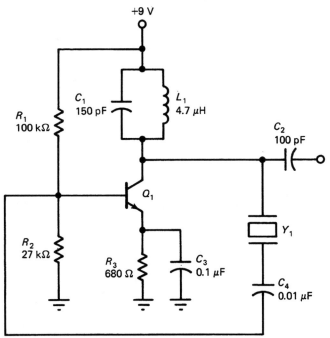

FIGURE 15-11 Oscillator operating at an overtone of the crystal.

EXAMPLE 15-3

A crystal has the following internal characteristics: $R = 500\ \Omega$, $L = 2$ H, $C_N = 0.01$ pF, and $C_M = 10$ pF. Determine (a) the series resonant frequency and (b) the Q factor.

Solution:

(a) The series resonant frequency can be found by

$$f_S = \frac{1}{2\pi\sqrt{(2\ \text{H})(0.01\ \text{pF})}}$$

$$= 1.1254\ \text{MHz}$$

(b) The Q factor of the crystal at this frequency is found by first determining the inductive reactance of the crystal.

$$X_L = 2\pi(1.1254\ \text{MHz})(2\ \text{H})$$

$$= 14.14\ \text{M}\Omega$$

The Q factor is the ratio of resistance to reactance:

$$Q = \frac{14.14\ \text{M}\Omega}{500\Omega}$$

$$= 28{,}280 \qquad \blacksquare$$

REVIEW QUESTIONS

26. What is the most common form of crystal used in electronics?
27. The thicker the wafer of crystal, the (higher or lower) its fundamental frequency.

28. The ability of a crystal to vibrate when mechanical pressure is applied to it is called the _____ effect.

29. The (series or parallel) resonant frequency of the crystal tends to be lower.

30. A high Q factor means that very little power is wasted. True or false?

31. When a crystal vibrates at a multiple of its fundamental frequency, it is vibrating at an _____.

15-5 SPEAKERS

Transducer: A device that converts one form of energy into another, for example, electricity into sound.
Acoustic: Able to produce or receive sound waves.

A speaker is an electroacoustic **transducer**. Transducers change, or convert, energy from one form to another. The speaker is electroacoustic in the sense that it changes electrical energy into sound or **acoustical** energy. Humans can hear sounds within a range of approximately 16 to 20 kHz.

A typical 3.5-in. speaker is shown in Fig. 15-12. The front outer surface is composed of a paper diaphragm or cone. A flexible material must be used since this surface vibrates to produce the sound. At the rear of the speaker is an encased voice coil that is surrounded by a permanent magnet. A coil develops a magnetic field when an ac signal is applied to it. The magnetic field varies with the frequency. The job of the permanent magnet is to interact with the magnetic field of the coil. A result of the interaction is that the voice coil (which is attached to the diaphragm) moves back and forth. The frequency at which the coil moves back and forth determines the frequency of vibration for the diaphragm. As the diaphragm vibrates, it produces acoustical waves in the air. These acoustical air waves vibrate our ear drum to produce the sound that we hear.

Speakers come in a variety of sizes and the size most often determines the frequency capabilities of the speaker. The woofer speaker is used in sound systems to reproduce the low audio frequencies (up to 300 Hz). They are the largest because low frequencies have a longer wavelength. This means that the diaphragm must vibrate a greater distance. Midrange speakers fill the middle of the frequency range (300 Hz to 3 kHz). At the upper end of the range is the tweeter. It is the smallest because of the shorter wavelength of higher frequencies (3 to 20 kHz).

Impedance Matching   The typical 3.5-in. speaker has an impedance of 8 Ω and can handle 0.25 W of power. In order to obtain the most power from a sound system, the impedances within the system should be matched. An output

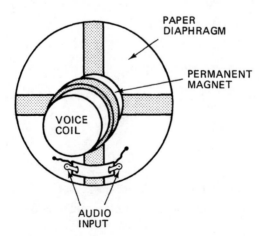

FIGURE 15-12 Speaker (rear view).

impedance of 8 Ω should be matched to an 8-Ω load. When impedances are matched, maximum transfer of power can occur.

Remember that impedance is the ac opposition to current flow. Therefore, if a speaker is rated at 8 Ω, it will offer 8 Ω of resistance in an active ac circuit. Don't expect to measure the speaker with an ohmmeter and read 8 Ω. The ohmmeter makes dc measurements and the speaker is essentially a coil, so the reading will be near zero.

A simple example demonstrates the logic of impedance matching and maximum power transfer. Suppose that the output impedance of an amplifier is 8 Ω and that it provides a 1-V ac signal. Any load connected to the output of the amplifier views it as a 1-V source with 8 Ω of internal impedance. Only a load of 8 Ω receives the maximum amount of power. Figure 15-13 shows a simplified drawing of the situation. Begin by assuming that R_L is 2 Ω. The circuit is treated as a simple series circuit in solving for the power of the load.

$$V_{R_L} = \frac{R_L}{R_S + R_L} V_S$$

$$= \frac{2\ \Omega}{8\ \Omega + 2\ \Omega} (1\ V)$$

$$= 0.2\ V$$

and

$$P_{R_L} = \frac{V_{R_L}^2}{R_L}$$

$$= \frac{(0.2\ V)^2}{2\ \Omega}$$

$$= 20\ mW$$

20 mW of power is transferred from the source to a 2-Ω load. Now let us match the impedances by replacing the load with an impedance of 8 Ω.

$$V_{R_L} = \frac{8\ \Omega}{8\ \Omega + 8\ \Omega} (1\ V)$$

$$= 0.5\ V$$

$$P_{R_L} = \frac{(0.5\ V)^2}{8\ \Omega}$$

$$= 31.25\ mW$$

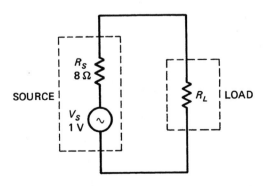

FIGURE 15-13 Maximum transfer of power occurs when R_S and R_L are matched.

An increase in the power of the load has occurred. It may seem that power increases with load resistance and that this is not the maximum amount of transferrable power. The contrary can be proven by increasing the load resistance above 8 Ω to a value of 25 Ω.

$$V_{R_L} = \frac{25\ \Omega}{8\ \Omega + 25\ \Omega}\ (1\ V)$$

$$= 0.758\ V$$

$$P_{R_L} = \frac{(0.758\ V)^2}{25\ \Omega}$$

$$= 23\ mW$$

The power transferred to the load decreases when the impedance of the load is made greater than the source impedance. Comparing all three proves that maximum power transfer occurs when impedances are matched, as shown in this table.

Source Impedance (Ω)	Load Impedance (Ω)	Load Power (mW)
8	2	20
8	8	31.25
8	25	23

REVIEW QUESTIONS

32. _____ convert energy from one form to another.
33. Speakers convert ac signals into _____ energy.
34. What part of the speaker actually vibrates and produces the sound waves?
35. Which speaker (woofer, midrange, or tweeter) is physically the largest?
36. The upper end of the audio spectrum is converted into sound by the (woofer, midrange, or tweeter).
37. (Higher or lower) frequencies have longer wavelengths.
38. Maximum power transfer occurs when _____ are matched.
39. An 8-Ω speaker measures 8 Ω on an ohmmeter. True or false?
40. What size load receives the maximum transfer of power from a source that provides a 200-mV signal and has 150 Ω of internal impedance?

15-6 RELAYS

A relay is basically a switch that is activated electrically rather than manually. The switching action is based on the magnetic principle of the solenoid. A solenoid is a coil that is used as a magnet. Figure 15-14 shows a block diagram of a relay. The solenoid is located between the two contact posts.

DC voltage is applied to the coil and causes a magnetic field to build. The magnetic field is strong enough to attract the metal armature downward toward the top of the coil. Once the armature makes contact at the left contact post, it creates a conductive path between the A and B terminals. These terminals are connected to the circuit being controlled.

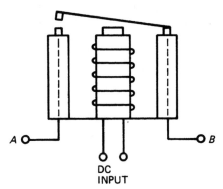

FIGURE 15-14 Block diagram of a relay.

Types of Relays   The relay of Fig. 15-14 is considered a single-pole single-throw (SPST) relay. Relays come in configurations similar to ordinary switches. The schematic diagram for the SPST relay is shown in Fig. 15-15. Other popular configurations include the single-pole double-throw (SPDT) and the double-pole double-throw (DPDT).

The SPDT shows that terminals A and B are connected when the solenoid is not magnetized. Upon receiving the proper voltage at the dc input, a conductive path between terminals B and C is created. Terminals A and B open at this time. The DPDT is similar to the SPDT except that two switches are controlled by the one solenoid.

The voltage necessary to close a relay depends on the type of relay. Here is a sample listing:

Operating Voltage (VDC)	Internal Resistance (Ω)	Current (A)
5	120	1
6	32	10
12	185	5
24	280	25
48	186	30

The operating voltage must be applied to the dc input of the relay to close it. The necessary current can be calculated by dividing the operating voltage by the internal resistance of the solenoid. The current listing refers to the amount of current that the terminals of the relay can handle once they are closed. As we can see, the current-handling capabilities of the relay range from 1 to 30 A. Industrial relays can control hundreds of amps.

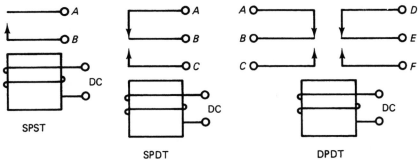

FIGURE 15-15   Schematic symbols for various relays.

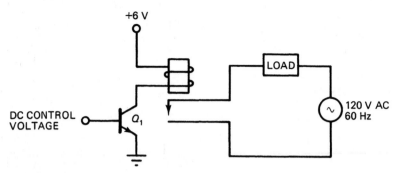

+6 V

DC CONTROL VOLTAGE

Q_1

LOAD

120 V AC
60 Hz

FIGURE 15-16 **A small dc voltage and current are used to control a large ac voltage and current via a relay.**

Uses of Relays Anywhere an electrically controlled switch is needed, a relay can be used. Relays have been replaced by thyristors in many circuits. The relay is limited with regard to its switching speed. Reed relays are able to switch at higher speeds (about 1 kHz). Solid-state devices replace relays in high-speed switching applications since there are no moving parts.

The most practical use of a relay is in situations where a dc circuit must be **interfaced** with an ac circuit. Relays allow a relatively small dc voltage and current to control a large ac voltage and current. Figure 15-16 is an example of how a relay might be used to accomplish this.

A dc control voltage at the base of the transistor is used to turn Q_1 on and off. A positive voltage on the base turns on Q_1 causing it to conduct, thus building the magnetic field of the solenoid. The terminals of the relay are connected to an ac circuit that provides 120 VAC at 60 Hz to a load. When the relay is closed, the load receives the necessary ac voltage. The ac and dc circuits are isolated from one another, with the relay acting as an interface between the two circuits.

Interface: To connect with; to provide a means for two or more circuits to interact.

EXAMPLE 15-4 _____

Explain the operation of the DPDT relay illustrated in Fig. 15-15.

Solution: When a dc voltage is applied to the dc input of the solenoid, it causes a magnetic field to develop. The magnetic field attracts the B and E armatures, connecting them to C and F, respectively. ■

REVIEW QUESTIONS _____

41. A relay is activated externally by a _____ .

42. A coil that is used as a magnet is called a _____ .

43. What does SPDT represent?

44. Which terminals of the DPDT relay in Fig. 15-15 are connected as a result of energizing the solenoid?

45. How much current does a 5-V relay with 120 Ω of internal resistance require to operate?

46. Relays are limited in their switching speed. True or false?

47. The relays discussed in this section are activated by (dc or ac) voltage.

48. The closed terminals of the relay can provide a complete path for (ac, dc, or both ac and dc).

Motors can be found anywhere, from consumer electronics to the disk drive of a computer. A motor is designed to convert electrical energy into mechanical energy. DC motors occupy our attention in this section. We begin with the inexpensive dc motors that can be purchased in most electronics or hobby shops and then move on to the more expensive stepper motor.

DC Motor   The dc motor relies on magnetic fields for its operation. Basically, a voltage is applied to the leads of the motor that causes it to spin. The action inside the motor can be explained through the block diagram of Fig. 15-17.

The external dc voltage is applied to the external leads of the motor. These leads connect to a pair of brushes. The brushes provide an electrical connection with the split-ring commutator.

A negative polarity is applied to one-half of the ring and a positive to the other half by the dc applied voltage. These polarities are passed on to the armature coil located within the magnet. The resulting current causes a magnetic field to develop within the armature coil and a corresponding set of north and south polarities result, as indicated on the armature coil. The arrows on the armature coil indicate the direction of spin.

Unlike polarities attract and like polarities repel. Therefore, the armature coil continues to turn in an attempt to align its north to the south of the magnet and its south to the north of the magnet. But remember that the commutator is split, and as soon as the opposite magnetic fields attempt to line up with each other, the polarities from the applied voltage are applied to the opposite halves of the split ring. The commutator provides a reversal of polarity to the armature coil. A change in polarities to the coil causes a change in its magnetic field. The north and south of the armature coil switch to the opposite polarities of those shown and the spinning continues due to the attraction and repulsion of the magnetic fields.

Each time the polarities of the armature and magnet seem to reach their final destination, the commutator provides a switch of polarities. This keeps the motor

DC motor.

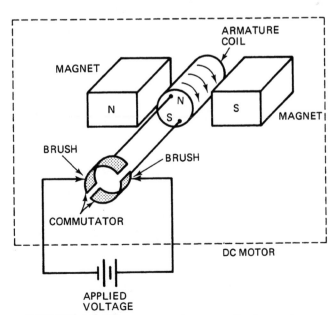

FIGURE 15-17   Block diagram of a dc motor.

spinning. The greater the voltage that is applied to the motor, the greater the current and magnetic field, and, therefore, the faster the motor spins. The speed of the motor is controlled by the applied voltage.

DC Motor Control   There are many methods of controlling the speed of a dc motor. Figure 15-18 illustrates one that uses the 555 timer to control the duty cycle of the output, which, in turn, controls the speed of the motor. The importance of the duty cycle is brought out in this circuit. It was mentioned earlier in the book that the duty cycle is the time during a cycle in which work is done. In this circuit, the work is performed when current flows in the dc motor and this occurs when the output of the timer is high.

The 555 timer is connected to operate as an oscillator in the astable mode. Frequency, however, is not the focus of this circuit. The key aspect of this circuit is the percentage of time that the output is pulsed high (P_{WH}), which is called the duty cycle. It is directly controlled by potentiometer R_2, which is connected as a rheostat. Resistor R_1 is used to prevent overloading the timer should the rheostat be adjusted to zero. If R_2 is adjusted to 5 kΩ, then the time that the output pulses high is determined by the charge path of the capacitor, which includes R_1, R_2, and R_3.

$$P_{WH} = 0.693 \times (R_1 + R_2 + R_3) \times C_2$$
$$= 0.693(1 \text{ k}\Omega + 5\text{k}\Omega + 1 \text{ k}\Omega) \times 0.1 \text{ } \mu\text{F}$$
$$= 485 \text{ } \mu\text{s}$$

The output pulses low for the remainder of the cycle. The low pulse width (P_{WL}) is determined by the RC time constant of the discharge path and includes only R_3 and C_2.

$$P_{WL} = 0.693 \times R_3 \times C_2$$
$$= 0.693 \times 1 \text{ k}\Omega \times 0.1 \text{ } \mu\text{F}$$
$$= 69.3 \mu\text{s}$$

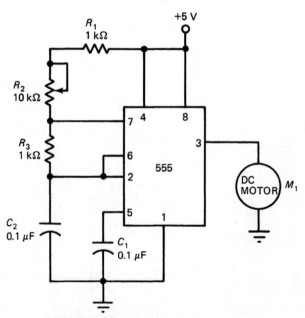

FIGURE 15-18   DC motor control.

The time of the cycle is equal to the sum of these pulse widths.

$$T = P_{WH} + P_{WL}$$
$$= 485 \ \mu s + 69.3 \ \mu s$$
$$= 554 \ \mu s$$

Finally, the duty cycle is the percentage of time that the output pulses high.

$$D = \frac{P_{WH}}{T} \times 100$$
$$= \left(\frac{485 \ \mu s}{554 \ \mu s}\right) \times 100$$
$$= 87.5\%$$

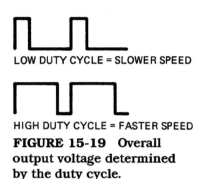

LOW DUTY CYCLE = SLOWER SPEED

HIGH DUTY CYCLE = FASTER SPEED

FIGURE 15-19 Overall output voltage determined by the duty cycle.

The duty cycle increases as the resistance is increased. An increased duty cycle means that more voltage and current are available to the motor and, therefore, its speed increases. Figure 15-19 compares two duty cycles. We can plainly see how more voltage and current are available within a given period of time when the duty cycle is longer.

EXAMPLE 15-5 _____

What is the duty cycle in the circuit of Fig. 15-18 if R_2 were adjusted to 2 kΩ?

Solution: The duty cycle is determined by the following steps.

$$P_{WH} = 0.693 \times (1 \ k\Omega + 2 \ k\Omega \ 1 \ k\Omega) \times 0.1 \ \mu F$$
$$= 277 \ \mu s$$

$$P_{WL} = 0.693 \times 1 \ k\Omega \times 0.1 \ \mu F$$
$$= 69.3 \ \mu s$$

$$T = 277 \ \mu s + 69.3 \ \mu s$$
$$= 346 \ \mu s$$

$$D = \frac{277 \ \mu s}{346 \ \mu s} \times 100$$
$$= 80\%$$ ■

Practice Problem: What is the value of the duty cycle when R_2 is adjusted to 7 kΩ in the circuit of Fig. 15-18?   *Answer:* $D = 90\%$.

Stepper Motor   Robotics and disk drives are two areas where you are likely to find stepper motors. The stepper motor provides a controlled rotation in exact increments or decrements. Precision has its price, so stepper motors are much more expensive than an ordinary free-spinning dc motor.

A block diagram of a four-phase stepper motor is shown in Fig. 15-20. Each of the blocks labeled $\phi 1$ through $\phi 4$ (phase 1 through phase 4) contains a coil. When the coil is activated by a positive voltage at the input, current develops a magnetic field that attracts a cog toward itself. The $\phi 1$ input is now active and so the cog aligns itself with the $\phi 1$ coil.

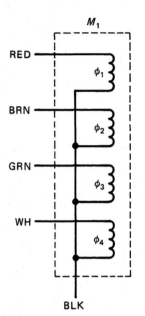

FIGURE 15-21 Schematic representation of a stepper motor.

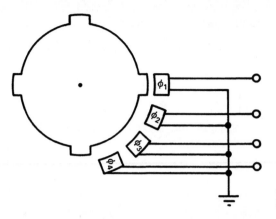

FIGURE 15-20 Block diagram of a four-phase stepper motor.

The cog can be caused to rotate clockwise one step by energizing the $\phi 2$ coil and deenergizing the $\phi 1$ coil. The clockwise action is continued by energizing the $\phi 3$ coil and deenergizing the $\phi 2$ coil. Eventually, the cog reaches the $\phi 4$ coil and a further step brings the next cog into alignment with $\phi 1$. A counterclockwise stepping action can be caused by energizing and deenergizing the coils in reverse order.

The stepper motor requires a separate line for each phase input and a line for ground. Figure 15-21 shows a typical schematic of a stepper motor. The manufacturer would indicate which color belongs to which phase. All stepper motors do not necessarily use this color-coding scheme.

REVIEW QUESTIONS

49. A motor converts electrical energy into _____ energy.
50. The _____ provide an electrical connection with the commutator.
51. The _____ coil develops magnetic polarities that cause it to spin.
52. (Like or unlike) polarities repel and (like or unlike) polarities attract.
53. The _____ provides a reversal of polarity to the armature coil.
54. The time during a cycle in which work is done is called the _____ cycle.
55. What would be the duty cycle in the circuit of Fig. 15-18 if R_2 were adjusted to its maximum of 10 kΩ?
56. The speed of the motor in the circuit of Fig. 15-18 (increases or decreases) as the duty cycle increases?
57. A _____ motor can be controlled in exact increments or decrements.

SUMMARY

1. Solar cells are photovoltaic cells that produce electricity from light.
2. Light is transmitted in packets of energy called photons.
3. The solar cell is a PN junction device whose output voltage, current, and power are directly related to the surface area of the cell itself.
4. The optical fiber is a strand of glass or plastic surrounded by cladding that aids in the reflection of the light beam along the cable.
5. Information is transmitted over the optical fiber by modulating light.

6. Amplitude modulation (AM) involves changing the voltage level to represent different information.

7. Frequency modulation (FM) changes the frequency to represent different information.

8. The LCD (liquid-crystal display) is made from transparent nematic crystal fluid that becomes opaque when an electric field is applied.

9. An electric field is applied between the segment and backplane to activate that particular segment.

10. Crystals are minerals such as tourmaline, Rochelle salt, and quartz that are used to provide accurate frequencies.

11. Thickness and angle of cut are important in determining the frequency of a crystal.

12. The Q factor for a crystal represents the ratio of reactance to resistance wherein the lower the resistance, the less power wasted.

13. Overtones are multiples of the resonant frequency.

14. A speaker is an electroacoustic transducer, which means it changes electrical energy into sound (acoustical energy).

15. Magnetic fields are used to vibrate the voice coil, which is connected to the diaphragm, and the diaphragm produces acoustical waves in the air.

16. Matching the impedance of a speaker to that of the amplifier provides for a maximum transfer of power.

17. A relay is a magnetic switch that is activated by an applied dc voltage.

18. Motors convert electrical energy into mechanical energy.

19. DC motors are activated by an applied dc voltage.

20. Stepper motors provide a controlled rotation in exact increments and decrements by applying a series of dc pulses to the phase 1 through phase 4 inputs.

GLOSSARY

Acoustic: Able to produce or receive sound waves.

Amplitude Modulation (AM): Modulation that varies voltage levels to convey information.

Frequency Modulation (FM): Modulation that varies frequencies to convey information.

Interface: To connect with; to provide a means for two or more circuits to interact.

Photon: A packet of light.

Phototransistor: A transistor that conducts when exposed to light.

Photovoltaic Process: The process of converting light to voltage.

Transducer: A device that converts one form of energy into another, for example, electricity into sound.

PROBLEMS

SECTION 15-1

15-1. What is the schematic symbol for a solar cell?

15-2. On what principle does the solar cell operate?

15-3. What is a photon?

15-4. Explain the results of using series and parallel connections of solar cells.

15-5. What is the purpose of Ni-Cd batteries?

15-6. What are some of the factors that infuence the voltage and current capabilities of a solar cell?

15-7. Design a small circuit that delivers 3 V at 600 mA to a load using the solar cells specified in Section 15-1.

SECTION 15-2

15-8. What is an optical fiber?

15-9. How is information transmitted using fiber optics?

15-10. What are the advantages of optical fiber over copper wire?

15-11. Define modulation.

15-12. What is the difference between AM and FM?

15-13. Explain how a phototransistor and an infrared LED work.

15-14. What is the relationship between the frequency of light, bandwidth, and information-carrying capabilities?

SECTION 15-3

15-15. What is an LCD?

15-16. List the differences between an LCD and an LED.

15-17. How are characters formed using the LCD?

15-18. What is a backplane?

15-19. What type of crystal is used in the LCD and how does it work?

SECTION 15-4

15-20. What are crystals?

15-21. Name three types of crystal.

15-22. What is the piezoelectric effect?

15-23. Define the fundamental frequency of a crystal.

15-24. How does the cut and thickness of a crystal affect its characteristics?

15-25. What is the approximate frequency range of crystals?

15-26. Explain the difference between the series and parallel resonances of a crystal.

15-27. What is an overtone?

SECTION 15-5

15-28. What is sound?

15-29. How does magnetism relate to the operation of a speaker?

15-30. What is the purpose of a speaker's diaphragm?

15-31. What is meant by the impedance of a speaker?

15-32. List the different types of speakers and their applications.

15-33. What is the approximate range of frequencies that can be heard by the human ear?

SECTION 15-6

15-34. On what principle does a relay work?

15-35. What is the difference between a relay and an ordinary switch?

15-36. Explain the difference between SPST, SPDT, and DPDT relays.

15-37. How do thyristors compare with relays?

15-38. Design a circuit in which a small dc voltage turns on a large ac appliance using a relay.

SECTION 15-7

15-39. What forms of energy are involved with dc motors?

15-40. Explain the purpose of the brushes, commutator, permanent magnet, and armature coil within a dc motor.

15-41. How is the speed of a motor controlled?

15-42. Explain the relationship between the duty cycle and motor speed.

15-43. How does a stepper motor differ from a free-running continuous motor?

15-44. Describe how a stepper motor can be stepped in either direction; use Fig. 15-20 as a reference.

ANSWERS TO REVIEW QUESTIONS

1. photovoltaic	**2.** light	**3.** A packet of light
4. True	**5.** Decrease	**6.** add
7. 0.5 V at 400 mA	**8.** Nickel–cadmium	**9.** Light
10. Glass or plastic	**11.** modulation	**12.** Amplitude modulation
13. False	**14.** Infrared LED	**15.** 5 V, lights
16. A phototransistor	**17.** Receiving	**18.** Pressed
19. Liquid-crystal display	**20.** False	**21.** Nematic crystal
22. Backplane	**23.** AC	**24.** +5 V
25. True	**26.** Quartz	**27.** lower
28. piezoelectric	**29.** series	**30.** True
31. overtone	**32.** Transducers	**33.** acoustical
34. The diaphragm	**35.** woofer	**36.** tweeter
37. lower	**38.** impedances	**39.** False
40. 150 Ω	**41.** voltage	**42.** solenoid
43. Single-pole double-throw	**44.** *B* and *C*, and *E* and *F*	**45.** 41.7 mA
46. True	**47.** dc	**48.** both ac and dc
49. mechanical	**50.** brushes	**51.** armature
52. Like, unlike	**53.** commutator	**54.** duty
55. $D = 92.3\%$	**56.** increases	**57.** stepper

Periodic Table of the Elements

PERIODS	IA	IIA	IIIB	IVB	VB	VIB	VIIB	VIII			IB	IIB	IIIA	IVA	VA	VIA	VIIA	VIIIA
1	1.008 **H** 1																	4.003 **He** 2
2	6.941 **Li** 3	9.012 **Be** 4											10.81 **B** 5	12.011 **C** 6	14.007 **N** 7	15.999 **O** 8	18.998 **F** 9	20.179 **Ne** 10
3	22.990 **Na** 11	24.305 **Mg** 12											26.982 **Al** 13	28.086 **Si** 14	30.9738 **P** 15	32.06 **S** 16	35.453 **Cl** 17	39.948 **Ar** 18
4	39.102 **K** 19	40.08 **Ca** 20	44.956 **Sc** 21	47.90 **Ti** 22	50.941 **V** 23	51.996 **Cr** 24	54.938 **Mn** 25	55.847 **Fe** 26	58.933 **Co** 27	58.71 **Ni** 28	63.546 **Cu** 29	65.37 **Zn** 30	69.72 **Ga** 31	72.59 **Ge** 32	74.922 **As** 33	78.96 **Se** 34	79.904 **Br** 35	83.80 **Kr** 36
5	85.468 **Rb** 37	87.62 **Sr** 38	88.906 **Y** 39	91.22 **Zr** 40	92.9064 **Nb** 41	95.94 **Mo** 42	98.906 **Tc** 43	101.07 **Ru** 44	102.906 **Rh** 45	106.4 **Pd** 46	107.868 **Ag** 47	112.40 **Cd** 48	114.82 **In** 49	118.69 **Sn** 50	121.75 **Sb** 51	127.60 **Te** 52	126.904 **I** 53	131.30 **Xe** 54
6	132.906 **Cs** 55	137.34 **Ba** 56	138.906 *****La** 57	178.49 **Hf** 72	180.948 **Ta** 73	183.85 **W** 74	186.2 **Re** 75	190.2 **Os** 76	192.22 **Ir** 77	196.09 **Pt** 78	196.967 **Au** 79	200.59 **Hg** 80	204.37 **Tl** 81	207.2 **Pb** 82	208.981 **Bi** 83	(209) **Po** 84	(210) **At** 85	(222) **Rn** 86
7	(223) **Fr** 87	226.025 **Ra** 88	(227) ******Ac** 89	(261) **[Rf]** 104	(260) **[Ha]** 105	(263) **[ ]** 106												

GROUPS

TRANSITION ELEMENTS

PERIODIC TABLE OF THE ELEMENTS

*Lanthanides

140.12 **Ce** 58	140.908 **Pr** 59	144.24 **Nd** 60	(145) **Pm** 61	150.4 **Sm** 62	151.96 **Eu** 63	157.25 **Gd** 64	158.925 **Tb** 65	162.50 **Dy** 66	164.930 **Ho** 67	167.26 **Er** 68	168.934 **Tm** 69	173.04 **Yb** 70	174.97 **Lu** 71

**Actinides

232.038 **Th** 90	231.031 **Pa** 91	238.029 **U** 92	237.048 **Np** 93	(244) **Pu** 94	(243) **Am** 95	(247) **Cm** 96	(247) **Bk** 97	(251) **Cf** 98	(254) **Es** 99	(253) **Fm** 100	(256) **Md** 101	(253) **No** 102	(257) **Lr** 103

Numbers below the symbol of the element indicate the atomic numbers. Atomic masses, above the symbol of the element, are based on the assigned relative atomic mass of ^{12}C = exactly 12; ( ) indicates the mass number of the isotope with the longest half-life. [ ] indicates not officially approved or named.

Model of an Atom

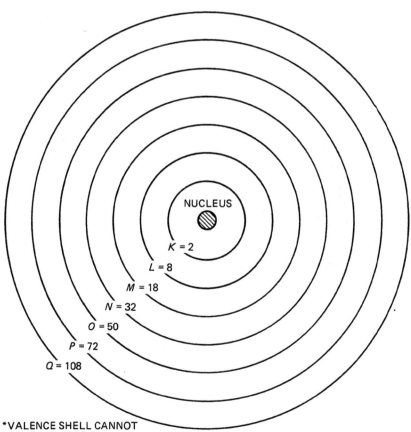

NUCLEUS

$K = 2$
$L = 8$
$M = 18$
$N = 32$
$O = 50$
$P = 72$
$Q = 108$

*VALENCE SHELL CANNOT
CONTAIN MORE THAN
8 ELECTRONS

*INNER SHELL NEXT TO VALENCE
SHELL CANNOT CONTAIN MORE
THAN 18 ELECTRONS

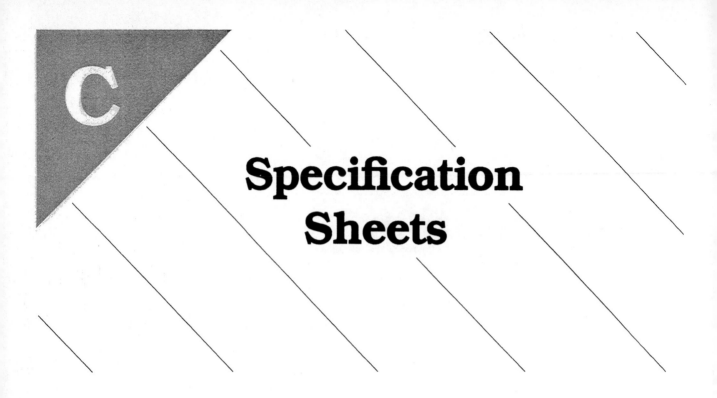

Specification Sheets

INTERPRETING SPECIFICATION SHEETS

Manufacturers provide specification sheets to aid engineers and technicians in designing with or substituting their components. Reference guides can be found at any electronic parts store. In this section, we examine how to read important information from the spec sheet that is applicable to our studies.

OP-AMP SPEC SHEET

The uA741C op amp listed in Fig. C-2 through Fig. C-5 (see pages 436–438) contains some maximum values that are important for experimentation. The maximum supply voltages are listed as +18 V above the midpoint.

Supply voltage V_{CC+} (see Note 1)	18 V
Supply voltage V_{CC-} (see Note 1)	−18 V

Note 1 tells us that these voltage values are with respect to the midpoint between the two supply voltages. If the supply voltages are equal, as in this text, then the midpoint would be 0 V.

The maximum input voltage on any input is listed as 15 V above or below the midpoint of the supply voltages. Once again, if the midpoint is 0 V, as in our examples, then you must not exceed either a positive or negative 15 V at either input of the op amp.

Input voltage any input (see Notes 1 and 3)	±15 V

In Fig. C-3, the typical input and output impedances for the uA741C are listed as 2 MΩ and 75Ω, respectively.

r_i	input resistance	2 MΩ
r_o	output resistance	75 Ω

The common-mode rejection ratio (CMRR) is also listed in the figure. Recall that this is the ability of an op amp to reject noise while amplifying the desired signal. The typical value for the uA741C op amp is 90 dB, which means that it amplifies the desired signal 90 dB more than the noise.

CMRR	90 dB

Figure C-5 illustrates the open-loop voltage gain vs. frequency characteristics of the op amp (the spec sheet's Fig. 8). This is similar to the graph used in Chapter 11, "Operational Amplifiers." At a frequency of 1 MHz, the gain is 1, whereas a frequency of 10 Hz allows a gain of 100,000.

555 TIMER SPEC SHEET

Figures C-6 through C-15 contain data about the 555 timer (see pages 439–443). The top of Fig. C-7 shows that the trigger voltage required to cause the output to pulse high is less than $\frac{1}{3}V_{DD}$ (supply voltage) with the reset input held high. The output is low when the trigger voltage is greater than $\frac{1}{3}V_{DD}$ and the threshold voltage is greater than $\frac{2}{3}V_{DD}$.

RESET	TRIGGER VOLTAGE	THRESHOLD VOLTAGE	OUTPUT
High	$<\frac{1}{3}V_{DD}$	Irrelevant	Low
High	$>\frac{1}{3}V_{DD}$	$>\frac{2}{3}V_{DD}$	High

The supply voltage for the NE555 should be between 4.5 and 16 V according to the lower table in Fig. C-7. The input voltage should not exceed the supply voltage. The recommended maximum output current is ±200 mA.

	NE555		
	MIN	MAX	UNIT
Supply voltage, V_{CC}	4.5	16	V
Input voltage (control voltage, reset, threshold, trigger)		V_{CC}	V
Output current		±200	mA

Figures C-10 to C-15 provide typical applications for the 555 timer along with descriptions and formulas.

POWER TRANSISTOR SPEC SHEET

Figures C-16 through C-22 contain data on power transistors (see pages 444–447). The maximum ratings for the TIP29A power transistor are found in Fig. C-16. Notice the ratings are given for a case temperature (not room temperature!) of 25°C (or 77°F) unless otherwise noted. Keep in mind that as a transistor conducts, its temperature increases and the ratings decrease (derate).

The maximum collector–emitter voltage for the TIP29A at cutoff ($I_B = 0$) is 60 V and the maximum continuous collector current is 1 A. The base current should not exceed 0.4 A. Finally, when soldering, the lead temperature 3.2 mm from the case for 10 seconds should not exceed 250°C.

	TIP29A
Collector–emitter voltage ($I_B = 0$)	60 V
Continuous collector current	1 A
Continuous base current	0.4 A
Lead temperature 3.2 mm from case for 10 seconds	250°C

The lower chart in Fig. C-17 contains some parameters that need explanation. Note the test conditions under which these values are attained.

$V_{(BR)CEO}$	Collector–emitter breakdown voltage with base open
I_{CEO}	DC collector cutoff current with base open
I_{CES}	DC collector cutoff current with base shorted to emitter
I_{EBO}	DC emitter cutoff current with collector open
h_{FE}	DC current gain for common-emitter
V_{BE}	DC base–emitter voltage
$V_{CE(MAT)}$	DC collector–emitter saturation voltage
h_{fe}	AC current gain for common-emitter

Some of the parameters are shown graphically in Fig. C-21. The top graph illustrates how the dc current gain varies depending on the collector current. The bottom graph shows the maximum collector current and the collector–emitter voltage for the TIP29A as roughly 0.3 A at 50 V.

DIODE SPEC SHEET

The diode spec sheet is shown in Fig. C-23 (see page 448). Once again, it is helpful to know what each of the abbreviations at top of the chart represents.

V_{RPM}	Repetitive peak reverse voltage
I_{RRM}	Repetitive peak reverse current
V_{FM}	Maximum total forward voltage

I_F	DC forward current
t_{rr}	Reverse recovery time

PART NO.	V_{PRM} (V)	I_{RRM} (nA)	V_{FM} @ I_F (V)	(mA)	t_{rr} (ns) MAX
1N4148	100	25	1.0	10	4.0

Examining the information on the 1N4148 diode shows that this diode should not be used in a rectifier circuit where the reverse voltage as a result of an applied ac voltage is greater than 100 V. Also, it should not conduct more than 10 mA of current. These characteristics are important in choosing a diode to substitute in a bridge rectifier circuit.

ZENER DIODE SPEC SHEET

The zener diode spec sheet is shown in Fig. C-24 (see page 449). The abbreviations are

V_z	DC zener regulator voltage
Z_z	Zener regulator impedance (for a small signal at I_z)
I_z	DC zener regulator current
P_D	maximum power dissipation

DEVICE NO.	V_z (V) NOM	Z_z (Ω) @ I_z MAX	mA	P_D (mW) (25°C)
1N751A	5.1	17	20	500

Examining the information on the 1N751A zener diode reveals a zener voltage of 5.1 V and the maximum power that it should dissipate is 500 mW (or 0.5 W).

GENERAL-PURPOSE TRANSISTORS SPEC SHEET

The NPN General Purpose Transistor spec sheet is shown in Fig. C-25 (see page 450). The abbreviations are

V_{ceo}	Collector–emitter voltage, open base
V_{cbo}	Collector–base voltage, open emitter
H_{fe}	DC current gain
I_c/V_{ce}	Ratio of collector current to collector–emitter voltage
Ft	Transition frequency

The most useful bit of information to us from this chart is the range of dc beta. The range of beta for 2N2222 is between 100 and 300. Now you can see why the voltage-divider bias that eliminated beta from the calculations is a preferred method given the range of beta for most transistors.

LINEAR INTEGRATED CIRCUITS

TYPES uA741M, uA741C
GENERAL-PURPOSE OPERATIONAL AMPLIFIER

D920, NOVEMBER 1970 – REVISED AUGUST 1983

- Short-Circuit Protection
- Offset-Voltage Null Capability
- Large Common-Mode and Differential Voltage Ranges
- No Frequency Compensation Required
- Low Power Consumption
- No Latch-up
- Designed to be Interchangeable with Fairchild μA741M, μA741C

description

The uA741 is a general-purpose operational amplifier featuring offset-voltage null capability.

The high common-mode input voltage range and the absence of latch-up make the amplifier ideal for voltage-follower applications. The device is short-circuit protected and the internal frequency compensation ensures stability without external components. A low potentiometer may be connected between the offset null inputs to null out the offset voltage as shown in Figure 2.

The uA741M is characterized for operation over the full military temperature range of −55°C to 125°C; the uA741C is characterized for operation from 0°C to 70°C.

symbol

uA741M . . . J PACKAGE
(TOP VIEW)

NC	1	14	NC
NC	2	13	NC
OFFSET N1	3	12	NC
IN−	4	11	VCC+
IN+	5	10	OUT
VCC−	6	9	OFFSET N2
	7	8	

uA741M . . . JG PACKAGE
uA741C . . . D, P, OR JG PACKAGE
(TOP VIEW)

OFFSET N1	1	8	NC
IN−	2	7	VCC+
IN+	3	6	OUT
VCC−	4	5	OFFSET N2

uA741M . . . U FLAT PACKAGE
(TOP VIEW)

NC	1	10	NC
OFFSET N1	2	9	VCC+
IN−	3	8	VCC+
IN+	4	7	OUT
VCC−	5	6	OFFSET N2

uA741M . . . FH, FK PACKAGE
(TOP VIEW)

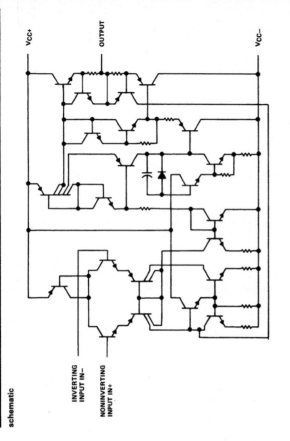

NC – No internal connection

TYPES uA741M, uA741C
GENERAL-PURPOSE OPERATIONAL AMPLIFIERS

schematic

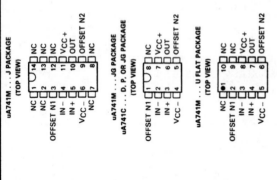

absolute maximum ratings over operating free-air temperature range (unless otherwise noted)

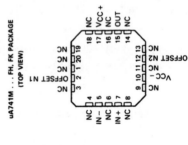

	uA741M	uA741C	UNIT
Supply voltage VCC+ (see Note 1)	22	18	V
Supply voltage VCC− (see Note 1)	−22	−18	V
Differential input voltage (see Note 2)	±30	±30	V
Input voltage any input (see Notes 1 and 3)	±15	±15	V
Voltage between either offset null terminal (N1/N2) and VCC−	±0.5	±0.5	V
Duration of output short-circuit (see Note 4)	unlimited	unlimited	
Continuous total power dissipation at (or below) 25°C free-air temperature (see Note 5)	500	500	mW
Operating free-air temperature range	−55 to 125	0 to 70	°C
Storage temperature range	−65 to 150	−65 to 150	°C
Lead temperature 1.6 mm (1/16 inch) from case for 60 seconds FH, FK, J, JG, or U package	300		°C
Lead temperature 1.6 mm (1/16 inch) from case for 10 seconds D, N or P package		260	°C

NOTES: 1. All voltage values, unless otherwise noted, are with respect to the midpoint between VCC+ and VCC−.
2. Differential voltages are at the noninverting input terminal with respect to the inverting input terminal.
3. The magnitude of the input voltage must never exceed the magnitude of the supply voltage or 15 volts, whichever is less.
4. The output may be shorted to ground or either power supply. For the uA741M only, the unlimited duration of the short-circuit applies at (or below) 125°C case temperature or 75°C free-air temperature.
5. For operation above 25°C free-air temperature, refer to Dissipation Derating Curves, Section 2. In the J and JG packages, uA741M chips are alloy mounted; uA741C chips are glass mounted.

FIGURE C-2

FIGURE C-1

Figures C-1 through C-15 are reproduced by permission of Texas Instruments, © 1983 by Texas Instruments Incorporated.

electrical characteristics at specified free-air temperature, V_{CC+} = 15 V, V_{CC-} = −15 V

PARAMETER		TEST CONDITIONS[†]		uA741M			uA741C			UNIT
				MIN	TYP	MAX	MIN	TYP	MAX	
V_{IO}	Input offset voltage	$V_O = 0$	25°C		1	5		1	6	mV
			Full range			6			7.5	
$\Delta V_{IO(adj)}$	Offset voltage adjust range	$V_O = 0$	25°C		±15			±15		mV
I_{IO}	Input offset current	$V_O = 0$	25°C		20	200		20	200	nA
			Full range			500			300	
I_{IB}	Input bias current	$V_O = 0$	25°C		80	500		80	500	nA
			Full range			1500			800	
V_{ICR}	Common-mode input voltage range		25°C	±12	±13		±12	±13		V
			Full range	±12			±12			
V_{OM}	Maximum peak output voltage swing	$R_L = 10\ k\Omega$	25°C	±12	±14		±12	±14		V
		$R_L \geq 10\ k\Omega$	Full range	±12			±12			
		$R_L = 2\ k\Omega$	25°C	±10	±13		±10	±13		
		$R_L \geq 2\ k\Omega$	Full range	±10			±10			
A_{VD}	Large-signal differential voltage amplification	$R_L \geq 2\ k\Omega$	25°C	50	200		20	200		V/mV
		$V_O = \pm10$ V	Full range	25			15			
r_i	Input resistance		25°C	0.3	2		0.3	2		MΩ
r_o	Output resistance	$V_O = 0$, See Note 6	25°C		75			75		Ω
C_i	Input capacitance		25°C		1.4			1.4		pF
CMRR	Common-mode rejection ratio	$V_{IC} = V_{ICR}$ min	25°C	70	90		70	90		dB
			Full range	70			70			
k_{SVS}	Supply voltage sensitivity ($\Delta V_{IO}/\Delta V_{CC}$)	$V_{CC} = \pm9$ V to ±15 V	25°C		30	150		30	150	μV/V
			Full range			150			150	
I_{OS}	Short-circuit output current		25°C		±25	±40		±25	±40	mA
I_{CC}	Supply current	No load, $V_O = 0$	25°C		1.7	2.8		1.7	2.8	mA
			Full range			3.3			3.3	
P_D	Total power dissipation	No load, $V_O = 0$	25°C		50	85		50	85	mW
			Full range			100			100	

[†]All characteristics are measured under open-loop conditions with zero common-mode input voltage unless otherwise specified. Full range for uA741M is −55°C to 125°C and for uA741C is 0°C to 70°C.

NOTE 6: This typical value applies only at frequencies above a few hundred hertz because of the effects of drift and thermal feedback.

FIGURE C-3

TYPICAL CHARACTERISTICS

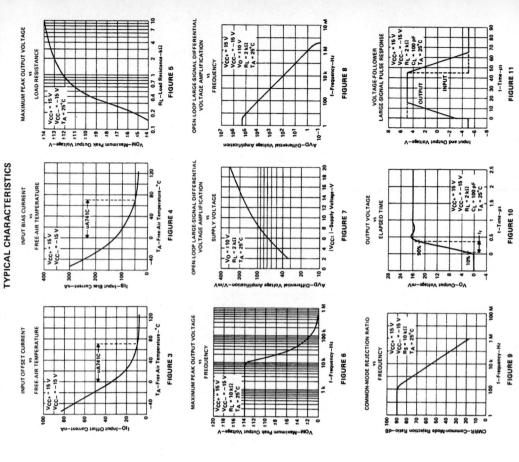

FIGURE C-5

TYPES uA741M, uA741C
GENERAL-PURPOSE OPERATIONAL AMPLIFIERS

operating characteristics, V_{CC+} = 15 V, V_{CC-} = −15 V, T_A = 25°C

			uA741M			uA741C			UNIT
	PARAMETER	TEST CONDITIONS	MIN	TYP	MAX	MIN	TYP	MAX	
t_r	Rise time	V_I = 20 mV, R_L = 2 kΩ, C_L = 100 pF, See Figure 1		0.3			0.3		μs
	Overshoot factor			5%			5%		
SR	Slew rate at unity gain	V_I = 10 V, R_L = 2 kΩ, C_L = 100 pF, See Figure 1		0.5			0.5		V/μs

PARAMETER MEASUREMENT INFORMATION

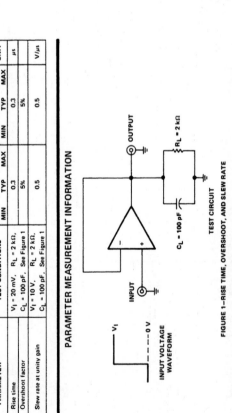

INPUT VOLTAGE WAVEFORM

TEST CIRCUIT

FIGURE 1–RISE TIME, OVERSHOOT, AND SLEW RATE

TYPICAL APPLICATION DATA

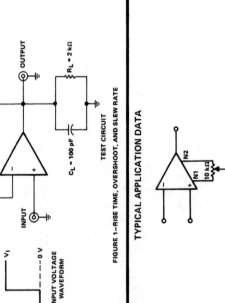

FIGURE 2 – INPUT OFFSET VOLTAGE NULL CIRCUIT

FIGURE C-4

- Timing from Microseconds to Hours
- Astable or Monostable Operation
- Adjustable Duty Cycle
- TLL-Compatible Output Can Sink or Source up to 200 mA
- Functionally Interchangeable with the Signetics SE555, SE555C, SA555, NE555; Have Same Pinout

SE555C FROM TI IS NOT RECOMMENDED FOR NEW DESIGNS

description

These devices are monolithic timing circuits capable of producing accurate time delays or oscillation. In the time-delay or monostable mode of operation, the timed interval is controlled by a single external resistor and capacitor network. In the astable mode of operation, the frequency and duty cycle may be independently controlled with two external resistors and a single external capacitor.

The threshold and trigger levels are normally two-thirds and one-third, respectively, of V_CC. These levels can be altered by use of the control voltage terminal. When the trigger input falls below the trigger level, the flip-flop is set and the output goes high. If the trigger input is above the trigger level and the threshold input is above the threshold level, the flip-flop is reset and the output is low. The reset input can override all other inputs and can be used to initiate a new timing cycle. When the reset input goes low, the flip-flop is reset and the output goes low. Whenever the output is low, a low-impedance path is provided between the discharge terminal and ground.

The output circuit is capable of sinking or sourcing current up to 200 milliamperes. Operation is specified for supplies of 5 to 15 volts. With a 5-volt supply, output levels are compatible with TTL inputs.

The SE555 and SE555C are characterized for operation over the full military range of −55°C to 125°C. The SA555 is characterized for operation from −40°C to 85°C, and the NE555 is characterized for operation from 0°C to 70°C.

NE555, SE555, NE555 . . . JG DUAL-IN-LINE PACKAGE
SA555, NE555 . . . D, JG, OR P DUAL-IN-LINE PACKAGE
(TOP VIEW)

GND	1	8 VCC
TRIG	2	7 DISCH
OUT	3	6 THRES
RESET	4	5 CONT

SE555, SE555C . . . FK CHIP CARRIER PACKAGE
(TOP VIEW)

NC — No internal connection

functional block diagram

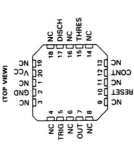

Reset can override Trigger, which can override Threshold.

FIGURE C-6

FUNCTION TABLE

RESET	TRIGGER VOLTAGE†	THRESHOLD VOLTAGE†	OUTPUT	DISCHARGE SWITCH
Low	Irrelevant	Irrelevant	Low	On
High	< 1/3 V_DD	Irrelevant	High	Off
High	> 1/3 V_DD	> 2/3 V_DD	Low	On
High	> 1/3 V_DD	< 2/3 V_DD	As previously established	

†Voltage levels shown are nominal.

absolute maximum ratings over operating free-air temperature range (unless otherwise noted)

Supply voltage, V_CC (see Note 1)	18 V
Input voltage (control voltage, reset, threshold trigger)	V_CC
Output current	±225 mA
Continuous total dissipation at (or below) 25°C free-air temperature (see Note 2)	600 mW
Operating free-air temperature range: SE555, SE555C	−55°C to 125°C
SA555	−40°C to 85°C
NE555	0°C to 70°C
Storage temperature range	−65°C to 150°C
Lead temperature 1,6 mm (1/16 inch) from case for 60 seconds: FK or JG package	300°C
Lead temperature 1,6 mm (1/16 inch) from case for 10 seconds: D or P package	260°C

NOTES: 1. All voltage values are with respect to network ground terminal.
2. For operation above 25°C free-air temperature, refer to Dissipation Derating Curves, Section 2. In the JG package, SE555 and SE555C chips are alloy mounted. SA555 and NE555 chips are glass mounted.

recommended operating conditions

	SE555		SE555C		SA555		NE555		UNIT
	MIN	MAX	MIN	MAX	MIN	MAX	MIN	MAX	
Supply voltage, V_CC	4.5	18	4.5	16	4.5	16	4.5	16	V
Input voltage (control voltage, reset, threshold, trigger)		V_CC		V_CC		V_CC		V_CC	V
Output current		±200		±200		±200		±200	mA
Operating free-air temperature, T_A	−55	125	−55	125	−40	85	0	70	°C

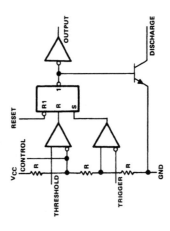

FIGURE C-7

TYPICAL CHARACTERISTICS†

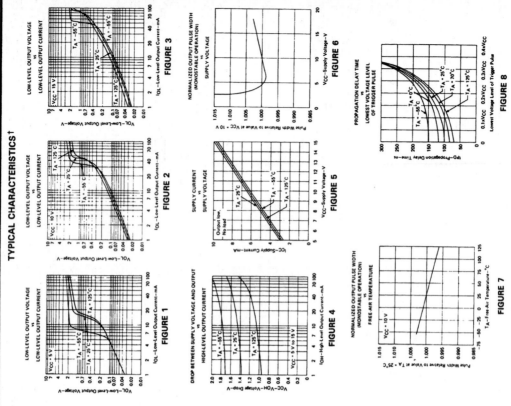

FIGURE C-9

†Data for temperatures below 0°C and above 70°C are applicable for SE555 circuits only.

SE555, SE555C, SA555, NE555
PRECISION TIMERS

electrical characteristics at 25°C free-air temperature, VCC = 5 V to 15 V (unless otherwise noted)

PARAMETER	TEST CONDITIONS	SE555 MIN	TYP	MAX	SE555C, SA555 NE555 MIN	TYP	MAX	UNIT
Threshold voltage level	VCC = 15 V	9.4	10	10.6	8.8	10	11.2	V
	VCC = 5 V	2.7	3.3	4	2.4	3.3	4.2	V
Threshold current (see Note 3)			30	250		30	250	nA
Trigger voltage level	VCC = 15 V	4.8	5	5.2	4.5	5	5.6	V
	VCC = 5 V	1.45	1.67	1.9	1.1	1.67	2.2	V
Trigger current	Trigger at 0 V		0.5	0.9		0.5	2.2	µA
Reset voltage level		0.4	0.7	1	0.4	0.7	1	V
Reset current	Reset at VCC		0.1	0.4		0.1	0.4	mA
	Reset at 0 V		-0.4	-1		-0.4	-1	
Discharge switch off-state current			20	100		20	100	nA
Control voltage (open circuit)	VCC = 15 V	9.6	10	10.4	9	10	11	V
	VCC = 5 V	2.9	3.3	3.8	2.6	3.3	4	
Low-level output voltage	IOL = 10 mA, VCC = 15 V		0.1	0.16		0.1	0.25	V
	IOL = 50 mA		0.4	0.5		0.4	0.75	
	IOL = 100 mA		2	2.2		2	2.5	
	IOL = 200 mA		2.5			2.5		
	IOL = 5 mA, VCC = 5 V		0.05	0.2		0.05	0.35	
	IOL = 8 mA		0.1	0.25		0.25	0.4	
High-level output voltage	IOH = -100 mA, VCC = 15 V	13	13.3		12.75	13.3		V
	IOH = -200 mA		12.5			12.5		
	IOH = -100 mA, VCC = 5 V	3	3.3		2.75	3.3		
Supply current	Output low, No load, VCC = 15 V		10	12		10	15	mA
	VCC = 5 V		3	5		3	6	
	Output high, No load, VCC = 15 V		9	10		9	13	
	VCC = 5 V		2	4		2	5	

NOTE 3: This parameter influences the maximum value of the timing resistors R_A and R_B in the circuit of Figure 12. For example, when V_{CC} = 5 V the maximum value is R = $R_A + R_B$ = 3.4 MΩ and for V_{CC} = 15 V the maximum value is 10 MΩ.

operating characteristics, VCC = 5 V and 15 V

PARAMETER	TEST CONDITIONS†	SE555 MIN	TYP	MAX	SE555C, SA555 NE555 MIN	TYP	MAX	UNIT
Initial error of timing interval‡	Each timer, monostable§ TA = 25°C		0.5	1.5		1	3	%
	Each timer, astable¶		1.5			2.25		
Temperature coefficient of timing interval	Each timer, monostable§ TA = MIN to MAX		30	100		50		ppm/°C
	Each timer, astable¶		90			150		
Supply voltage sensitivity of timing interval	Each timer, monostable§ TA = 25°C		0.05	0.2		0.1	0.5	%/V
	Each timer, astable¶		0.15			0.3		
Output pulse rise time	CL = 15 pF, TA = 25°C		100	200		100	300	ns
Output pulse fall time	TA = 25°C		100	200		100	300	

† For conditions shown as MIN or MAX, use the appropriate value specified under recommended operating conditions.
‡ Timing interval error is defined as the difference between the measured value and the nominal value computed by the formula: t_w = 1.1 $R_A C$.
§ Values specified are for a device in a monostable circuit similar to Figure 9, with component values as follow: R_A = 2 kΩ to 100 kΩ, C = 0.1 µF.
¶ Values specified are for a device in an astable circuit similar to Figure 1, with component values as follow: R_A = 1 kΩ to 100 kΩ, C = 0.1 µF.

FIGURE C-8

TYPICAL APPLICATION DATA

monostable operation

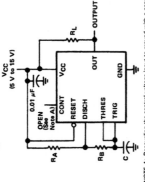

FIGURE 9. CIRCUIT FOR MONOSTABLE OPERATION

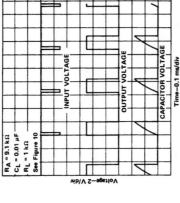

$R_A = 9.1\ k\Omega$
$C_L = 0.01\ \mu F$
$R_L = 1\ k\Omega$
See Figure 10

Time—0.1 ms/div

FIGURE 10. TYPICAL MONOSTABLE WAVEFORMS

For monostable operation, any of these timers may be connected as shown in Figure 9. If the output is low, application of a negative-going pulse to the trigger input sets the flip-flop ($\overline{Q}$ goes low), drives the output high, and turns off Q1. Capacitor C is then charged through R_A until the voltage across the capacitor reaches the threshold voltage of the threshold input. If the trigger input has returned to a high level, the output of the threshold comparator will reset the flip-flop ($\overline{Q}$ goes high), drive the output low, and discharge C through Q1.

Monostable operation is initiated when the trigger input voltage falls below the trigger threshold. Once initiated, the sequence will complete only if the trigger input is high at the end of the timing interval. Because of the threshold level and saturation voltage of Q1, the output pulse width is approximately $t_w = 1.1\ R_A C$. Figure 11 is a plot of the time constant for various values of R_A and C. The threshold levels and charge rates are both directly proportional to the supply voltage, VCC. The timing interval is therefore independent of the supply voltage, so long as the supply voltage is constant during the time interval.

Applying a negative-going trigger pulse simultaneously to the reset and trigger terminals during the timing interval will discharge C and re-initiate the cycle, commencing on the positive edge of the reset pulse. The output is held low as long as the reset pulse is low. When the reset input is not used, it should be connected to VCC to prevent false triggering.

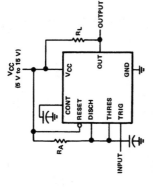

t_w — Output Pulse Duration—s

C—Capacitance—μF

$R_A = 10\ M\Omega$
$R_A = 1\ M\Omega$
$R_A = 100\ k\Omega$
$R_A = 10\ k\Omega$
$R_A = 1\ k\Omega$

**FIGURE 11. OUTPUT PULSE
DURATION vs CAPACITANCE**

FIGURE C-10

TYPICAL APPLICATION DATA

astable operation

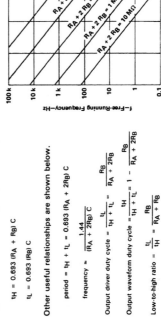

V_{CC}
(5 V to 15 V)

0.01 μF

OPEN
(See
Note A)

NOTE A: Decoupling the control voltage input to ground with a capacitor may improve operation. This should be evaluated for individual applications.

FIGURE 12. CIRCUIT FOR ASTABLE OPERATION

Addition of a second resistor, R_B, to the circuit of Figure 9, as shown in Figure 12, and connection of the trigger input to the threshold input will cause the timer to self-trigger and run as a multivibrator. The capacitor C will charge through R_A and R_B then discharge through R_B only. The duty cycle may be controlled, therefore, by the values of R_A and R_B.

This astable connection results in capacitor C charging and discharging between the threshold-voltage level ($\approx 0.67 \cdot V_{CC}$) and the trigger-voltage level ($\approx 0.33 \cdot V_{CC}$). As in the monostable circuit, charge and discharge times (and therefore the frequency and duty cycle) are independent of the supply voltage.

Figure 13 shows typical waveforms generated during astable operation. The output high-level duration t_H and low-level duration t_L may be found by:

$$t_H = 0.693\ (R_A + R_B)\ C$$

$$t_L = 0.693\ (R_B)\ C$$

Other useful relationships are shown below.

$$period = t_H + t_L = 0.693\ (R_A + 2R_B)\ C$$

$$frequency = \frac{1.44}{(R_A + 2R_B)\ C}$$

$$Output\ driver\ duty\ cycle = \frac{t_L}{t_H + t_L} = \frac{R_B}{R_A + 2R_B}$$

$$Output\ waveform\ duty\ cycle = \frac{t_H}{t_H + t_L} = 1 - \frac{R_B}{R_A + 2R_B}$$

$$Low\text{-}to\text{-}high\ ratio = \frac{t_L}{t_H} = \frac{R_B}{R_A + R_B}$$

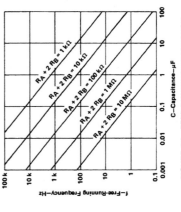

$R_A = 5\ k\Omega$
$R_B = 3\ k\Omega$
$C = 0.15\ \mu F$
$R_L = 1\ k\Omega$
See Figure 13

OUTPUT VOLTAGE

CAPACITOR VOLTAGE

Time—0.5 ms/div

FIGURE 13. TYPICAL ASTABLE WAVEFORMS

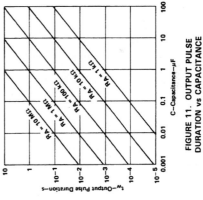

f—Free-Running Frequency—Hz

C—Capacitance—μF

$R_A + 2\ R_B\ 1\ k\Omega$
$R_A + 2\ R_B\ 10\ k\Omega$
$R_A + 2\ R_B\ 100\ k\Omega$
$R_A + 2\ R_B\ 1\ M\Omega$
$R_A + 2\ R_B\ 10\ M\Omega$

FIGURE 14. FREE-RUNNING FREQUENCY

FIGURE C-11

TYPICAL APPLICATION DATA

missing-pulse detector

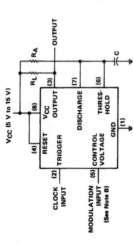

Vcc (5 V to 15 V)

RESET OUTPUT Vcc
(4) (3) (8)
RL
RA
OUTPUT

TRIGGER DISCHARGE
(2) (7)
C

INPUT

CONTROL VOLTAGE THRES-HOLD
(5) (6)
GND (1)

0.01 µF

A5T3644

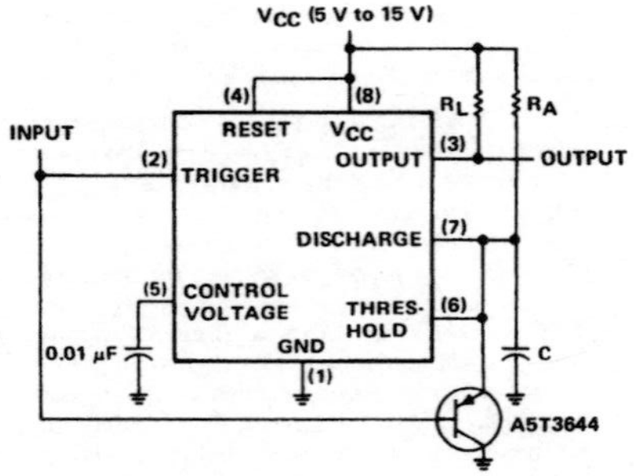

**FIGURE 15. CIRCUIT FOR
MISSING-PULSE DETECTOR**

Vcc = 5 V
RA = 1 kΩ
C = 0.1 µF
See Figure 16

Voltage—2 V/div

INPUT VOLTAGE

OUTPUT VOLTAGE

CAPACITOR VOLTAGE

Time—0.1 ms/div

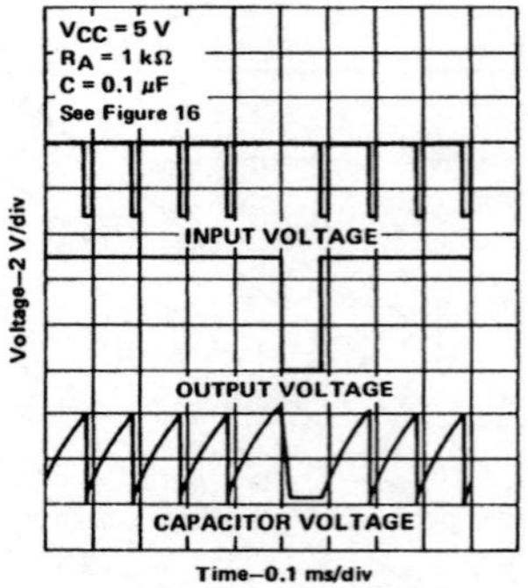

**FIGURE 16. MISSING-PULSE
DETECTOR WAVEFORMS**

The circuit shown in Figure 15 may be utilized to detect a missing pulse or abnormally long spacing between consecutive pulses in a train of pulses. The timing interval of the monostable circuit is continuously retriggered by the input pulse train as long as the pulse spacing is less than the timing interval. A longer pulse spacing, missing pulse, or terminated pulse train will permit the timing interval to be completed, thereby generating an output pulse as illustrated in Figure 16.

frequency divider

By adjusting the length of the timing cycle, the basic circuit of Figure 9 can be made to operate as a frequency divider. Figure 17 illustrates a divide-by-3 circuit that makes use of the fact that retriggering cannot occur during the timing cycle.

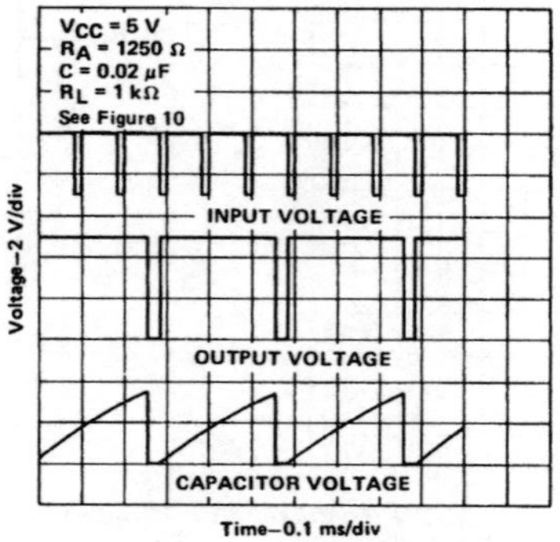

Vcc = 5 V
RA = 1250 Ω
C = 0.02 µF
RL = 1 kΩ
See Figure 10

Voltage—2 V/div

INPUT VOLTAGE

OUTPUT VOLTAGE

CAPACITOR VOLTAGE

Time—0.1 ms/div

**FIGURE 17. DIVIDE-BY-THREE
CIRCUIT WAVEFORMS**

FIGURE C-12

TYPICAL APPLICATION DATA

pulse-width modulation

Vcc (5 V to 15 V)

RESET OUTPUT Vcc
(4) (3) (8)
RL
RA
OUTPUT

CLOCK INPUT TRIGGER DISCHARGE
(2) (7)
C

MODULATION INPUT
(See Note B)

CONTROL VOLTAGE THRES-HOLD
(5) (6)
GND (1)

NOTE B: The modulating signal may be direct or capacitively coupled to the control voltage terminal. For direct coupling, the effects of modulation source voltage and impedance on the bias of the timer should be considered.

**FIGURE 18. CIRCUIT FOR PULSE-WIDTH
MODULATION**

The operation of the timer may be modified by modulating the internal threshold and trigger voltages. This is accomplished by applying an external voltage (or current) to the control voltage pin. Figure 18 is a circuit for pulse-width modulation. The monostable circuit is triggered by a continuous input pulse train and the threshold voltage is modulated by a control signal. The resultant effect is a modulation of the output pulse width, as shown in Figure 19. A sine-wave modulation signal is illustrated, but any wave-shape could be used.

RA = 3 kΩ
C = 0.02 µF
RL = 1 kΩ
See Figure 19

Voltage—2 V/div

MODULATION INPUT VOLTAGE

CLOCK INPUT VOLTAGE

OUTPUT VOLTAGE

CAPACITOR VOLTAGE

Time—0.5 ms/div

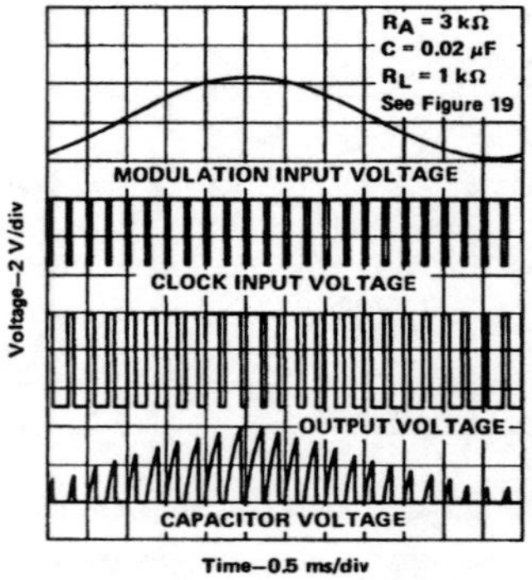

**FIGURE 19. PULSE-WIDTH MODULATION
WAVEFORMS**

FIGURE C-13

TYPICAL APPLICATION DATA

sequential timer

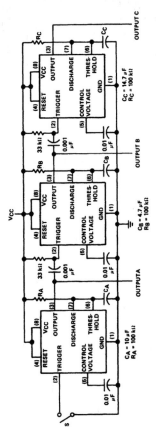

S closes momentarily at t = 0.

FIGURE 22. SEQUENTIAL TIMER CIRCUIT

Many applications, such as computers, require signals for initializing conditions during start-up. Other applications such as test equipment require activation of test signals in sequence. These timing circuits may be connected to provide such sequential control. The timers may be used in various combinations of astable or monostable circuit connections, with or without modulation, for extremely flexible waveform control. Figure 22 illustrates a sequencer circuit with possible applications in many systems and Figure 23 shows the output waveforms.

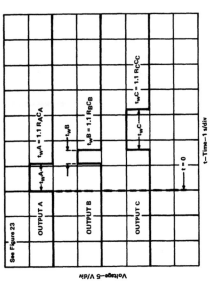

FIGURE 23. SEQUENTIAL TIMER WAVEFORMS

FIGURE C-15

TYPICAL APPLICATION DATA

pulse-position modulation

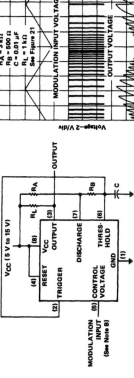

NOTE B: The modulating signal may be direct or capacitively coupled to the control voltage terminal. For direct coupling, the effects of modulation source voltage and impedance on the bias of the timer should be considered.

FIGURE 20. CIRCUIT FOR PULSE-POSITION MODULATION

FIGURE 21. PULSE POSITION-MODULATION WAVEFORMS

Any of these timers may be used as a pulse-position modulator as shown in Figure 20. In this application, the threshold voltage, and thereby the time delay, of a free-running oscillator is modulated. Figure 21 shows such a circuit, with a triangular-wave modulation signal, however, any modulating wave-shape could be used.

FIGURE C-14

- Designed for Complementary Use With TIP30 series
- 30 W at 25°C Case Temperature
- 1 A Continuous Collector Current
- 3 A Peak Collector Current
- Minimum f_T of 3 MHz at 10 V, 0.2 A
- Customer-Specified Selections Available

device schematic

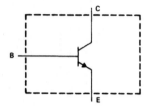

TO-220AB PACKAGE

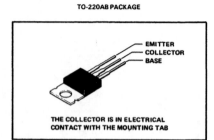

EMITTER
COLLECTOR
BASE

THE COLLECTOR IS IN ELECTRICAL
CONTACT WITH THE MOUNTING TAB

absolute maximum ratings at 25°C case temperature (unless otherwise noted)

	TIP29	TIP29A	TIP29B	TIP29C
Collector-base voltage	80 V	100 V	120 V	140 V
Collector-emitter voltage (I_B = 0)	40 V	60 V	80 V	100 V
Emitter-base voltage	5 V			
Continuous collector current	1 A			
Peak collector current (see Note 1)	3 A			
Continuous base current	0.4 A			
Safe operating areas at 25°C case temperature	See Figure 4			
Continuous device dissipation at 25°C case temperature (see Note 2)	30 W			
Continuous device dissipation at (or below) 25°C free-air temperature (see Note 3)	2 W			
Unclamped inductive load energy (see Note 4)	32 mJ			
Operating collector junction and storage temperature range	− 65°C to 150°C			
Lead temperature 3,2 mm (0.125 inch) from case for 10 seconds	250°C			

NOTES: 1. This value applies for t_w ≤ 0.3 ms, duty cycle ≤ 10%.
2. Derate linearly to 150°C case temperature at the rate of 0.24 W/°C.
3. Derate linearly to 150°C free-air temperature at the rate of 16 mW/°C.
4. This rating is based on the capability of the transistor to operate safely in the circuit in Figure 2.

FIGURE C-16

Figures C-16 through C-22 are reproduced by permission of Texas Instruments, first printed July 1968, revised October 1984.

electrical characteristics at 25°C case temperature (unless otherwise noted)

PARAMETER	TEST CONDITIONS	TIP29D MIN	TYP	MAX	TIP29E MIN	TYP	MAX	TIP29F MIN	TYP	MAX	UNIT
$V_{(BR)CEO}$	$I_C = 30$ mA, See Note 5			120			140			160	V
I_{CEO}	$V_{CE} = 90$ V, $I_B = 0$			0.3			0.3			0.3	mA
I_{CES}	$V_{CE} = 160$ V, $V_{BE} = 0$; $V_{CE} = 180$ V, $V_{BE} = 0$; $V_{CE} = 200$ V, $V_{BE} = 0$			0.2			0.2			0.2	mA
I_{EBO}	$V_{EB} = 5$ V, $I_C = 0$			1			1			1	mA
h_{FE}	$V_{CE} = 4$ V, $I_C = 0.2$ A, See Notes 5 and 6	40			40			40			
	$V_{CE} = 4$ V, $I_C = 1$ A, See Notes 5 and 6	15			15			15			
V_{BE}	$V_{CE} = 4$ V, See Notes 5 and 6			1.3			1.3			1.3	V
$V_{CE(sat)}$	$I_B = 125$ mA, $I_C = 1$ A, See Notes 5 and 6			0.7			0.7			0.7	V
h_{fe}	$V_{CE} = 10$ V, $I_C = 0.2$ A, $f = 1$ MHz	20			20			20			
$|h_{fe}|$	$V_{CE} = 10$ V, $I_C = 0.2$ A, $f = 1$ MHz	3			3			3			

NOTES: 5. These parameters must be measured using pulse techniques, $t_w = 300$ µs, duty cycle ≤ 2%.
6. These parameters are measured with voltage-sensing contacts separate from the current-carrying contacts.

thermal characteristics

PARAMETER	MIN	TYP	MAX	UNIT
$R_{\theta JC}$			4.17	°C/W
$R_{\theta JA}$			62.5	°C/W

resistive-load switching characteristic at 25°C case temperature (unless otherwise noted)

PARAMETER	TEST CONDITIONS†	MIN	TYP	MAX	UNIT
t_{on}	$I_C = 1$ A, $I_{B1} = 0.1$ A, $I_{B2} = -0.1$ A,			0.5	µs
t_{off}	$V_{BE(off)} = -4.3$ V, $R_L = 30 Ω$, See Figure 1			2	µs

† Voltage and current values shown are nominal; exact values vary slightly with transistor parameters.

FIGURE C-18

---

absolute maximum ratings at 25°C case temperature (unless otherwise noted)

	TIP29D	TIP29E	TIP29F	UNIT
Collector-base voltage	160 V	180 V	200 V	V
Collector-emitter voltage ($I_B = 0$)	120 V	140 V	160 V	V
Emitter-base voltage	5 V			V
Continuous collector current	1 A			A
Peak collector current (see Note 1)	3 A			A
Continuous base current	0.4 A			A
Safe operating areas at 25°C case temperature	See Figure 4			
Continuous device dissipation at 25°C case temperature (see Note 2)	30 W			W
Continuous device dissipation at (or below) 25°C free-air temperature (see Note 3)	2 W			W
Unclamped inductive load energy (see Note 4)	32 mJ			mJ
Operating collector junction and storage temperature range	−65°C to 150°C			
Lead temperature 3.2 mm (0.125 inch) from case for 10 seconds	250°C			

NOTES: 1. This value applies for $t_w ≤ 0.3$ ms, duty cycle ≤ 10%.
2. Derate linearly to 150°C case temperature at the rate of 0.24 W/°C.
3. Derate linearly to 150°C free-air temperature at the rate of 16 mW/°C.
4. This rating is based on the capability of the transistor to operate safely in the circuit in Figure 2.

electrical characteristics at 25°C case temperature (unless otherwise noted)

PARAMETER	TEST CONDITIONS	TIP29 MIN	TYP	MAX	TIP29A MIN	TYP	MAX	TIP29B MIN	TYP	MAX	TIP29C MIN	TYP	MAX	UNIT
$V_{(BR)CEO}$	$I_C = 30$ mA, $I_B = 0$, See Note 5	40			60			80			100			V
I_{CEO}	$V_{CE} = 30$ V, $I_B = 0$; $V_{CE} = 60$ V, $I_B = 0$			0.3			0.3			0.3			0.3	mA
I_{CES}	$V_{CE} = 80$ V, $V_{BE} = 0$; $V_{CE} = 100$ V, $V_{BE} = 0$; $V_{CE} = 120$ V, $V_{BE} = 0$; $V_{CE} = 140$ V, $V_{BE} = 0$			0.2			0.2			0.2			0.2	mA
I_{EBO}	$V_{EB} = 5$ V, $I_C = 0$			1			1			1			1	mA
h_{FE}	$V_{CE} = 4$ V, $I_C = 0.2$ A, See Notes 5 and 6	40			40			40			40			
	$V_{CE} = 4$ V, $I_C = 1$ A, See Notes 5 and 6	15			15			15			15			
V_{BE}	$V_{CE} = 4$ V, See Notes 5 and 6			1.3			1.3			1.3			1.3	V
$V_{CE(sat)}$	$I_B = 125$ mA, $I_C = 1$ A, See Notes 5 and 6			0.7			0.7			0.7			0.7	V
h_{fe}	$V_{CE} = 10$ V, $I_C = 0.2$ A, $f = 1$ MHz	20			20			20			20			
$|h_{fe}|$	$V_{CE} = 10$ V, $I_C = 0.2$ A, $f = 1$ MHz	3			3			3			3			

NOTES: 5. These parameters must be measured using pulse techniques, $t_w = 300$ µs, duty cycle ≤ 2%.
6. These parameters are measured with voltage-sensing contacts separate from the current-carrying contacts.

FIGURE C-17

TIP29, TIP29A, TIP29B, TIP29C,
TIP29D, TIP29E, TIP29F
N-P-N SILICON POWER TRANSISTORS

PARAMETER MEASUREMENT INFORMATION

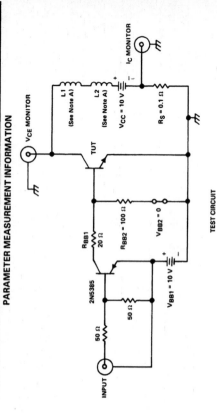

TEST CIRCUIT

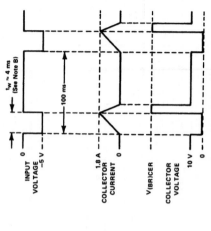

VOLTAGE AND CURRENT WAVEFORMS

FIGURE 2.   INDUCTIVE-LOAD SWITCHING

NOTES: A. L1 and L2 are 10 mH, 0.11 Ω, Chicago Standard Transformer Corporation C-2688, or equivalent.
        B. Input pulse duration is increased until I_{CM} = 1.8 A.

FIGURE C-20

TIP29, TIP29A, TIP29B, TIP29C,
TIP29D, TIP29E, TIP29F
N-P-N SILICON POWER TRANSISTORS

PARAMETER MEASUREMENT INFORMATION

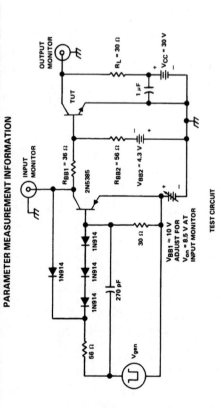

TEST CIRCUIT

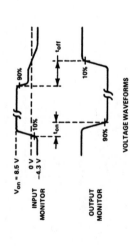

VOLTAGE WAVEFORMS

FIGURE 1.   RESISTIVE-LOAD SWITCHING

NOTES: A. V_{gen} is a −30-V pulse into a 50 Ω termination.
        B. The V_{gen} waveform is supplied by a generator with the following characteristics: t_r ≤ 15 ns, t_f ≤ 15 ns, Z_{out} = 50 Ω.
            t_w = 20 μs, duty cycle ≤ 2%.
        C. Waveforms are monitored on an oscilloscope with the following characteristics: t_r ≤ 15 ns, R_{in} ≥ 10 MΩ, C_{in} ≤ 11.5 pF.
        D. Resistors must be noninductive types.
        E. The d-c power supplies may require additional bypassing in order to minimize ringing.

FIGURE C-19

THERMAL INFORMATION

DISSIPATION DERATING CURVE

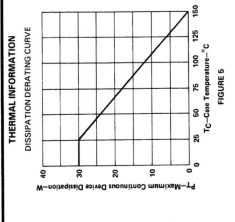

FIGURE 5

FIGURE C-22

TYPICAL CHARACTERISTICS

STATIC FORWARD CURRENT TRANSFER RATIO
vs
COLLECTOR CURRENT

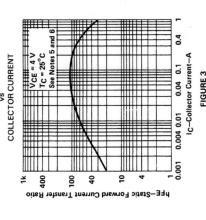

FIGURE 3

NOTES: 5. These parameters must be measured using pulse techniques, t_w = 300 µs, duty cycle ≤ 2%.
6. These parameters are measured with voltage-sensing contacts separate from the current-carrying contacts.

MAXIMUM SAFE OPERATING AREA

FORWARD-BIAS SAFE OPERATING AREA

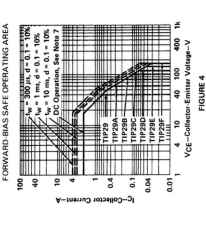

FIGURE 4

FIGURE C-21

NOTE 7. This combination of maximum voltage and current may be achieved only when switching from saturation to cutoff with a clamped inductive load.

Diode Products by Ascending V_{RRM} and t_{rr} (Continued)

Part No.	V_{RRM} (V)	I_{RRM} (nA)	V_{FM} @ I_F (V)	(mA)	t_{rr} (ns) Max	Package
1N659	60	5000	1.0	6.0		DO-35
BAV18	60	100	1.0	100	50	DO-35
FDLL659	60	5000	1.0	6.0		LL-34
FDH600	65	100	1.0	200	4.0	DO-35
BAV70	70	5000	1.1	50	6.0	TO-236
BAV99	70	2500	1.1	50	6.0	TO-236
BAW56	70	2500	1.1	50	6.0	TO-236
1N457	70	25	1.0	100		DO-35
1N457A	70	25	1.0	100		DO-35
1N462A	70	500	1.0	100		DO-35
FDLL457	70	25	1.0	20		LL-34
FDLL457A	70	25	1.0	100		LL-34
FDLL462A	70	500	1.0	100		LL-34
1N4153	75	50	0.66	20	2.0	DO-35
1N4151	75	50	1.0	50	2.0	DO-35
1N4305	75	100	0.85	10	2.0	DO-35
FDLL4153	75	50	0.88	20	2.0	LL-34
FDLL4151	75	50	1.0	50	2.0	LL-34
FDLL4305	75	100	0.85	10	2.0	LL-34
1N3600	75	100	1.0	200	4.0	DO-35
1N3064	75	100	1.0	10	4.0	DO-35
1N4150	75	100	1.0	200	4.0	DO-35
1N4454	75	100	1.0	10	4.0	DO-35
FDLL3600	75	100	1.0	200	4.0	LL-34
FDLL600	75	100	1.0	200	4.0	LL-34
FDLL3604	75	100	1.0	10	4.0	LL-34
FDLL4150	75	100	1.0	200	4.0	LL-34
FDLL4454	75	100	1.0	10	4.0	LL-34
BAS16	75	1000	1.1	50	6.0	TO-236
FDH1000	75	50	1.0	500	100	DO-35
BA128	75	100	1.0	50		DO-35
BAW76	75	100	1.0	100	2.0	DO-35
1N5194	80	25	1.0	100		DO-35
1N5282	80	100	1.3	500	2.0	DO-35
1N483B	80	25	1.0	100		DO-35
FDLL483B	80	25	1.0	100		LL-34
1N914	100	25	1.0	10	4.0	DO-35
1N914A	100	25	1.0	20	4.0	DO-35
1N914B	100	25	1.0	100	4.0	DO-35
1N916	100	25	1.0	10	4.0	DO-35
1N916A	100	25	1.0	20	4.0	DO-35
1N916B	100	25	1.0	30	4.0	DO-35
1N4148	100	25	1.0	10	4.0	DO-35
1N4149	100	25	1.0	10	4.0	DO-35
1N4446	100	25	1.0	20	4.0	DO-35
1N4447	100	25	1.0	20	4.0	DO-35
1N4448	100	25	1.0	100	4.0	DO-35
1N4449	100	25	1.0	30	4.0	DO-35
FDLL914	100	25	1.0	10	4.0	LL-34
FDLL914A	100	25	1.0	20	4.0	LL-34
FDLL914B	100	25	1.0	100	4.0	LL-34
FDLL916	100	25	1.0	10	4.0	LL-34
FDLL916B	100	25	1.0	30	4.0	LL-34
FDLL4148	100	25	1.0	10	4.0	LL-34
FDLL4149	100	25	1.0	10	4.0	LL-34
FDLL4446	100	25	1.0	20	4.0	LL-34
FDLL4447	100	25	1.0	20	4.0	LL-34

FIGURE C-23

Figures C-23 through C-26 are reproduced by permission of National Semiconductor.

Zener Diodes by Ascending Zener Voltage

Device No.	V_z (V) Nom	Z_z (Ω) @ I_z Max	mA	P_D (mW) (25°C)	Package
1N5226B	3.3	28	20	500	DO-35
1N746A	3.3	28	20	500	DO-35
1N4728A	3.3	10	7.6	1000	DO-41
1N5227B	3.6	24	20	500	DO-35
1N747A	3.6	24	20	500	DO-35
1N4729A	3.6	10	69	1000	DO-41
1N5228B	3.9	23	20	500	DO-35
1N748A	3.9	23	20	500	DO-35
1N4730A	3.9	9.0	64	1000	DO-41
1N5229B	4.3	22	20	500	DO-35
1N749A	4.3	22	20	500	DO-35
1N4731A	4.3	9.0	58	1000	DO-41
1N5230B	4.7	19	20	500	DO-35
1N750A	4.7	19	20	500	DO-35
1N4732A	4.7	8.0	53	1000	DO-41
1N5231B	5.1	17	20	500	DO-35
1N751A	5.1	17	20	500	DO-35
1N4733A	5.1	7.0	49	1000	DO-41
1N5232B	5.6	11	20	500	DO-35
1N752A	5.6	11	20	500	DO-35
1N4734A	5.6	5.0	45	1000	DO-41
1N5233B	6.0	7.0	20	500	DO-35
1N5234B	6.2	7.0	20	500	DO-35
1N753A	6.2	7.0	20	500	DO-35
1N4735A	6.2	2.0	41	1000	DO-41
1N5235B	6.8	5.0	20	500	DO-35
1N754A	6.8	5.0	20	500	DO-35
1N957B	6.8	4.5	18.5	500	DO-35
1N4736A	6.8	3.5	37	1000	DO-41
1N5236B	7.5	6.0	20	500	DO-35
1N755A	7.5	6.0	20	500	DO-35
1N958B	7.5	5.5	16.5	500	DO-35
1N4737A	7.5	4.0	34	1000	DO-41
1N5237B	8.2	8.0	20	500	DO-35
1N756A	8.2	8.0	20	500	DO-35
1N959B	8.2	6.5	15	500	DO-35
1N4738A	8.2	4.5	34	1000	DO-41
1N5238B	8.7	8.0	20	500	DO-35
1N5239B	9.1	10	20	500	DO-35
1N757A	9.1	10	20	500	DO-35
1N960B	9.1	7.5	14	500	DO-35
1N4739A	9.1	5.0	8	1000	DO-41
1N5240B	10	17	20	500	DO-35
1N758A	10	17	20	500	DO-35
1N961B	10	8.5	12.5	500	DO-35
1N4740A	10	7.0	25	1000	DO-41
1N5241B	11	22	20	500	DO-35
1N962B	11	9.5	11.5	500	DO-35
1N4741A	11	8.0	23	1000	DO-41
1N5242B	12	30	20	500	DO-35
1N759A	12	30	20	500	DO-35
1N963B	12	11.5	10.5	500	DO-35
1N4742A	12	9.0	21	1000	DO-41
1N5243B	13	13	9.5	500	DO-35
1N964B	13	13	9.5	500	DO-35

FIGURE C-24

NPN General Purpose Transistors by Ascending V_{ceo} (Continued)

Part Type	V_{ceo} (V) Min	V_{cbo} (V) Min	H_{fe} Min	H_{fe} Max	I_c/V_{ce} (mA/V)	Ft (MHz) Min	Package
MPS3393	25		90	180	2/4.5		TO-92
MPS3394	25		55	110	2/4.5		TO-92
MPS3395	25		150	500	2/4.5		TO-92
MPS3396	25		90	500	2/4.5		TO-92
MPS3397	25		55	500	2/4.5		TO-92
MPS3398	25		55	800	2/4.5		TO-92
MPS5172	25	25	100	500	10/10		TO-92
MPS6515	25	40	250	500	2/10		TO-92
MPS6520	25		200	400	2/10		TO-92
MPS6521	25		200	600	2/10		TO-92
MPS6522	25		200	400	2/10		TO-92
MPS6560	25	25	50	200	500/1.0	60	TO-92
PE4010	25	30	200	100	1.0/10		TO-92
PE8050	25	30	65	200	100/1.0	100	TO-92
PN3565	25	30	150	600	1.0/10	40	TO-92
PN5135	25	30	50	600	10/10	40	TO-92
2N2218	30	60	40	120	150/10	250	TO-39
2N2219	30	60	100	300	150/10	250	TO-39
2N2221	30	60	40	120	150/10	250	TO-18
2N2222	30	60	100	300	150/10	250	TO-18
2N3704	30	50	100	300	50/2.0	100	TO-92
2N3705	30	50	50	150	50/2.0	100	TO-92
2N3858	30	30	60	120	2/4.5	250	TO-92
2N3859	30	30	100	200	2/4.5	250	TO-92
2N3860	30	30	150	300	2/4.5	250	TO-92
2N4123	30	40	50	150	2/1.0	250	TO-92
2N4951	30	60	60	200	150/10	250	TO-92
2N4952	30	60	100	300	150/10	250	TO-92
2N4953	30	60	200	600	150/10	250	TO-92
2N4954	30	40	60	600	150/10	250	TO-92
2N4970	30	50	100	350	150/10	200	TO-92
2N5088	30	35	300	900	0.1/5.0	50	TO-92
BSR13	30	60	100	300	150/10	250	TO-236
MMBT2218	30	60	40	120	150/10	250	TO-236
MMBT2219	30	60	100	300	150/10	250	TO-236
MMBT2221	30	60	40	120	150/10	250	TO-236
MMBT2222	30	60	100	300	150/10	250	TO-236
MMBT3566	30	40	150	600	10/10	40	TO-236
MMBT3641	30	60	40	120	150/10	150	TO-236
MMBT3643	30	60	100	300	150/10	250	TO-236
MMBT3704	30	50	100	300	50/1.0	100	TO-236
MMBT3705	30	50	50	150	50/20	100	TO-236
MMBT4123	30	40	5	150	2/1.0	250	TO-236
MMBT5088	30	35	300	900	0.1/5.0	50	TO-236
MPS3704	30	50	100	300	50/20	100	TO-92
MPS3705	30	50	50	150	50/20	100	TO-92
MPS6512	30	40	50	180	2/10		TO-92
MPS6513	30	40	90	180	2/10		TO-92
MPS6532	30	50	30		100/1.0		TO-92
PN2218	30	60	40	120	150/10	250	TO-92
PN2219	30	60	100	300	150/10	250	TO-92
PN2221	30	60	40	120	150/10	250	TO-92
PN2222	30	60	100	300	150/10	250	TO-92

FIGURE C-25

General Purpose N-Channel JFETs by BV$_{gss}$

Type No.	BV$_{gss}$ (V) Min	V$_p$ (V) @ V$_{ds}$ Min	Max	(V)	I$_d$ (nA)	G$_{fs}$ @ (mS) Min	Max	C$_{iss}$ (pF) Max	C$_{rss}$ (pF) Max	e$_n$ @	F (Hz)	Package
MPF110	20	0.5	10	10	1	0.5						TO-92
MPF111	20	0.5	10	10	1000	0.5						TO-92
2N5103	25	0.5	4	15	1	2	8	5	1	100	10	TO-72
2N5104	25	0.5	4	15	1	3.5	7.5	5	1	50	10	TO-72
2N5105	25	0.5	4	15	1	5	10	5	1			TO-72
2N5457	25	0.5	6	15	10	2	5	7	3			TO-92
2N5458	25	1	7	15	10	1.5	5.5	7	3			TO-92
2N5459	25	2	8	15	10	2	6	7	3			TO-92
J210	25	1	3	15	1	4	12	15	11.5	110	999	TO-92
J211	25	2.5	4.5	15	1	7	12	15	11.5	110	999	TO-92
J212	25	4	6	15	1	7	12	15	11.5	110	999	TO-92
MPF103	25		6	15	1	1	5	7	3			TO-92
MPF104	25		7	15	1	1.5	5.5	7	3			TO-92
MPF105	25		8	15	1	2	6	7	3			TO-92
MPF109	25	0.2	8	15	10	0.8	6	7	3	115	999	TO-92
MPF112	25	0.5	10	10	1000	1	7.5					TO-92
PN5163	25	0.4	8	15	1000	2	9	12	3	50	999	TO-92
TIS58	25	0.5	5	15	20	1.3	4	6	3			TO-92
TIS59	25	1	9	15	20	1.3		6	3			TO-92
2N3967	30	2	5	20	1	2.5		5	1.3	84	100	TO-72
2N3967A	30	2	5	20	1	2.5		5	1.3	160	10	TO-72
2N3968	30		3	20	1	2		5	1.3	84	100	TO-72
2N3968A	30		3	20	1	2		5	1.3	160	10	TO-72
2N3969	30		1.7	20	1	1.3		5	1.3	84	100	TO-72
2N3969A	30		1.7	20	1	0.3		5	1.3	160	10	TO-72
2N4220	30		4	15	0.1	1	4	6	2			TO-72
2N4220A	30		4	15	0.1	1	4	6	2	115	100	TO-72
2N4221	30		6	15	0.1	2	5	6	2			TO-72
2N4221A	30		6	15	0.1	2	5	6	2	115	100	TO-72
2N4222	30		8	15	0.1	2.5	6	6	2			TO-72
2N4222A	30		8	15	0.1	2.5	6	6	2	115	100	TO-72
2N5556	30	0.2	4	15	1	1.5	6.5	6	3	35	10	TO-72
2N5557	30	0.8	5	15	1	1.5	6.5	6	3	35	10	TO-72
2N5558	30	1.5	6	15	1	1.5	6.5	6	3	35	10	TO-72
PN4220	30		4	15	1	1	4	6	2			TO-92
PN4221	30		6	15	1	2	5	6	2			TO-92
PN4222	30		8	15	1	2.5	6	6	2			TO-92
PN4302	30		4	20	10	1		6	3	100	999	TO-92
PN4303	30		6	20	10	2		6	3	100	999	TO-92
PN4304	30		10	20	10	1		6	3	125	999	TO-92
2N3369	40		6.5	20	1000	0.6	2.5	20	3			TO-18
2N3370	40		3.2	20	1000	0.3	2.5	20	3			TO-18
2N5358	40	0.5	3	15	100	1	3	6	2	115	100	TO-72
2N5359	40	0.8	4	15	100	1.2	3.6	6	2	115	100	TO-72
2N5360	40	0.8	4	15	100	1.4	4.2	6	2	115	100	TO-72
2N5361	40	1	6	15	100	1.5	4.5	6	2	115	100	TO-72
2N5362	40	2	7	15	100	2	5.5	6	2	115	100	TO-72
2N5363	40	2.5	8	15	100	2.5	6	6	2	115	100	TO-72
2N5364	40	2.5	8	15	100	2.7	6.5	6	2	115	100	TO-72
J201	40	0.3	1.5	20	10	0.5		15	12	110	999	TO-92
J202	40	0.8	4	20	10	1		15	12	110	999	TO-92
J203	40	2	10	20	10	1.5		15	12	110	999	TO-92
2N3458	50		7.8	20	1000	2.5	10	18	5	225	20	TO-18
2N3459	50		3.4	20	1000	1.5	6	18	5	155	20	TO-18
2N3460	50		1.8	20	1000	0.8	4.5	18	5	155	20	TO-18

FIGURE C-26

Answers to Selected Problems

CHAPTER 1

3. This is the shell that gains or loses electrons.

4. Electron: a particle with a negative charge. Proton: a particle with a positive charge. Neutron: a particle with a neutral charge.

5. The net charge of an atom is determined by the number of electrons and protons. More electrons than protons creates a net negative charge and vice versa.

6. A conductor has many free electrons available for current flow.

7. The atomic number refers to the number of electrons the atom contains.

8. Light, heat, voltage, and chemical action can affect the valence electrons of an atom.

9. A good conductor has few valence electrons located as far as possible from the nucleus.

10. Conduction occurs when the valence electron takes on a new energy level and breaks free of the valence shell.

11. An electron is excited by the application of a voltage. The kinetic and potential energies of the electron are increased.

12. Coulomb's first law of electrostatics states that like charges repel and unlike charges attract.

13. A hole is created when an electron vacates the valence shell. A hole is a positively charged particle and has no mass.

14. Insulation confines current to a determined path.

15. The valence shell of an insulator is full.

16. It is difficult to break the valence electrons of an insulator from their orbit.

17. The forbidden band receives its name because electrons are forbidden to exist in this energy band.

18. A semiconductor has four electrons in its valence shell.

19. Silicon has 14 electrons and germanium has 32 electrons.

20. Covalent bonding occurs when atoms share their valence electrons. Ionic bonding occurs when atoms gain or lose an electron.

21. Intrinsic means having no impurities.

22. Doping is the process in which impurities are added to form P-type and N-type materials.

23. Doping occurs when trivalents or pentavalents are added to a silicon or germanium wafer.

24. An N-type material contains more electrons than holes.

25. A pentavalent is an atom with five valence electrons.

26. A hole only exists when an electron breaks free of its covalent bond.

27. Electrons are the majority carriers and holes are the minority carriers for N-type materials.

28. A P-type material contains more holes than electrons.

29. A trivalent is an atom with three valence electrons.

30. Boron, gallium, and indium are examples of trivalents.

31. Holes are the majority carriers and electrons are the minority carriers for P-type materials.

32. A PN junction is the point at which P-type and N-type materials meet.

33. Ions are formed when atoms gain or lose an electron.

34. The depletion region is an area of the wafer that is depleted of current carriers.

35. Germanium = 0.3 V and silicon = 0.7 V.

36. Biasing consists of setting up a device for proper operation.

37. Forward bias is provided by applying a positive charge to the P-type material and a negative charge to the N-type material.

38. Forward biasing reduces the depletion region that forms at the PN junction and reverse biasing expands the depletion region.

39. A positive potential repels the holes in the P-type material and a negative potential repels the electrons in the N-type material. The barrier voltage is overcome and current flows.

40. Positive and negative potentials attract the respective N-type and P-type regions, causing the depletion region to expand.

41. An I–V curve demonstrates the relationship between current and voltage.

42. Peak inverse voltage is the maximum reverse voltage allowable before a diode begins to conduct.

CHAPTER 2

2. The barrier voltage is the internal voltage of the diode that must be overcome in order for conduction to occur.

6. The diode varies its resistance in accordance with the current that flows through it.

8. Rectification is the process of changing ac to dc.

11. $V_{dc\ out} = 6.46$ VDC.

14. The filter capacitor maintains the peak voltage.

15. Ripple voltage is the small change in output voltage that remains after filtering.

17. A larger filter capacitor increases dc output voltage by reducing the ripple voltage.

18. The PRV of the half-wave rectifier is double the peak secondary voltage.

19. $V_{dc\ out} = 23.3$ VDC.

21. The full-wave rectifier conducts on both alternations of the input signal.

22. The center-tapped transformer divides the secondary voltage in half.   **23.** $V_{dc\ out} = 7.5$ VDC.

28. $v_s = 33.9$ V pk.   **29.** Ripple frequency = 120 Hz.

30. $V_{dc\ out} = 16$ VDC.   **31.** $V_{dc\ out} = 11.3$ VDC.

35. $V_{dc\ out} = 15$ VDC.   **37.** PRV = 11.3 V.

38. The output is positive.

39. Ripple frequency = 2 kHz.   **40.** $v_{rip} = 4.96$ mV p–p.

41. $V_{dc\ out} = 9.91$ VDC

42. The clipper is used to eliminate a portion of a wave.

48. The forward resistance of a diode should be low and the reverse resistance should be high.

51. A shorted diode results in a blown fuse.

52. A shorted filter cap results in damage to the diodes and a blown fuse.

CHAPTER 3

2. A zener diode is designed to operate in reverse bias at which time it develops its zener voltage.

7. $V_{RS} = 6$ V.   **8.** $I_{ZD} = 1.5$ mA.   **9.** $V_{out} = 5$ V.

10. $I_{ZD} = 4$ mA.

14. Reducing the resistance causes the LED to become brighter.   **19.** Z, M, X, W, and K.

20. The common-anode display requires ground on the segment to light it. The common-cathode display requires a positive voltage on the segment to light it.

23. $V_D = 4.05$ V.

24. $V_{RS} = 2.48$ V (estimated current of 7.5 μA).

26. Laser light is coherent, monochromatic, and collimated.

CHAPTER 4

1. Bipolar devices rely on both majority and minority carriers for current flow.   **4.** Base, emitter, and collector.

7. Electron current flow enters the emitter and exits the base and collector.

8. The base–emitter voltage drop is the result of forward biasing the B–E PN junction.

12. $I_B = 151$ μA and $V_C = 3.71$ V.

16. $I_{C(SAT)} = 24$ mA and $V_{C(SAT)} = 0$ V.

17. $I_{C(SAT)} = 22.1$ mA and $V_{C(SAT)} = 0$ V.

20. $I_B = 338$ μA and $V_C = 4.93$ V.

21. $V_{C(SAT)} = 4.55$ V and $R_{B(SAT)} = 13.1$ kΩ.

22. $I_B = 192$ μA and $V_C = 5.68$ V.

23. $V_{C(SAT)} = 3.2$ V and $R_{B(SAT)} = 30.4$ kΩ.

26. $V_C = 8.16$ V and $I_B = 228$ μA.

27. $V_C = 6.25$ V and $I_B = 207$ μA.

28. $V_C = 10$ V and $I_B = 0$ A.

33. $V_B = 4.31$ V and $V_C = 4.32$ V.   **34.** $V_{C(SAT)} = 3.84$ V.

35. $V_B = 6.07$ V and $V_C = 8.29$ V.   **36.** $V_{C(SAT)} = 6.67$ V.

38. $V_C = V_{CC}$.   **40.** $V_C = V_{CC}$.   **41.** $V_C = V_{C(SAT)}$.

42. $V_C = 0$ V.   **43.** $V_C = V_{CC}$.

CHAPTER 5

1. $V_B = 3.19$ V.   **2.** $V_C = 6.27$ V.   **3.** $A_v = 149$.

4. $A_p = 14,900$.   **5.** $v_{out} = 2.98$ V p-p.

6. $z_{in} = 13.8$ kΩ.   **7.** $z_{out} = 2.7$ kΩ.   **8.** $V_{C(SAT)} = 4$ V.

11. $V_{CQ1} = 7.7$ V.   **12.** $A_{vQ1} = 172$.

13. $v_{CQ1} = 2.58$ V p-p.   **14.** $V_{CQ2} = 7.83$ V.

15. $A_{vQ2} = 1.11$.   **16.** $v_{CQ2} = 2.87$ V p-p.

17. $V_{CQ3} = 7.59$ V.   **18.** $v_{CQ3} = 2.87$ V p-p.

19. Total $A_V = 191$.   **20.** $V_B = 5$ V.   **21.** $V_E = 4.3$ V.

22. A_v is approximately 1.   **23.** $A_i = 100$.

24. $v_{out} = 3$ V p-p.   **25.** $z_{in} = 4.52$ kΩ.

26. $z_{out} = 45$ Ω (using a power supply impedance of 5 kΩ).

28. $V_B = 7.5$ V.   **29.** $V_{EQ1} = 6.8$ V.

30. Total A_v is approximately 1.   **31.** Total $A_i = 10,000$.

32. $z_{in} = 500$ kΩ **33.** $z_{out} < 50$ Ω **34.** $V_{EQ2} = 6.1$ V.

35. $V_C = 9.66$ V.   **36.** $A_v = 93.7$.   **37.** $A_i = 1$.

38. $z_{in} = 28.8$ Ω.   **39.** $z_{out} = 2.7$ kΩ.

CHAPTER 6

1. $I_{VD} = 429$ μA.   **2.** $V_B = 1.42$ V.   **3.** $V_E = 716$ mV.

4. $I_E = 2.65$ mA.   **5.** $V_{CEQ} = 6.63$ V.

6. $V_{CQ} = 7.35$ V.   **7.** $P_Q = 17.6$ mW.

8. $P_{dc} = 30.8$ mW.   **9.** $r_c = 286$ Ω.   **10.** $A_v = 30.3$.

11. $A_i = 110$.   **12.** $A_p = 3,340$.   **13.** $v_{out} = 3.79$ V p–p.
14. $P_L = 4.48$ mW.   **15.** % efficiency = 14.6%.
21. $I_{VD} = 860$ μA.   **22.** $V_{B1} = 5.7$ V.   **23.** $V_{B2} = 4.3$ V.
24. $V_E = 5$ V.   **25.** $I_{C(ave)} = 31.8$ mA.
27. $P_{dc} = 327$ mW.   **28.** $A_v = 1$.   **29.** $A_i = 135$.
30. $A_p = 135$.   **31.** 8 V p–p.   **32.** $P_L = 200$ mW.
33. % efficiency = 61.2%.   **38.** $V_{C1} = 1.8$ V.
39. $V_C = 10$ V.   **40.** $f_r = 22.5$ kHz.
41. $v_{out} = 10$ V p–p.
42. The best input frequency for this circuit is the resonant frequency of 22.5 kHz.

CHAPTER 7

1. $V_{OUT} = 10$ V.   **2.** $I_T = 152$ mA.   **3.** $I_z = 142$ mA.
4. $P_Z = 1.42$ W.   **5.** $R_{MIN} = 78.7 \Omega$.
6. A 2-W resistor is a safe value for R_1.
7. A resistance of 3.9Ω is needed.   **8.** $V_{OUT} = 8.4$ V.
9. $I_E = 16.8$ mA.   **10.** $V_{CE} = 6.6$ V.
11. $P_{Q1} = 111$ mW.   **12.** $V_{R1} = 5.9$ V.
16. $I_L = 84$ mA.   **17.** $P_{Q1} = 554$ mW.
18. $V_{out(MIN)} = 0.7$ V.   **19.** $V_{out(MAX)} = 5.7$ V.
20. $I_L = 5.7$ mA.   **21.** $P_{Q1} = 35.9$ mW.
23. $P_{Q1} = 7.91$ mW.   **24.** $V_{R2} = 7$ V.
25. V_{C2} is equal to 0.7 V less than the output voltage.
26. $V_{OUT} = 3.2$ V.   **27.** $P_{Q1} = 282$ mW.
30. $V_{out(MIN)} = 6$ V.   **31.** $V_{out(MAX)} = 12$ V.
32. $I_L = 12$ mA.   **33.** $V_{R3} = 39.6$ mV.   **34.** No.
35. $P_{Q1} = 35.5$ mW.   **36.** $P_{Q2} = 217$ μW.
39. Current to the pass transistor is limited to 212 mA.
40. Q_4 will begin to conduct if R_L is less then 69.8Ω.
41. $R_3 = 5.83 \Omega$.   **42.** $P = 25$ μs.   **43.** $W = 16$ μs.
44. $D = 0.64$.   **45.** $V_{OUT} = 15.4$ V.
46. $V_{OUT} = 19.2$ V.
50. $V_{OUT} = 12$ V.   **51.** $V_{OUT} = 20.5$ V.

CHAPTER 8

5. $Q_{B2} =$ High.   **6.** $X =$ Low.
8. V_{C2} is approximately 1.4 V.   **9.** Q_1 and Q_3.
10. V_{B2} is approximately 0 V.   **11.** $X = 3.4$ V.   **16.** Q_2.
17. V_{C2} is equal to V_{CC} minus the small voltage drop across R_3.
19. $X =$ high.   **23.** No.   **24.** No.
25. The output of Gate A is low.   **27.** No.
28. The output of Gate B is high.   **30.** No.   **31.** Yes.
32. The output of Gate C is low.   **33.** No.   **34.** High.
35. Q_{B1} and Q_{B2} must both be high.   **37.** $Q =$ high.
39. $S =$ low and $C =$ high.
43. The LED lights when both A and B are high.
44. No.   **45.** Yes.   **46.** $Q_{C2} =$ high.   **47.** Yes.
48. No.   **50.** No.

CHAPTER 9

4. The depletion region is formed by reverse biasing the gate–source. Drain current decreases as the depletion region increases.

5. The high input impedance of the JFET is the result of reverse biasing the gate–source.   **6.** N-channel.
7. $I_{DSS} = 8$ mA.   **8.** $V_{GS(OFF)} = -6$ V.   **9.** $I_G = 0$ A.
10. $V_P = 5$ V.   **11.** Characteristic drain curve.
14. $I_G = 0$ A.
15. Common-source amplifier using self-bias.
16. $V_{GS} = -3$ V.
17. As I_D attempts to increase, V_{GS} increases.
18. $V_D = 6.6$ V.   **19.** $V_{DS} = 2.4$ V.   **20.** $V_S = 3$ V.
22. $V_G = 6.03$ V.   **23.** $V_S = 8.53$ V.
24. $I_D = 1.81$ mA.   **25.** $V_D = 16.4$ V.
26. $V_{DS} = 7.84$ V.
29. The input ac signal causes V_{GS} to vary between -1.5 and -4.5 V.
30. Drain current fluctuates between 2.44 and 1.15 mA.
31. $g_m = 1470$ μS.   **32.** $A_v = 3.97$.
33. $v_{out} = 11.9$ V p–p.   **34.** $z_{in} = 15$ MΩ.
36. Common source.   **37.** Common drain.
38. V_{GS} varies between -1.5 and -2.5 V.
39. I_D varies between 3.43 and 1.75 mA.
40. $g_m = 1680$ μS.   **41.** $A_v = 0.904$.
42. $v_{out} = 904$ mV p–p.
44. The disadvantage of the common-gate amp is the low input impedance.   **47.** $I_D = 1.75$ mA.
52. $I_D = 24.5$ mA.   **55.** $I_D = 12$ mA.   **56.** $V_{DS} = 5$ V.
57. $R_D = 1.58$ kΩ.   **58.** $g_m = 3000$ μS.
59. $A_v = -4.74$.   **60.** $z_{in} = 7.5$ MΩ.
67. Common source.   **70.** Common drain.
71. R_4 acts as a volume control.

CHAPTER 10

1. It is a relaxation oscillator. It causes the lamp to blink at a predetermined rate.   **2.** $R_{MIN} = 7.25$ kΩ.
3. $R_{MAX} = 66.7$ kΩ.   **4.** $T = 705$ mS.   **5.** $f = 3.5$ Hz.
9. A SCR is turned on by exceeding the breakover voltage and is turned off by a low current dropout.
10. $I_{R3} = 795$ μA.   **11.** $R_W = 1.26$ kΩ.   **12.** 6.5 V.
13. $V_{R1} = 7.95$ mV when SCR is off and 5.5 V when SCR is on.
14. $I_{R1} = 795$ μA when SCR is off and 55 mA when SCR is on.   **15.** $R_{MIN} = 275 \Omega$.
19. A diac is turned on by exceeding the breakover voltage.
21. v_C lags v_{in}.
22. A triac is turned on by exceeding its breakover voltage.
24. A diac is turned off by reducing current below the value of the holding current.
25. A diac is turned off by reducing current below the value of the holding current.   **29.** $V_P = 6.2$ V.

CHAPTER 11

1. Differential amplifier.   **2.** $I_E = 56.5$ μA.
3. $V_{OUT} = 6.35$ V.   **4.** $A_V = 113$.
10. $10^6 = 140$ dB, $10^3 = 80$ dB, and $1 = 0$ dB.
13. Inverting amplifier.   **14.** $R_F = 33$ kΩ.

15. $A_V = -7.02$.   **16.** $R_O = 4.11$ kΩ.
17. $V_{OUT} = -10.5$ V.   **18.** BW = 95 kHz.
19. $z_{in} = 4.7$ kΩ and $z_{out} = 5.27$ mΩ.
25. Noninverting amplifier.   **26.** $I_F = 66.7$ μA.
27. $A_V = 32.3$.   **28.** $V_{OUT} = -3.23$ V.
29. BW = 31 kHz.   **30.** $z_{in} = 6.19$ GΩ, $z_{out} = 24.3$ mΩ
31. $R_2 = 1.45$ kΩ.   **35.** $A_V = 1$.
37. $z_{in} = 1$ TΩ (tera = 1×10^{12})   **38.** $z_{out} = 400$ uΩ.
39. Summing amplifier.
40. $A_{VA} = -2.13$, $A_{VB} = -3.03$, $A_{VC} = -1.79$,
and $A_{VP} = -2.56$.
41. $V_{OUT(A)} = -4.26$ V. $V_{OUT(B)} = -3.03$ V,
$V_{OUT(C)} = -2.69$ V, and $V_{OUT(D)} = -1.28$ V.
42. $V_{OUT} = -11.3$ V.
46. $V_{R_F} = -11.3$ V.   **47.** Difference amplifier.
48. V_{OUT} = noninverting input − inverting input.
49. $A_{V(-)} = 1$ and $A_{V(+)} = 2$.   **50.** $V_{OUT} = 1$ V.
51. 2.5 V.   **54.** $R_F = 3$ kΩ.   **56.** $V_{OUT} = 3.5$ V.
61. The resulting waveform should be a square wave
due to positive and negative saturations.
65. An open loop describes a condition when no feedback
is used.
66. A Schmitt trigger uses positive feedback and a
comparator uses an open-loop configuration.
70. Voltage follower.   **71.** Difference amplifier.
72. Comparator.   **74.** $V_{REF} = -3.24$ V.
75. The LED lights when Request Access = +5 and
Clear to Access = 0 V. All other combinations result in an
unlit LED.

CHAPTER 12

1. High pass.   **2.** $f_{CO} = 106$ kHz.   **3.** $X_C = 796$ Ω.
4. $Z = 810$ Ω.   **5.** $V_{out} = 4.91$ V.
11. First-order low-pass active filter.   **13.** $A_v = 7$.
14. $A_v = 16.9$ dB.   **15.** $f_{CO} = 3.39$ khz.
16. $v_{out} = 13.4$ V p–p.   **17.** Passed.
21. Second-order high-pass active filter.
22. $f_{CO} = 32.2$ kHz.   **23.** $R_4 = 1406$ Ω.
26. Band-pass filter.
27. The first stage is a low-pass and the second stage is a
high-pass filter.
28. It passes a narrow band of frequencies.
29. $f_{CO1} = 154$ kHz.   **30.** $f_{CO2} = 124$ kHz.
31. BW = 30 kHz.

32. All frequencies between 124 kHz and 154 kHz
are passed.
33. $A_{v1} = 1.58$ and $A_{v2} = 1.58$.   **34.** $A_v = 2.5$.

CHAPTER 13

8. The frequency of oscillation is determined by the
frequency-determining components.
10. An overall phase shift of 180° is provided by the *RC*
phase-shift network.   **11.** $f_O = 5.12$ kHz.
12. $A_v = 30.8$.   **14.** $R_4 = 3.45$ kΩ.   **20.** $C_T = 558$ pF.
21. $f_O = 98.3$ kHz.   **22.** $A_v = 11$.   **29.** $L_T = 2.59$ μH.
30. $f_O = 989$ kHz.   **31.** $A_v = 6.2$.   **37.** $A_v = 67.7$.
38. $V_{out} = 5.64$ V pk.   **43.** $f_O = 106$ kHz.
45. $R_{4(MAX)} = 1.5$ kΩ.   **46.** $V_{out} = 9$ V rms.
52. $\beta = 0.167$.   **53.** $f_O = 105$ kHz.

CHAPTER 14

2. $I_{C1+} = 2$ V.   **3.** $I_{C1(OUT)} = 0$ V.   **4.** $I_{C2-} = 4$ V.
5. Clear.   **6.** Yes.   **7.** 0 V.
8. Pulse the trigger input low.
13. A trigger voltage of less than 3 V activates the output.
14. $P_W = 24.2$ ms.
15. A threshold voltage greater than 4 V is necessary.
18. The voltage at pin 5 is 6 V.   **23.** $P_{WH} = 32.4$ μs.
24. $P_{WL} = 20.8$ μs.   **25.** $f = 18.8$ kHz.
26. $D = 60.9\%$.   **32.** $P_W = 103$ ms.
38. $P_{WH} = 9.12$ μs.   **39.** $P_{WL} = 10.4$ μs.
40. $f = 51.2$ kHz.   **46.** $P_{WL} = 457$ μs.   **47.** $f = 131$ Hz.
48. $f = 261$ Hz.

CHAPTER 15

3. A photon is a packet of light.
5. Ni–Cd stores energy.
12. AM varies the amplitude of the signal to represent different information, whereas FM varies the frequency.
16. The LED emits light when activated, whereas the LCD
causes its crystal segments to become opaque.
21. Quartz, Rochelle salt, and tourmaline.
30. The diaphragm of the speaker is used to vibrate the
surrounding air in order to create sound.
34. A relay works on the principle of magnetism.
41. The speed of a motor can be controlled by varying the
voltage or current that is applied to it.

Index